Developments in
Strategic Materials

Developments in Strategic Materials

*A Collection of Papers Presented at the
32nd International Conference on Advanced
Ceramics and Composites
January 27–February 1, 2008
Daytona Beach, Florida*

Editors

Hua-Tay Lin
Kunihito Koumoto
Waltraud M. Kriven
Edwin Garcia
Ivar E. Reimanis
David P. Norton

Volume Editors
Tatsuki Ohji
Andrew Wereszczak

A John Wiley & Sons, Inc., Publication

Published by John Wiley & Sons, Inc., Hoboken, New Jersey.
Published simultaneously in Canada.

For general information on our other products and services or for technical support, please contact our
Customer Care Department within the United States at (800) 762-2974, outside the United States at
(317) 572-3993 or fax (317) 572-4002.

Wiley also publishes its books in a variety of electronic formats. Some content that appears in print may
not be available in electronic format. For information about Wiley products, visit our web site at
www.wiley.com.

Library of Congress Cataloging-in-Publication Data is available.

ISBN 978-0-470-34500-9

Printed in the United States of America.

10 9 8 7 6 5 4 3 2 1

Contents

*This paper was presented at the 31st International Conference on Advanced Ceramics and Composites, held January 21-26, 2007 and was mistakenly excluded from the proceedings. It is being included in this 2008 proceedings.

MULTIFUNCTIONAL CERAMICS

SCIENCE OF CERAMIC INTERFACES

MATERIALS FOR SOLID STATE LIGHTING

Preface

This proceedings issue, *Developments in Strategic Materials,* contains a collection of 28 papers presented during the 32nd International Conference on Advanced Ceramics and Composites, Daytona Beach, FL, January 27-February 1, 2008. Papers are included from five symposia as listed below

- Geopolymers
- Basic Science of Multifunctional Materials
- Science of Ceramic Interfaces
- Materials for Solid State Lighting

The first paper provides an overview on developments in oxynitride glasses. This paper was presented by Stuart Hampshire, University of Limerick, during the plenary session of the conference.

The editors thank the assistance of all the organizers and session chairs and to the authors and reviewers for their contribution and hard work. The successful international gathering of geopolymer researchers was directly due to the generous financial support for speakers, provided by the US Air Force Office of Scientific Research (AFOSR) through Dr. Joan Fuller, Program Director of Ceramic and Non-Metallic Materials, Directorate of Aerospace and Materials Science.

Hua-Tay Lin
Oak Ridge National Laboratory

Kunihito Koumoto
Nagoya University

Waltraud M. Kriven
University of Illinois at Urbana-Champaign

Edwin Garcia
Purdue University

Ivar E. Reimanis
Colorado School of Mines

David P. Norton
University of Florida

Introduction

Organized by the Engineering Ceramics Division (ECD) in conjunction with the Basic Science Division (BSD) of The American Ceramic Society (ACerS), the 32nd International Conference on Advanced Ceramics and Composites (ICACC) was held on January 27 to February 1, 2008, in Daytona Beach, Florida. 2008 was the second year that the meeting venue changed from Cocoa Beach, where ICACC was originated in January 1977 and was fostered to establish a meeting that is today the most preeminent international conference on advanced ceramics and composites

The 32nd ICACC hosted 1,247 attendees from 40 countries and 724 presentations on topics ranging from ceramic nanomaterials to structural reliability of ceramic components, demonstrating the linkage between materials science developments at the atomic level and macro level structural applications. The conference was organized into the following symposia and focused sessions:

Symposium 1	Mechanical Behavior and Structural Design of Monolithic and Composite Ceramics
Symposium 2	Advanced Ceramic Coatings for Structural, Environmental, and Functional Applications
Symposium 3	5th International Symposium on Solid Oxide Fuel Cells (SOFC): Materials, Science, and Technology
Symposium 4	Ceramic Armor
Symposium 5	Next Generation Bioceramics
Symposium 6	2nd International Symposium on Thermoelectric Materials for Power Conversion Applications
Symposium 7	2nd International Symposium on Nanostructured Materials and Nanotechnology: Development and Applications
Symposium 8	Advanced Processing & Manufacturing Technologies for Structural & Multifunctional Materials and Systems (APMT): An International Symposium in Honor of Prof. Yoshinari Miyamoto
Symposium 9	Porous Ceramics: Novel Developments and Applications

Symposium 10   Basic Science of Multifunctional Ceramics
Symposium 11   Science of Ceramic Interfaces: An International Symposium
                  Memorializing Dr. Rowland M. Cannon
Focused Session 1 Geopolymers
Focused Session 2 Materials for Solid State Lighting

Peer reviewed papers were divided into nine issues of the 2008 Ceramic Engineering & Science Proceedings (CESP); Volume 29, Issues 2-10, as outlined below:

- Mechanical Properties and Processing of Ceramic Binary, Ternary and Composite Systems, Vol. 29, Is 2 (includes papers from symposium 1)
- Corrosion, Wear, Fatigue, and Reliability of Ceramics, Vol. 29, Is 3 (includes papers from symposium 1)
- Advanced Ceramic Coatings and Interfaces III, Vol. 29, Is 4 (includes papers from symposium 2)
- Advances in Solid Oxide Fuel Cells IV, Vol. 29, Is 5 (includes papers from symposium 3)
- Advances in Ceramic Armor IV, Vol. 29, Is 6 (includes papers from symposium 4)
- Advances in Bioceramics and Porous Ceramics, Vol. 29, Is 7 (includes papers from symposia 5 and 9)
- Nanostructured Materials and Nanotechnology II, Vol. 29, Is 8 (includes papers from symposium 7)
- Advanced Processing and Manufacturing Technologies for Structural and Multifunctional Materials II, Vol. 29, Is 9 (includes papers from symposium 8)
- Developments in Strategic Materials, Vol. 29, Is 10 (includes papers from symposia 6, 10, and 11, and focused sessions 1 and 2)

The organization of the Daytona Beach meeting and the publication of these proceedings were possible thanks to the professional staff of ACerS and the tireless dedication of many ECD and BSD members. We would especially like to express our sincere thanks to the symposia organizers, session chairs, presenters and conference attendees, for their efforts and enthusiastic participation in the vibrant and cutting-edge conference.

ACerS and the ECD invite you to attend the 33rd International Conference on Advanced Ceramics and Composites (http://www.ceramics.org/daytona2009) January 18–23, 2009 in Daytona Beach, Florida.

TATSUKI OHJI and ANDREW A. WERESZCZAK, Volume Editors
July 2008

Oxynitride Glasses

DEVELOPMENTS IN OXYNITRIDE GLASSES: FORMATION, PROPERTIES AND CRYSTALLIZATION

Stuart Hampshire
Materials and Surface Science Institute
University of Limerick, Limerick, Ireland

ABSTRACT

Oxynitride glasses are effectively alumino-silicates in which nitrogen substitutes for oxygen in the glass network. They are found at triple point junctions and as intergranular films in silicon nitride based ceramics. The properties of silicon nitride, especially fracture behaviour and creep resistance at high temperatures are influenced by the glass chemistry, particularly the concentrations of modifyer, usually Y or a rare earth (RE) ion, and Al, and their volume fractions within the ceramic. This paper provides an overview of the preparation of M-Si-Al-O-N glasses and outlines the effects of composition on properties. As nitrogen substitutes for oxygen, increases are observed in glass transition (T_g) and dilatometric softening (T_{ds}) temperatures, viscosities, elastic moduli and microhardness. If changes are made to the cation ratios or different rare earth elements are substituted, properties can be modified. The effects of these changes on mechanical properties of silicon nitride based ceramics are discussed.

This paper also outlines new research on M-Si-Al-O-N-F glasses. It was found that fluorine expands the glass forming region in the Ca-Sialon system and facilitates the solution of nitrogen into glass melts. T_g and T_{ds} decreased with increasing fluorine substitution levels, whilst increasing nitrogen substitution resulted in increases in values for these thermal properties. Nitrogen substitution for oxygen caused increases in Young's modulus and microhardness whereas these two properties were virtually unaffected by fluorine substitution for oxygen.

Oxynitride glasses may be crystallized to form glass-ceramics containing oxynitride phases and a brief outline is presented.

INTRODUCTION

Oxynitride glasses were first discovered as intergranular phases in silicon nitride based ceramics[1,2] in which the composition, particularly Al content as well as N content, and volume fraction of such glass phases determine the properties of the silicon nitride. Oxynitride glasses can be formed when a nitrogen containing compound, such as Si_3N_4 (or AlN), dissolves in either a silicate or alumino-silicate liquid at ~1600-1700°C which then cools to form a M-Si-O-N or M-Si-Al-O-N glass (M is usually a di-valent [Mg, Ca] or tri-valent [Y, Ln] cation). In particular, the chemistry of these oxynitride glasses has been shown to control high temperature mechanical properties and ambient fracture behaviour of silicon nitride based ceramics[1-4]. The desire to understand the nature of these grain boundary phases has resulted in a number of investigations on oxynitride glass formation and properties[5-12].

EXPERIMENTAL PROCEDURE

The extent of the glass forming regions in various M-Si-Al-O-N systems (M = Mg, Y, Ca, etc.) has been studied previously[5,7,8] and represented using the Jänecke prism with compositions expressed in equivalent percent (e/o) of cations and anions[5,7] instead of atoms or gram-atoms. One equivalent of any element always reacts with one equivalent of any other element or species. For a system containing three types of cations, A, B and C with valencies of v_A, v_B, and v_C, respectively, then:

Equivalent concentration of $A = (v_A [A])/(v_A [A] + v_B[B] + v_C[C])$,
where [A], [B] and [C] are, respectively, the atomic concentrations of A, B and C, in this case, Si^{IV}, Al^{III} and the metal cation, M, with its normal valency.

If the system also contains two types of anions, C and D with valencies v_C and v_D, respectively, then:

Equivalent concentration of $C = (v_C [C])/(v_C [C] + v_D[D])$,
where [C] and [D] are, respectively, the atomic concentrations of C and D, i.e. O^{II} and N^{III}.

Fig. 1 shows the glass forming region in the Y-Si-Al-O-N system which was studied by exploring glass formation as a function of Y:Si:Al ratio on vertical planes in the Jänecke prism representing different O:N ratios. The region is seen to expand initially as nitrogen is introduced and then diminishes when more than 10 e/o N is incorporated until the solubility limit for nitrogen is exceeded at ~28 e/o N.

Preparation of glasses involves mixing appropriate quantities of silica, alumina, the modifying oxide and silicon nitride powders by wet ball milling in isopropanol for 24 hours, using sialon milling media, followed by evaporation of the alcohol before pressing into pellets. Batches of 50-60g are melted in boron nitride lined graphite crucibles at 1700-1725°C for 1h under 0.1MPa nitrogen pressure in a vertical tube furnace, after which the melt is poured into a preheated graphite mould. The glass is annealed at a temperature close to the glass transition temperature (T_g) for one hour to remove stresses and slowly cooled.

Bulk densities were measured by the Archimedes principle using distilled water as the working fluid. X-ray analysis was used to confirm that the glasses were totally amorphous. Scanning electron microscopy allowed confirmation of this and assessment of homogeneity.

Differential thermal analysis (DTA) was carried out in order to measure the glass transition temperature, T_g, which is observed as the onset point of the endothermic drift on the DTA curve, corresponding to the beginning of the transition range.

The viscosity results presented were obtained from a high temperature "deformation-under-load" (compressive creep) test on cylinders of 10 mm diameter in air between 750 and 1000°C. These have also been compared with results from three point bending tests (bars of dimensions: 25mm x 4mm (width) x 3mm (height) with a span of 21 mm. Viscosity, η, is derived from the relationships between (i) the stress/strain relations in an elastic solid and

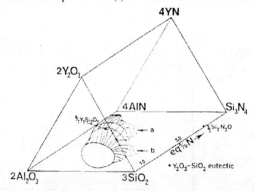

Fig. 1 Glass forming region of the Y-Si-Al-O-N system on cooling from 1700°C[5,7]

(ii) those that relate to a viscous fluid:

$$\eta = \sigma / [2(1+\upsilon)\dot{\varepsilon}] \tag{1}$$

where σ and $\dot{\varepsilon}$ are the applied stress and the creep rate on the outer tensile fibre and υ is Poisson's ratio (taken as 0.5). The results from both types of test show good agreement[5,7,12].

RESULTS AND DISCUSSION

EFFECTS OF NITROGEN ON PROPERTIES

The first systematic studies on the effect of replacing oxygen by nitrogen on properties of oxynitride glasses with fixed cation compositions were reported by Drew, Hampshire and Jack[5,7]. Fig. 2 shows that for all Ca-, Mg-, Nd- and Y- Si-Al-O-N glasses with a fixed cation composition (in e/o) of 28Y: 56Si: 16Al (standard cation composition), incorporation of nitrogen resulted in increases in glass transition temperature (T_g). They also reported that nitrogen increases microhardness, viscosity, resistance to devitrification, refractive index, dielectric constant and a.c. conductivity. In a more extensive study of the Y-Si-Al-O-N system[8], it was confirmed that glass transition temperature (T_g), viscosity, microhardness and elastic moduli all increase systematically while coefficient of thermal expansion (CTE) decreases with increasing nitrogen:oxygen ratio for different series of glasses.

As shown in Fig. 3, values of Young's modulus increase by 15 to 25% as ~17-20 e/o N is substituted for oxygen at fixed cation ratios[8,9]. The coefficient of thermal expansion (α) was found to decrease as N content increased[8] at fixed Y:Si:Al ratios.

Fig. 4 shows the effects of nitrogen content on viscosity for a series of glasses[13] with composition (in e/o) of 28Y:56Si:16Al:(100-x)O:xN (x=0, 5, 10, 18). It can be seen that viscosity increases by much more than 2 orders of magnitude as 18 e/o oxygen is replaced by nitrogen. Similar trends have been reported for other Y-Si-Al-O-N glasses with different cation ratios[12].

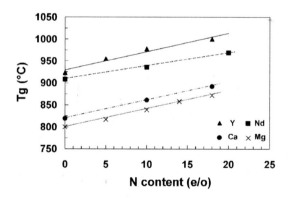

Fig. 2. Effect of N content (e/o) on the glass transition temperature, T_g, of Mg-, Ca-, Nd- and Y-Si-Al-O-N glasses with fixed M:Si:Al: ratio = 28:56:16 (after ref. 5).

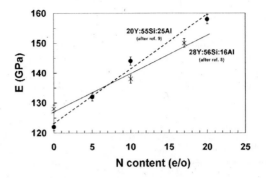

Fig. 3. Effect of N (e/o) on Young's modulus (E) for glasses with fixed Y:Si:Al ratios (data from refs. 8 and 9).

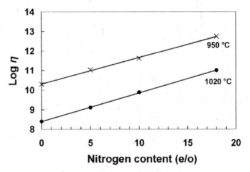

Fig. 4 Effect of N (e/o) on viscosity for glass with fixed Y:Si:Al ratio = 28:56:16 at 950 and 1020 °C (data from ref. 13).

All of these increases in properties are known to be due to the increased cross-linking within the glass structure as 2-coordinated bridging oxygen atoms are replaced by 3-coordinated nitrogen atoms[5-9]. In certain cases, some nitrogen atoms may be bonded to less than three Si atoms, as in:

(i) $\equiv Si - N^- - Si \equiv$

or

(ii) $\equiv Si - N^{2-}$

The local charge on the so-called "non-bridging" nitrogen ions is balanced by the presence of interstitial modifying cations (Y, etc.) in their local environment. In the case of silicate glasses, non-bridging oxygen atoms replace bridging oxygen atoms at high modifier contents. In (i) above, while the N atom links two silicon atoms rather than three, it still effectively "bridges" the network ions.

EFFECTS OF LANTHANIDE CATIONS ON PROPERTIES

Fig. 5 demonstrates the effects of different rare earth lanthanide cations on viscosity of Ln-Si-Al-O-N glasses[10] with fixed cation ratio of 28Ln:56Si:16Al. Viscosity changes by ~3 orders of magnitude in the series: Eu<Ce<Sm<Y<Dy <Ho<Er. As found also for other properties, viscosity increases almost linearly with increase in cation field strength, CFS (where CFS = v/r^2, v is valency and r is ionic radius) of the Ln ion. Viscosities of Ln-Si-Al-O-N liquids, containing Sm, Ce, Eu, where the ionic radii are larger than that of Y, are less than those of the equivalent Y-Si-Al-O-N liquids and this will have implications for easier densification of silicon nitride ceramics. However, there will also be consequences for high temperature properties, particularly creep resistance. Liquids and glasses containing Ln cations with ionic radii smaller than Y (Lu, Er, Ho, Dy) have been shown to have higher viscosities than the Y-containing glasses and, in silicon nitride these particular cations will form grain boundary glasses with higher softening temperatures and, hence, better creep resistance.

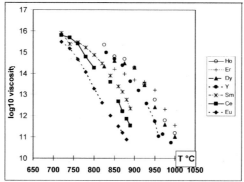

Fig. 5   Effects of Ln cation on viscosity of glasses with cation ratio (in e/o) of 28Ln:56Si:16Al (Ln = Y, Eu, Ce, Sm, Dy, Er, Ho; fixed N = 17 e/o).

The effects on properties of changes in grain boundary glass chemistry, as a result of changes in sintering additives to silicon nitride, can be summarised as follows[12,14]:

(i)   As up to 20 e/o N is substituted for oxygen at a fixed cation ratio, viscosity increases by >2 orders of magnitude.

(ii)   At a fixed N content, increasing the Y:Al ratio of the glass results in further increases in viscosity.

(iii)   Changing the rare earth cation from a larger ion, such as La or Ce, to a smaller cation, such as Er or Lu, increases viscosity by a further 3 orders of magnitude.

Overall, a change of almost 6 orders of magnitude in viscosity can be achieved by increasing N and modifying the cation ratio and the type of rare earth ion. The implications for silicon nitride ceramics are that intergranular glasses containing more N and less Al and smaller RE cations will provide enhanced creep resistance.

CRYSTALLIZATION OF OXYNITRIDE GLASSES

The glass-ceramic transformations in a glass of composition (in e/o) 28Y:56Si:16Al:83O:17N have been studied[15] using both classical and differential thermal analysis techniques and these two methods were found to be in close agreement. Optimum nucleation and crystallisation temperatures were determined in relation to the glass transition temperature. The major crystalline phases present are mixtures of different forms of yttrium di-silicate and silicon oxynitride. Bulk nucleation was observed to be the dominant nucleation mechanism. The activation energy for the crystallisation process was found to be 834kJ/mol.

For a glass of composition (in e/o) 35Y:45Si:20Al:77O:23N, crystallization[8] results in formation of B-phase[16] (Y_2SiAlO_2N), Iw-phase ($Y_2Si_3Al_2(O,N)^{10e/o}$) and wollastonite ($YSiO_2N$) at temperatures below 1200°C while α-yttrium di-silicate ($Y_2Si_2O_7$), apatite ($Y_2Si_3O_{12}N$) and YAG ($Y_3Al_2O_{12}$) are formed at higher temperatures. At relatively low heat treatment temperatures of ~950-1100°C, the nucleation and growth of N-wollastonite ($YSiO_2N$) and the intermediate phases B and Iw are kinetically favoured over that of the more stable equilibrium phases YAG and Si_2N_2O. Further studies on the crystallisation of B and Iw phases in these Y-Si-Al-O-N glasses have been reported[17,18]. The properties of the glass-ceramics exceed those of the parent glasses with values of elastic modulus greater than 200GPa.

PREPARATION AND PROPERTIES OF OXYFLUORONITRIDE GLASSES

Current work has explored a new generation of oxynitride glasses containing fluorine and aims to develop an initial understanding of the effects of composition on glass formation, structure and properties. Fig. 6 shows the glass forming region found for the Ca-Si-Al-O-N-F system[19] with 20 e/o N and 5 e/o F at 1650°C. In the surrounding regions the different crystalline phases observed are also shown. All glasses were dense except for a region of Si-rich compositions where porous glasses were observed.

In the porous glass area of Fig. 6, there are bubbles on the surface in addition to the bubbles (pores) within the bulk of the glasses which is due to SiF_4 loss. Formation of SiF_4 is favored perhaps due to high Si:F ratios (>3) and low Al (6-15 e/o) and Ca (13-25 e/o) contents. In some areas of this glass forming region inhomogeneous and phase separated glasses were found. Effectively, fluorine extends glass formation in the previously known Ca-Si-Al-O-N system. The effect of fluorine addition on the structure of silicate or aluminosilicate glasses has been previously invetsigated[20] and it has been shown that fluorine can bond to silicon as Si-F, to Al as Al-F, and to Ca as Ca-F. Fluorine loss occurs under conditions where Si-F bonds are favoured. The bonding of fluorine to Al prevents fluorine loss as SiF_4 from the glass melt and explains the reduction in the glass transition temperature[20].

The liquidus temperatures for these oxyfluoronitride compositions were compared with data for Ca-Si-Al-O glasses and it was found that the addition of both nitrogen and fluorine reduces the liquidus temperatures of the high silica and alumina compositions by 100-250°C. At higher Ca contents, much greater reductions in liquidus temperatures of about 800°C were found. Fluorine also facilitates the solution of higher amounts of nitrogen (up to 40 e/o N) into glasses compared with the Ca-Si-Al-O-N system[19]. Fluorine has the effect of lowering glass transition temperature in these glasses but has no effect on the elastic modulus or microhardness[21].

CONCLUSIONS

Oxynitride glasses which occur as intergranular amorphous phases in silicon nitride ceramics have been studied to assess the effects of composition on properties such as viscosity which increases by more nearly three orders of magnitude as 18 e/o N is substituted for oxygen. Viscosity generally increases as more Si or Y is substituted for Al but this is a smaller effect than that of nitrogen. A further increase in viscosity of two to three orders of magnitude is achieved by substituting smaller rare earth cations in place of larger ones. The implications for silicon nitride

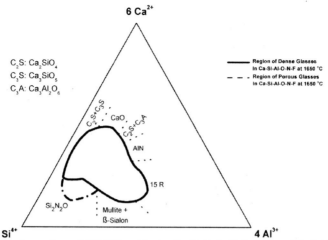

Fig. 6. The glass forming region at 1650°C found in the Ca-Si-Al-O-N-F system at 20 e/o N and 5 e/o F and the adjacent crystalline regions (after ref. 19).

ceramics are that intergranular glasses containing more N and less Al and smaller RE cations will provide enhanced creep resistance.

A new generation of oxynitride glasses containing fluorine have also been investigated. The glass forming region in the Ca-Si-Al-O-N-F system at 20 eq.% N and 5 eq.% F is larger than the glass forming region at 20 eq.% N in the Ca-Si-Al-O-N system. Fluorine expands the range of glass formation in this oxynitride system. Considerable reduction of liquidus temperatures by about 800°C at higher calcium contents occurs. Fluorine facilitates the solution of much higher amounts of nitrogen into the melt than are possible in the Ca-Si-Al-O-N system.

ACKNOWLEDGMENTS

The author wishes to thank the Engineering Ceramics Division of the American Ceramic Society for the Bridge Building Award 2008. I wish to acknowledge Science Foundation Ireland for financial support of research on Oxynitride Glasses and to thank Professor M. J. Pomeroy, Dr. Annaik Genson and Mr. Amir Hanifi of the Materials and Surface Science Institute for their work in this area.

REFERENCES

[1]F. L. Riley, "Silicon Nitride and Related Materials," *J. Amer. Ceram. Soc.*, **86** [2], 245-265 (2000).

[2]F. F. Lange, "The Sophistication of Ceramic Science through Silicon Nitride Studies," J. Ceram. Soc. Japan **114** (1335), 873-879 (2006).

[3]E. Y. Sun, P. F. Becher, K. P. Plucknett, C.-H. Hsueh, K. B. Alexander, S. B. Waters, K. Hirao, and M. E. Brito, "Microstructural Design of Silicon Nitride with Improved Fracture Toughness: II, Effects of Yttria and Alumina Additives," *J. Am. Ceram. Soc.*, **81** [11] 2831–40 (1998).

[4]E. Y. Sun, P. F. Becher, C.-H. Hsueh, G. S. Painter, S. B. Waters, S-L. Hwang, M. J. Hoffmann, "Debonding behavior between beta-Si_3N_4 whiskers and oxynitride glasses with or without an epitaxial beta-SiAlON interfacial layer," Acta Mater., **47** [9] 2777-85 (1999).

[5]R. A. L. Drew, S. Hampshire and K. H. Jack, "Nitrogen Glasses," Proc. Brit. Ceram. Soc., **31**, 119-132 (1981).

[6]R. E. Loehman, "Preparation and Properties of Oxynitride Glasses," J. Non-Cryst. Sol., **56**, 123-134 (1983).

[7]S. Hampshire, R. A. L. Drew and K. H. Jack, "Oxynitride Glasses," Phys. Chem. Glass., **26** [5], 182-186 (1985).

[8]S. Hampshire, E. Nestor, R. Flynn, J.-L. Besson, T. Rouxel, H. Lemercier, P. Goursat, M. Sebai, D. P. Thompson and K. Liddell, "Yttrium oxynitride glasses: properties and potential for crystallisation to glass-ceramics," J. Euro. Ceram. Soc., **14**, 261-273 (1994).

[9]E. Y. Sun, P. F. Becher, S-L. Hwang, S. B. Waters, G. M. Pharr and T. Y. Tsui, "Properties of silicon-aluminum-yttrium oxynitride glasses," J. Non-Cryst. Solids, **208**, 162-169 (1996).

[10]R. Ramesh, E. Nestor, M. J. Pomeroy and S. Hampshire, "Formation of Ln-Si-Al-O-N Glasses and their Properties," J. Euro. Ceram. Soc., **17**, 1933-9 (1997).

[11]S. Hampshire, "Oxynitride Glasses, Their Properties and Crystallization - A Review," J. Non-Cryst. Sol., **316**, 64-73 (2003).

[12]P. F. Becher and M. K. Ferber, "Temperature-Dependent Viscosity of SiREAl-Based Glasses as a Function of N:O and RE:Al Ratios (RE = La, Gd, Y, and Lu)," J. Am. Ceram. Soc., **87** [7], 1274–1279 (2004).

[13]S. Hampshire, R. A. L. Drew and K. H. Jack, "Viscosities, Glass Transition Temperatures and Microhardness of Y-Si-Al-O-N Glasses," J. Am. Ceram. Soc., **67** [3], C46-47 (1984).

[14]S. Hampshire and M. J. Pomeroy, "Effect of composition on viscosities of rare earth oxynitride glasses," J. Non-Cryst. Solids, **344**, 1-7 (2004).

[15]R. Ramesh, E. Nestor, M.J. Pomeroy and S. Hampshire, "Classical and DTA studies of the glass-ceramic transformation in a YSiAlON glass," J. Am. Ceram. Soc., **81** [5], 1285-97 (1998).

[16]M. F. Gonon, J. C. Descamps, F. Cambier, D. P. Thompson, "Determination and refinement of the crystal structure of M_2SiAlO_5N (M=Y, Er, Yb)," Ceram. Internat. **26**, 105-11 (2000).

[17]Y. Menke, L.K.L. Falk and S. Hampshire, "The Crystallisation of Er-Si-Al-O-N B-Phase Glass-Ceramics," J. Mater. Sci., **40** [24], 6499-512 (2005).

[18]Y. Menke, S. Hampshire and L.K.L. Falk, "Effect of Composition on Crystallization of Y/Yb–Si–Al–O–N B-Phase Glasses," J. Am. Ceram. Soc., **90**, 1566-73 (2007).

[19]A. R. Hanifi, A. Genson, M. J. Pomeroy and S. Hampshire, "An Introduction to the Glass Formation and Properties of Ca-Si-Al-O-N-F Glasses," Mater. Sci. Forum, **554**, 17-23 (2007).

[20]R. Hill, D. Wood and M. Thomas, "Trimethylsilylation Analysis of the Silicate Structure of Fluoro-Alumino-Silicate Glasses and the Structure Role of Fluorine," J. Mater. Sci., **34**, 1767-1774 (1999).

[21]A. Genson, A. R., Hanifi, A. Vande Put, M. J. Pomeroy and S. Hampshire, "Effect of Fluorine and Nitrogen anions on Properties and Crystallisation of Ca-Si-Al-O glasses," Mater. Sci. Forum, **554**, 31-35 (2007).

Thermoelectric Materials for Power Conversion Applications

THERMOELECTRIC PROPERTIES OF Ge DOPED In$_2$O$_3$

David Bérardan, Emmanuel Guilmeau, Antoine Maignan, and Bernard Raveau

Laboratoire CRISMAT, UMR 6508 CNRS-ENSICAEN, 6 Boulevard du Maréchal Juin, 14050 CAEN Cedex, France
E-mail: david.berardan@ensicaen.fr and emmanuel.guilmeau@ensicaen.fr

ABSTRACT

Chemical, structural and transport properties of a series of In$_2$O$_3$ based samples with germanium doping (from 0 to 15 atom%) have been studied. X-ray diffraction and scanning electron microscopy studies show that the solubility limit of Ge in In$_2$O$_3$ is very small and that additions of more than about 0.5 atom% Ge lead to the presence of In$_2$Ge$_2$O$_7$ inclusions. The electrical conductivity is strongly enhanced by Ge doping with best values exceeding 1200 S.cm^{-1} at room temperature. On the other hand, the thermopower decreases with Ge addition, but the thermoelectric power factor remains higher than that of undoped In$_2$O$_3$ and is close to 1 mW.m^{-1}.K^{-2} at 1100K in In$_{1.985}$Ge$_{0.015}$O$_3$. The thermal conductivity is strongly reduced by Ge additions. The dimensionless figure of merit ZT reaches 0.1 at 1273K in In$_2$O$_3$ and exceeds 0.45 at 1273K in composite compounds with nominal composition In$_{2-x}$Ge$_x$O$_3$.

INTRODUCTION

In the past decade, a great effort of research has been devoted to the development of novel materials for applications in thermoelectric energy conversion (see for example ref. 1). Thermoelectric generation systems can directly convert heat energy into electrical power without moving parts or carbon dioxide production. The efficiency of a thermoelectric material used for power generation increases with the so-called dimensionless figure of merit ZT defined as ZT = $S^2T\sigma/\lambda$, where S is the Seebeck coefficient or thermopower, σ the electrical conductivity, and λ the thermal conductivity. The thermal conductivity can be divided into two parts, the first one λ_L originating from the heat carrying phonons, and the second one λ_{elec} which is linked to the electrical conductivity by the Wiedmann-Franz law (valid for metallic-like conductivity) $\lambda_{elec}=L_0T\sigma$ with L_0 the Lorentz number 2.45.10^{-8} V^2.K^{-2}. Compared with conventional thermoelectric materials (see for example ref. 2), which are mainly intermetallic compounds, oxides are very suitable for high temperature applications due to their high chemical stability under air. Several families of n-type oxide materials have been studied up to now, but all of them exhibit rather poor thermoelectric performances even in single crystals, with best ZT values hardly exceeding 0.3-0.35 at 900-1000°C [3-5] in Al-doped ZnO or textured (ZnO)$_m$In$_2$O$_3$. As In$_2$O$_3$ based compounds are known to exhibit very good electrical conductivity and are widely used as transparent conducting oxides (TCO) (see for example ref. 6), we decided to investigate the thermoelectric properties of In$_2$O$_3$ with Ge doping.

EXPERIMENTAL

All samples, belonging to the In$_{2-x}$Ge$_x$O$_3$ series, were prepared using a standard solid reaction route. Starting powders, In$_2$O$_3$ (Neyco, 99.99 %) and GeO$_2$ (Alfa Aesar, 99.99%), were weighed in stoichiometric amounts and ground together by ball milling using agate balls and vial. The resulting powders were pressed uniaxially under 300 MPa, using polyvinyl alcohol binder to form parallelepipedic 2x3x12 mm^3 or cylindrical $\varnothing$20x3 mm samples. Then they were sintered at 1300°C during 48h under air on platinum foils to avoid any contamination from the alumina crucible. X-ray powder diffraction (XRD) was employed for structural characterization using a Philips X'Pert Pro diffractometer with X'Celerator using Cu-K$_\alpha$ radiation in a 2θ range 10°-90°. The XRD patterns were analyzed using the Rietveld method with the help of the FullProf software [7]. The scanning electron microscopy (SEM) observations were made using a FEG Zeiss Supra 55.

The cationic compositions were determined by energy dispersive X-ray spectroscopy (EDX) EDAX. The electrical conductivity and thermopower were measured simultaneously using a ULVAC-ZEM3 device between 50°C and 800°C under helium. The thermal conductivity was obtained at 26°C, 300°C, 800°C and 1000°C from the product of the geometrical density, the heat capacity and the thermal diffusivity from a Netzsch model 457 MicroFlash laser flash apparatus.

RESULTS AND DISCUSSION

The thermal treatment presented above leads to well crystallized samples. Figure 1 shows the XRD patterns of the samples with different Ge contents. Several reflexions can be observed that corresponds to the $In_2Ge_2O_7$ phase for $x \geq 0.015$.

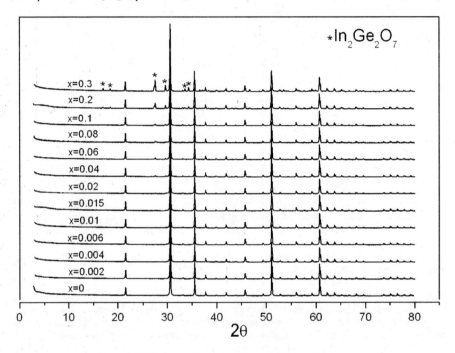

Figure 1: XRD patterns for the samples with different Ge contents.

Figure 2 shows the Rietveld refinement of a sample with nominal composition $In_{1.9}Ge_{0.1}O_3$. All Bragg peaks can be indexed taking into account a mixture of In_2O_3 (S.G. Ia-3 [8]) as main phase and $In_2Ge_2O_7$ (S.G. C2/m [9]) as secondary phase. No systematic evolution of the lattice parameter a occurs when the germanium fraction increases despite the very different ionic radii of In(+III) and Ge(+IV) [10], and the fractions of $In_2Ge_2O_7$ determined using Rietveld refinement of the X-ray diffraction patterns are very close to those expected without substitution (see inset of figure 2). Therefore, either Ge does not substitute in In_2O_3 either the solubility limit is very small. This result is confirmed by SEM observations coupled to EDX analyses (Figure 3), which show the presence of $In_2Ge_2O_7$ inclusions as a secondary phase (dark grains) and which does not allow to evidence the presence of germanium in the main phase within the detection limit of EDX spectroscopy. In the following discussions, $In_{2-x}Ge_xO_3$ will thus refer to a composite compound constituted of an In_2.

$_\delta$Ge$_\delta$O$_3$ phase (δ being too small to be detected) with In$_2$Ge$_2$O$_7$ inclusions. As can be seen in the SEM image, part of the In$_2$Ge$_2$O$_7$ inclusions form crown-like agglomerates inducing some porosity. Therefore, the presence of In$_2$Ge$_2$O$_7$ inclusions has a negative effect on the samples densification: the geometrical density decreases from 91% with x=0 to 82% with x=0.2 and 74% with x=0.3 (see figure 4). This effect on the densification probably originates from both the formation of the crown-like agglomerates, which might be linked to the ball-milling process which does not lead to a good mixing of the precursors, and from a decrease of the grain boundaries mobility that prevents a good filling of the initial porosity of the green pellets.

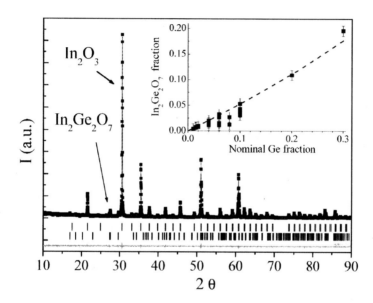

Figure 2: XRD pattern and Rietveld refinement for a sample with nominal composition In$_{1.9}$Ge$_{0.1}$O$_3$. Inset: In$_2$Ge$_2$O$_7$ fraction determined from rietveld refinement as a function of the nominal germanium fraction. The dashed curve corresponds to the theoretical In$_2$Ge$_2$O$_7$ fraction assuming that no substitution occurs.

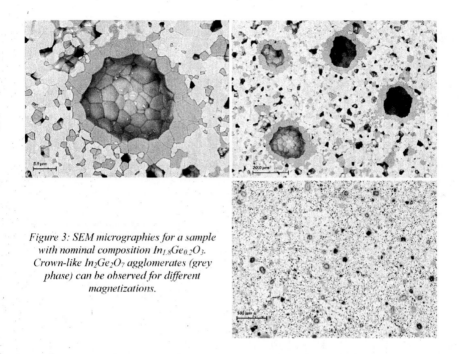

Figure 3: SEM micrographies for a sample with nominal composition In$_{1.8}$Ge$_{0.2}$O$_3$. Crown-like In$_2$Ge$_2$O$_7$ agglomerates (grey phase) can be observed for different magnetizations.

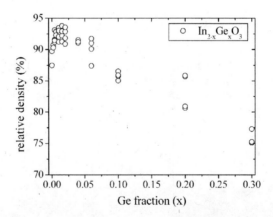

Figure 4: Influence of the germanium fraction on the geometrical density of the samples.

Figure 5 shows the temperature dependence of the electrical resistivity in the series with nominal composition In$_{2-x}$Ge$_x$O$_3$. All samples exhibit a metallic behavior with the electrical resistivity increasing with temperature except the composite In$_{1.8}$Ge$_{0.2}$O$_3$. The influence of the germanium fraction on the electrical resistivity at 1000K is shown in figure 6. Starting from undoped In$_2$O$_3$ with $\rho \sim 25$ mΩ.cm at 1000K, a very small addition of germanium, of the order 0.1 atom% (x=0.002), leads to a strong decrease of the electrical resistivity, by a factor of 5. Further additions lead to a decrease of the resistivity with a minimum reached at x=0.015 with $\rho \sim 1.5$ mΩ.cm at 1000K. It is noteworthy that this latter value is very close to the best electrical conductivities observed in bulk In$_2$O$_3$ based compounds (see for example [11] and ref. therein). Further additions lead to a slow increase of the resistivity. Therefore, we can propose that the solubility limit for Ge in In$_2$O$_3$ may be close to x=0.01-0.015 and that there is most probably a doping effect due to the substitution of Ge(+IV) for In(+III) which would lead to an increase of the electrons concentration in the system.

The behavior of the thermopower is similar (see Fig. 7). All samples are n-type with the absolute value of the thermopower increasing with temperature, which is consistent with the metal-like behavior of the electrical resistivity. The influence of the germanium fraction on the thermopower at 1000K is shown in figure 6. Starting from undoped In$_2$O$_3$ with S $\sim$ -225 μV.K^{-1} at 1000K, a very small addition of germanium leads to a decrease of the thermopower, which reaches a constant value S $\sim$ -110 μV.K^{-1} when the germanium fraction exceeds x=0.015.

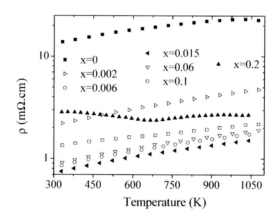

Figure 5: Temperature dependence of the electrical resistivity in the series with nominal composition In$_{2-x}$Ge$_x$O$_3$.

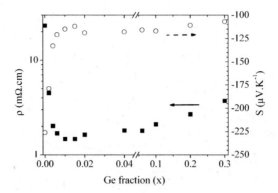

Figure 6: Influence of the germanium fraction on the electrical resistivity (filled symbols) and the thermopower (open symbols) at 1000K. Be careful to the break in the x-coordinate.

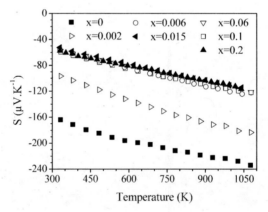

Figure 7: Temperature dependence of the thermopower in the series with nominal composition In$_2$-$_x$Ge$_x$O$_3$.

Thus, whereas the XRD and SEM studies show that the solubility of germanium in In$_2$O$_3$ is very small or even null, the decrease of the electrical resistivity and of the absolute value of the thermopower suggests strongly that small Ge fractions substitute for In leading to an increase of the electrons concentration. Then the increase of the electrical resistivity after the minimum may be connected to the increasing fraction of highly resistive In$_2$Ge$_2$O$_7$ phase or to the decreasing density. In$_2$Ge$_2$O$_7$ being highly resistive, the presence of inclusions hardly influences the thermopower values that remain constant above the solubility limit. Indeed, it has been shown that small inclusions of a highly resistive phase into a conductive matrix do not influence the magnititude of the thermopower which is mainly linked to the matrix [12].

Figure 8 shows the temperature dependence of the power factor S$^2\sigma$ calculated using the measured electrical resistivity and thermopower. The power factor increases with increasing temperature in the whole temperature range for all samples. It appears that Ge doping is very efficient to enhance the thermoelectric properties at high temperature. Starting from undoped In$_2$O$_3$ with PF ~ 2.2.10^{-4} W.m^{-1}.K^{-1} above 1000K, a very small addition of germanium, of the order x=0.002-0.006, leads to a strong increase of the power factor, by a factor of 4 approximately. Further additions of germanium, higher than x=0.02, lead to a slow decrease of the power factor, which is linked to the increase of the electrical resistivity. However, this decrease could probably be moderated by achieving a better densification of the samples.

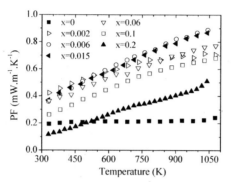

Figure 8: Temperature dependence of the power factor S$^2\sigma$ in the series with nominal composition In$_{2-x}$Ge$_x$O$_3$.

Figure 8 shows the influence of the germanium fraction on the thermal conductivity. Whatever the germanium content, the thermal conductivity decreases with increasing temperature as it is exemplified in the inset of figure 9 in the case of Ge = 0.06. At room temperature, the influence of the germanium additions is very important, with λ decreasing from about 10 W.m^{-1}.K^{-1} in undoped In$_2$O$_3$ to 3.3 W.m^{-1}.K^{-1} in the composite In$_{1.8}$Ge$_{0.2}$O$_3$. This trend is the same at higher temperature although the magnitude of the thermal conductivity decrease is smaller. It is noteworthy that λ is smaller in highly substituted samples than in the undoped one although the electrical resistivity is smaller too. The total thermal conductivity can be divided into two parts, one originating from the charge carriers and linked to the electrical conductivity by the Wiedmann-Franz law, and the other one originating from the heat carrying phonons. Figure 10 shows the calculated electronic and lattice part of the thermal conductivity at 1273K as a function of the germanium fraction. Starting from undoped In$_2$O$_3$ where the main part of the thermal conductivity is due to the phonons, small additions of Ge lead to a strong increase of the electronic part, which is linked to the strong decrease of the electrical resistivity due to germanium doping. Further additions of germanium above the solubility limit lead to a slow decrease of the electronic part due to the slow increase of the electrical resistivity. The lattice part of the thermal conductivity monotically decreases from about 3 W.m^{-1}.K^{-1} in undoped In$_2$O$_3$ to about 0.6 W.m^{-1}.K^{-1} in the composite In$_{1.8}$Ge$_{0.2}$O$_3$.

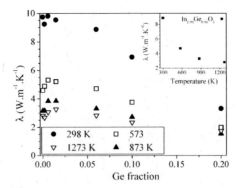

Figure 9: Influence of the germanium fraction on the thermal conductivity in the series with nominal composition In$_{2-x}$Ge$_x$O$_3$. Inset: Temperature dependence of the thermal conductivity of the composite In$_{1.94}$Ge$_{0.06}$O$_3$.

The efficiency of a thermoelectric material is linked to the dimensionless figure of merit ZT=S$^2\sigma$T/λ. Assuming that the Wiedmann-Franz law is valid in this system, the figure of merit is directly linked to the electronic and lattice part of the thermal conductivity according to the relation:

$$ZT = \frac{S^2}{L_0} \cdot \frac{1}{1+\dfrac{\lambda_L}{\lambda_{elec}}}$$

Therefore, ZT is maximized when the ratio of the electronic part of the thermal conductivity to its lattice part is maximized. The inset of the figure 9 shows the influence of the germanium fraction on the ratio λ_{elec}/λ_L. In undoped In$_2$O$_3$, the thermal conductivity is dominated by its lattice part. The first increase of λ_{elec}/λ_L for low germanium fractions is mainly linked to the decrease of the electrical resistivity. The ratio exhibits a local maximum for a germanium fraction that corresponds to the solubility limit. However, it is noteworthy that λ_{elec}/λ_L increases when x increases for high germanium fractions, although the electrical conductivity (and therefore λ_{elec}) decreases. It means that the germanium additions are more effective in decreasing the phonons mean free path that the charge carriers mobility and are therefore suitable for improving the thermoelectric efficiency. Nevertheless, as the presence of In$_2$Ge$_2$O$_7$ and the increase of porosity due to the crown-like agglomerates formation are strongly linked, it is not possible to determine whether this effect on the thermal conductivity mainly originates from the inclusions or from the pores. One way of separating these two possible contributions would be to study fully densified samples with well dispersed inclusions by using a different synthesis path that could induce a better mixing of the In$_2$O$_3$ and GeO$_2$ precursors.

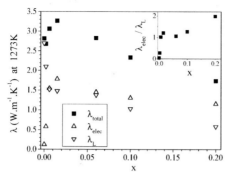

Figure 10: Calculated electronic and lattice part of the thermal conductivity at 1273K as a function of the germanium fraction. The inset shows the ratio of the electronic part to the lattice part of the total thermal conductivity at 1273K as a function of the germanium fraction.

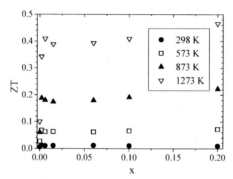

Figure 11: Calculated dimensionless figure of merit ZT as a function of the germanium fraction in the series with nominal composition $In_{2-x}Ge_xO_3$.

Figure 11 shows the calculated dimensionless figure of merit ZT as a function of the germanium fraction in the series with nominal composition $In_{2-x}Ge_xO_3$. ZT increases with the increasing temperature for all compositions. High temperature ZT outreaches 0.1 in In_2O_3, which is higher than in undoped ZnO [4] and almost one third of the values of the best n-type thermoelectric oxides. The figure of merit is strongly enhanced by Ge additions, with ZT=0.46 at 1273K in the composite $In_{1.8}Ge_{0.2}O_3$. This value is higher than that of the best previously reported bulk n-type oxides.

Based on our results, even if unquestionable responses have not been given to explain the influence of Ge additions on the transport properties, one can speculate that a small percentage of In^{3+} has been substituted by Ge^{4+} in the In_2O_3 phase, increasing the electrons concentration. With the view to modify the carriers concentration in the material, sintering treatments at 1350°C and 1400°C have been performed and thermoelectric properties compared to the samples sintered at 1300°C.

Figure 12a and 12b show the influence of the germanium fraction on, respectively, the electrical resistivity and the thermopower at 1000K according to different sintering temperatures. The decrease of the thermopower follows the decrease of the electrical resistivity which tends to indicate that the carriers concentration is significantly increased by increasing the sintering temperature. This result confirms, as it is well reported in the literature [13, 14], that the intrinsic donor concentration in In$_2$O$_3$ (i.e. oxygen vacancies) can vary strongly and the electron concentration and mobility can differ from a sample to another, depending on the preparation method and purity. In the present case, on can believe that the solubility of Ge in In$_2$O$_3$ has been modified and/or that the oxygen vacancies concentration was increased. Neutron diffraction experiments are planned in the next months to check, if possible, the oxygen content, and SIMS measurements (Secondary Ion Mass Spectroscopy) are underway to quantify the Ge content in In$_2$O$_3$ grains.

Based on electrical resistivity and thermopower measurements, the power factor has been calculated (figure 13). It appears that this latter is unchanged since the decrease of the thermopower is balanced by an increase of the electrical conductivity. It remains now to measure the thermal conductivity to estimate how its lattice and electron parts are changed by modifying the process conditions.

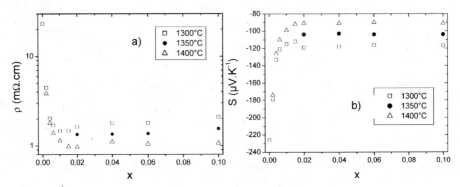

Figure 12: Influence of the germanium fraction (x) on (a) the electrical resistivity and (b) the thermopower at 1000K according to the sintering temperature.

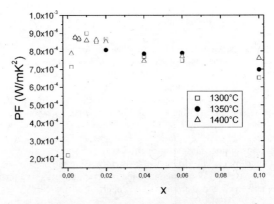

Figure 13: Influence of the germanium fraction (x) on the power factor S$^2\sigma$ according to the sintering temperature.

CONCLUSIONS

We have shown that Ge-doped In$_2$O$_3$ is a very promising thermoelectric oxide. Although the germanium solubility limit is very small, the electrical resistivity is strongly decreased by the doping, leading to high power factors, close to 1 mW.m^{-1}.K^{-1} around 1100K. As the thermal conductivity of undoped In$_2$O$_3$ is small at high temperature, the dimensionless figure of merit ZT is quite high, reaching 0.1. Moreover, further Ge additions appear to be very effective for decreasing the lattice thermal conductivity, with λ being lower than 2 W.m^{-1}.K^{-1} in the composite In$_{1.8}$Ge$_{0.2}$O$_3$ at high temperature. This low thermal conductivity, coupled with high power factors, leads to very promising thermoelectric figures of merit with ZT > 0.45 in the composite In$_{1.8}$Ge$_{0.2}$O$_3$ at 1273 K. This high values make these materials suitable for high temperature energy conversion in thermoelectric modules.

ACKNOWLEDGMENTS

EG gratefully acknowledges the French Ministère de la Recherche et de la Technologie and the Délégation Régionale à la Recherche et à la Technologie – région Basse Normandie – for financial support.
This work was supported by Corning SAS.

REFERENCES

[1] Recent trends in thermoelectric materials research I-III, in *Semiconductors and Semimetals* Vol. 69-71, edited by T.M. Tritt (Academic, San Diego, USA, 2000)
[2] G.D. Mahan, B.C. Sales and J. Sharp, Physics Today **50**, 42 (1997)
[3] B.T. Cong, T. Tsuji, P.X. Thao, P.Q. Thanh and Y. Yamamura, Physica B **352**, 18 (2004)
[4] Oxide thermoelectric materials: An overview with some historical and strategic perspectives, in *Oxide Thermoelectrics*, edited by K. Koumoto (Research Signpost, Trivandrum, India, 2002)
[5] H. Kaga, R. Asahi and T. Tani, Jpn J. Appl. Phys. **43**, 7133 (2004)
[6] K.L. Chopra, S. Major and D.K. Pandya, Thin Solid Films **102**, 1 (1983)
[7] J. Rodriguez-Carjaval, Physica B **192**, 55 (1993)
[8] M. Marezio, Acta Cryst. **20**, 723 (1966)
[9] R.D. Shannon and C.T. Prewitt, J. Solid State Chem. **2**, 199 (1970)
[10] R.D. Shannon, Acta Cryst. A **32**, 751 (1976)
[11] L. Bizo, J. Choinest, R. Retoux and B. Raveau, Solid state comm. **136**, 163 (2005)
[12] D.J. Bergman, O. Levy, J. Appl. Phys. **70**, 6821 (1991)
[13] J. H. W. De Wit, J. Solid State Chem. 13, 192 (1975)
[14] J. H. W. De Wit, J. Solid State Chem. 25, 101 (1978)

TRANSITION METAL OXIDES FOR THERMOELECTRIC GENERATION

J.P. Doumerc, M. Blangero, M. Pollet, D. Carlier, J. Darriet, C. Delmas, R. Decourt
ICMCB, CNRS, Université Bordeaux 1
87, Av. Dr. A. Schweitzer, 33608 Pessac Cedex
Pessac, France

ABSTRACT

In order to face the forthcoming energy crisis, search for alternative power sources is extensively growing. Of interest is the recovery of waste heat from either plants or thermal engines. Thermoelectric devices are reliable candidates as they contain no moving parts and require no maintenance. However, conventional materials such as bismuth telluride are not stable enough at high temperature. Although the ZT figure of merit still remains less than unity, intermetallic compounds seem promising, but transition metal oxides are expected to be more stable in air at high temperature.

In a first part chemical bonding in transition metal oxides and its relationship with crystal and electronic structures is discussed.

The main part of the paper is devoted to alkali cobaltites A_xCoO_2 and the following questions are considered.

How are chemical properties, crystal structure, magnetic and electrical behavior changing with the nature of the A alkali metal?

Up to which extent ordering of alkali atoms influences the electronic properties?

Comparing lithium, sodium and potassium cobaltites raises another question: why metallic phases can either show an enhanced conventional Pauli type behavior or a Curie-Weiss behavior denoting spin polarization? How does the strength of electronic correlations depend on the nature of the alkali metal?

Finally the case of delafossite-type oxides is briefly reviewed.

INTRODUCTION

According to a conventional approach, achieving large power factors requires a low carrier density and a large electronic mobility. This is the case for most of the thermoelectric materials commercially available nowadays, such as bismuth telluride.

However, Terasaki et al.[1] pointed out that some layered cobaltites of general formula Na_xCoO_2 (x=0.5) investigated more than twenty years ago by a Bordeaux group[2, 3, 4] had a power factor of the same order of magnitude as Bi_2Te_3 whereas the carrier density is much larger than in heavily doped semiconductors. In transition metal oxides, the conduction band that generally originates from d-orbitals is relatively narrow. Nevertheless highly conducting oxides such as tungsten bronzes Na_xWO_3, RuO_2 or even 2D $PdCrO_2$ with electrical conductivities nearly as large as those of elemental good conductors such as copper or silver do exist.

As oxides are generally stable in air at high temperatures, which in most cases are conditions in which waste heat can be converted into electricity using thermoelectric generators, in the present paper we review the properties of various families of conducting oxides in the scope of their relevance to thermoelectric generation.

The specificity of the Co^{3+} ions ($3d^6$) that can exhibit three different electronic configurations (S=0, S=1 and S=2) in oxides, depending on the interplay of exchange and crystal field energies, is also considered.

BASIC CONSIDERATIONS

A large figure of merit $ZT=(\alpha^2\sigma/\kappa)T$ where α is the thermoelectric power, σ the electrical conductivity, κ the thermal conductivity and T the absolute temperature requires a small value of the thermal conductivity. In the above expression of ZT, $\kappa=\kappa_e+\kappa_{ph}$ where e and ph subscripts indicate the electron and lattice contributions, respectively. Decreasing κ_{ph} brings back to usual strategies consisting in substitutions for heavier elements, creating disorder and introducing rattling atoms, which can be more or less easily applied to oxides. On the other hand, in conventional metals σ and κ_e cannot be varied independently as they are linked by the Wiedemann-Franz law ($\kappa_e/\sigma=(\pi^2k_B^2/3e^2)T$), which leads to a required minimum value of the Seebeck coefficient for achieving a given ZT value : $\alpha>156(ZT)^{1/2}\mu VK^{-1}$.

Large values of α can be theoretically and practically achieved in semiconducting oxides the electrical conductivity of which is thermally activated. However, at this point it is important to distinguish between the two main families of semiconducting oxides: on one hand, there are the so-called band-type semiconductors in which carriers are excited into states where they occupy Bloch states and can move with a given mean free path. On the other hand there are mixed valent oxides in which the metal-metal interaction (direct or through oxygen) is so weak that in the ground state carriers are trapped and can only hop from site to site thanks to a thermal activation.

In both cases the electrical conductivity increases exponentially with temperature contributing to a larger ZT at higher temperature, which is appropriate for waste heat recovery.

In the first case:
$$\sigma=e\mu n_0\exp(-\Delta E_n/k_BT)$$
where μ is the carrier mobility near the band edge, n_0 is the effective density of state that varies as $T^{3/2}$ (assuming a parabolic $E(\mathbf{k})$ dispersion) and other quantities have their usual meaning.

In the second case the carrier density is assumed temperature independent whereas the mobility is linked to the diffusion coefficient, D_0, by the Einstein relation ($\mu=eD_0/k_BT$), which results in:
$$\sigma=e^2n_0(D_0/k_BT)\exp(-\Delta E_\mu/k_BT).$$
Very often the T dependence of the pre-exponential term is hidden by the stronger dependence of the exponential term.

While in both cases the temperature dependence of σ lets expect a strong improvement of ZT at high temperature, the behavior of the thermoelectric power is extremely different. In the first case the absolute value of α decreases with temperature $|\alpha| \propto (k_B/e)(\Delta E_n/k_BT)$, which thwarts the effect of increasing σ all the more so α is squared in ZT.

On the contrary, for hopping transport, $|\alpha|$ increases with temperature and tends to a constant value at high temperature given by the well-known Heikes formula:
$$|\alpha|=|(-k/e)\ln\{[(1-c)/c]*\beta\}|.$$
Heikes formula can simply be derived considering that, as $T\rightarrow\infty$, α is given by the entropy per carrier. The first factor in the logarithm where $c=n/N$ corresponds to the distribution entropy of n electrons over N sites. The second factor, β, includes the spin and orbital degeneracies [5,6].

The case of metallic oxides is considered in another section.

THE METAL-OXYGEN CHEMICAL BONDING

Crystal structure

The metal-oxygen bond is iono-covalent. The ionic character accounts for the structural features in the sense that most of the oxide crystal structures can be described as a close packing of oxygen atoms the octahedral and/or tetrahedral vacancies of which are occupied by metallic atoms. For instance, in the rocksalt-type structure of CoO – where the packing sequence of oxygen layers is -

ABCABC- –, all the octahedral vacancies are occupied by Co atoms. In the so-called α-NaFeO₂ structure of $LiCoO_2$, the octahedral vacancies are occupied by both Co and Li ions in an ordered way giving rise to a packing of CoO_2 layers linked by alkali layers (figure 1). For all the intercalation oxides used in Li batteries a basic feature of the structure is the CoO_2 layers. These layers are also the basic units of thermoelectric cobaltites.

When the size of the alkali element or the occupancy rate varies, the type of oxygen layer packing may change and the alkali site is no longer octahedral, but prismatic. This is described in more detail below.

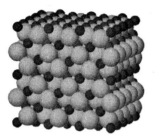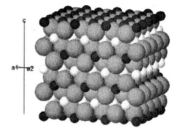

Figure 1. Close packing of oxygen atoms in the rocksalt structure of CoO (Co atoms are represented by the smaller and darker spheres, and the vertical direction corresponds to the [111] direction of the cubic (cfc) unit cell), on the left hand side, and in $LiCoO_2$ (Li atoms are shown as white small spheres), on the right hand side.

Perovskites can also be described as a packing of AO_3 layers in which one oxygen atom is replaced by a metal A atom. Depending whether the packing is …*ABCABC*… or …*ABAB*… one gets the cubic 3R or the hexagonal 2H polytype. Replacing three other oxygen atoms by an A' metal gives another type of layers. Combining both types of layers allows predicting a lot of new structures[7].

Electronic structure

The covalent character of the metal-oxygen bonding accounts well for the electronic structure that can be derived from the metal-oxygen-metal interactions. First, molecular orbitals can be formed between the metallic atom and the surrounding oxygen atoms; then they broaden into bands through the translation symmetry. The classical example of cubic tungsten bronzes is shown in figure 2.

Sometimes direct metal-metal interactions play also a part but they develop in low dimensional sub-systems such as pairs, trimers, chains or layers.

The same principles apply to most of the mixed valent systems of bronze type including the thermoelectric layered cobaltites.

For x>0.3 sodium tungsten bronzes are metallic, showing a Pauli type paramagnetism and a small thermoelectric power as conventional metals making them inappropriate for thermoelectric generation. Actually, for an electronic configuration between d^0 and d^1 electronic correlations generally are weak and no enhancement of the thermelectric power, as well as of the specific heat and magnetic susceptibility can be expected.

The above tungsten bronze example illustrates how a partly occupied d-shell may lead to a metallic state. Actually this is a necessary condition but not a sufficient condition for observing a metallic state. A non metallic state can result from various effects such as:

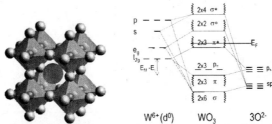

$W^{6+}(d^0)$ WO_3 $3O^{2-}$

Figure 2: In Na_xWO_3 tungsten bronzes, corner sharing WO_6 octahedra form a 3D network like in ABO_3 cubic perovskite. W-O-W interactions lead to the band structure shown on the right hand side. The A site is partially occupied by sodium atoms (red/large circle). Ionization of sodium provides x electron/formula unit to the π^* conduction band [8].

    - band splitting, opening a band gap at the Fermi level as in the Peierls-type dimerization. Such instabilities are announced by a nesting vector connecting two different parts of the Fermi surface,
    - electron-electron interactions accounting for Mott-Hubbard insulators as the U intra-atomic Coulomb repulsion overcomes the bandwidth,
    - electron-phonon interactions,
    - disorder, including a random distribution of inserted or substituted atoms in a crystalline network and leading to Anderson localization.

The Co(III) ion
    A unique feature of the Co^{3+} ions ($3d^6$) is the existence of three different electronic configurations (S=0, S=1 and S=2) in oxides, depending on the interplay of exchange and crystal field energies. The most investigated case is that of Co-atoms located in octahedra that are more or less elongated (D_{4h} symmetry) (figure 3).
    Cobaltites with this site symmetry such as those with a perovskite structure have never a very large thermopower and a large conductivity simultaneously.

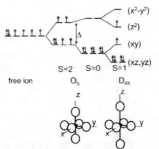

Figure 3: Co^{3+} electronic configurations in O_h and D_{4h} symmetries.

    In lamellar cobaltites such as those discussed below where CoO_6 edge sharing octahedra form a triangular network that implies a D_{3d} distortion, it is more convenient to choose the 3-fold symmetry axis as the quantization axis. With respect to the usual real d orbitals, a set of five d orbitals is the following [9] (coefficients are given for the undistorted O_h symmetry):

$$\sqrt{1/3}\,d_{x^2-y^2} + \sqrt{2/3}\,d_{xz} \qquad (e_{g1})$$

$$\sqrt{1/3}\,d_{xy} - \sqrt{2/3}\,d_{yz} \qquad (e_{g2})$$

$$d_{z^2-r^2} \qquad\qquad\qquad (a_1)$$

$$\sqrt{2/3}\,d_{x^2-y^2} - \sqrt{1/3}\,d_{xz} \qquad (e'_1)$$

$$\sqrt{2/3}\,d_{xy} + \sqrt{1/3}\,d_{yz} \qquad (e'_2)$$

A drawing of e'_1 and e'_2 orbitals is shown in figure 4.

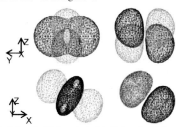

Figure 4. Drawing of the e'_1 (left hand side) and e'_2 (right hand side) orbitals for the two perpendicular axis orientations shown on the very left [10].

For a regular octahedron, the two degenerate e_g orbitals are less stable than the three degenerate a_1, e'_1 and e'_2 orbitals. As the octahedron is compressed along the three-fold axis, the two e_g and the two e' orbitals remain degerate, but the a_1 orbital is shifted upwards as illustrated in figure 5 [11]. This result cannot be obtained using a simple point charge model as easily as in the case of the D_{4h} symmetry. The reason is that symmetry now allows hybridization of e' and e_g orbitals [12].

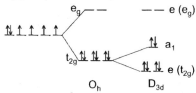

Figure 5: Orbital splitting for a low spin d^6 ion located in an octahedron compressed along a 3-fold axis.

2D ALKALI COBALTITES A_xCoO_2 (x=0.6; A= Li, Na, K)

Crystal structures: main features and nomenclature

As reminded above, the possible structures can be classified according to the packing of oxygen layers that can be labeled by the letters A, B and C.

When the packing is of the ...AB-CA-BC... type, the so-called α-NaFeO$_2$ structure is obtained. The alkali site is octahedral and the structural type is denoted O3 according to the nomenclature introduced by Delmas et al.[13] (the figure refers to the number of MO$_2$ sheets within the corresponding hexagonal cell). A monoclinic distortion of the unit cell is noted by a prime superscript (e. g. O'3).

When the packing is of the ...AB-BA-AB... type or of the ...AB-BA-AC-CA... type the alkali site is prismatic and the structure type is denoted P2 or P3, respectively,. These two possibilities of stacking also exist for the delafossite family (A=Cu, Ag, Pd, Pt), giving rise to the so-called 2H and 3R polytypes, respectively. The main difference with respect to alkali metallates is the linear twofold coordination of monovalent copper or silver and that non-stoichiometric occupancy of the A site has never been reported.

The prismatic sites are not exactly identical in the P2 and P3 structures. In P2 an alkali atom can have either two, or six M nearest neighbors depending whether the prism is sharing faces (in which case the site will be labeled A_f) or edges (site labeled A_c) with the closest MO_6 octahedra (figure 6). In A_f site an alkali ion undergoes a stronger electrostatic repulsion from the neighboring cobalt ions than for A_e sites. Therefore a general trend would be occupying mainly the A_e sites. However A^+-A^+ repulsions must also be taken into account. Then, in order to maximize the A-A average distances both type of sites can be occupied and the alkali ion can be shifted away from the ideal position at the prism center.

In P3 structure type, a single possibility exists as the prism is sharing a face with an octaheron of one adjacent oxygen double layer and edges with three octahedra of the other adjacent layer. We shall use a f subscript for labeling oxygen atoms (O_f) belonging to a face common to alkali and Co polyhedra and a e subscript for oxygen atoms (O_c) belonging to a common edge.

Figure 6. Various sites occupied by alkali ions in lamellar cobaltites. From left to right: the so-called A_e and A_f sites encountered in the P2 structure type, the alkali site in the P3 structure type, and the twofold linear site of monovalent copper (or silver) in the delafossite type structure.

Comparing the crystal structures

For $Li_{0.6}CoO_2$, the symmetry is rhombohedral (lattice parameters of the hexagonal cell: a= 2.810 Å and c= 14,287 Å) and the space group is R-3m. The stacking of the layers is of O3 type. The cobalt-oxygen distance amounts to 1.896 Å.

$Na_{0.6}CoO_2$ exhibits a monoclinic distortion (space group: C2/m) that remains weak. In fact, β angle is 106.07° against 106.55° for an undistorted hexagonal cell and the ratio $a/\sqrt{3}b$ is close to 1.001. This distortion likely results from an ordering of sodium ions and vacancies[2]. However, the nature of this ordering has still not been determined.

Sodium ions occupy a prismatic site of A_{ef} type that is sharing three edges and one face with octaedra of the two sandwiching CoO_2 layers. Sodium ion is strongly shifted from the prism center with Na-O distances ranging from 2.225 to 2.664 Å. Na-O distances also differ depending on whether the oxygen atoms are located on a face (O_f type) or an edge (O_c type) common to a CoO_6 octahedron and a NaO_6 prism. The average Na-O_f distance (2.448 Å) is larger than the average Na-O_c distance (2.383 Å). A shift along the z axis can be attributed to a Na-Co coulomb repulsion stronger in the case of a common face than in that of common edges.

The coulomb repulsion between nearest neighboring Na^+ ions determines the arrangement of sodium ions in the (a,b) plane : shifts along x and y axis permit a lowering of the coulomb repulsion thanks to an increase of the average distance separating Na^+ ions.

The unit cell of $K_{0.6}CoO_2$ is hexagonal (space group $P6_3/mmc$, a = 2.843 Å et c = 12.350 Å). The structure is of P2 type and Co-O distances are equal to 1.902 Å. Potassium ions are distributed in two prismatic sites, K_f et K_e, with a ratio 1:2. As in the case of sodium based P2 phases with x < 0.74[14], K_e atoms occupy 6h positions strongly shifted from the prism center in order to minimize the electrostatic repulsion.

Increasing the cation size, through substituting potassium for lithium, results in an increase of the CoO_2 slab spacing and alkali-oxygen distances are in good agreement with the sum of ionic radii. The 2D character of the structure is then expected to increase going from potassium to lithium cobaltites.

A second effect of this alkali substitution is an increase of the magnitude of the D_{3d} distortion of the CoO_6 octahedra that are compressed along their 3-fold axis (that is taken parallel to the z axis). This distortion can be measured considering the value of the θ angle between the 3-fold (z) axis and the direction of the Co-O bonds. Keeping a pure hexagonal symmetry, θ angle depends on the Co-Co distance (equal to the a lattice constant) and on the Co-O distance:

$$\theta = \text{Arcsin}\left(\frac{1}{\sqrt{3}} \frac{d_{Co-Co}}{d_{Co-O}} \right).$$

For a regular octahedron (O_h symmetry), $d_{Co-O} = \sqrt{2} \times d_{Co-Co}$ a then $\theta = 54.74°$.

When the alkali rate remains close to 0.6 the Co-O distance does not vary much and remains close to 1.90 Å, in agreement with the sum of ionic radii, whereas the Co-Co distance significantly increases passing from lithium to potassium, mainly due to steric effects, and, as a result θ angle increases. The simultaneous increase of both the θ-value and the slab spacing involves only a very weak change in the O-O distance corresponding to an edge shared by two CoO_6 octahedra (2.547 Å for Li compound and 2.528 Å for the K compound). Keeping constant this contact distance explains the large change of the a lattice constant of the hexagonal cell of delafossite type oxides $A'MO_2$ (A'=Cu, Ag) with the size of the trivalent M element[15, 16]. When x decreases, θ increases, which accounts for the fact that d_{Co-O} decreases more rapidly than d_{Co-Co} as the Co(+IV) rate increases.

Alkali ordering

Whether the alkali atoms are ordered or randomly distributed within their planar sublattice is a permanent issue widely debated. Although a few experimental evidences of long range ordering, either commensurate or incommensurate, have been given using neutron or electron diffraction, most of the reported works are devoted to theoretical considerations using various method of calculation[17, 18, 19]. To our knowledge, the first direct evidence through usual X-ray diffraction of potassium ordering was given for $K_4Co_7O_{14}$[20].

The ordered $K_4Co_7O_{14}$ phase has been prepared using the same experimental conditions as random $K_{0.6}CoO_2$. Actually, investigating the effect of thermodynamic as well as kinetic parameters did not allow us to determine precisely the conditions appropriate for a reproducible synthesis of the one or the other. Their compositions are very close (x=0.57 vs. 0.61) and the evolution of K_2O is making complex the control of the alkali rate in the final product.

In the X-ray diffraction pattern, main peaks are indexed assuming a P2 type structure (space group $P6_3/mmc$, $a_0 = 2.841(1)$ Å and $c_0 = 12.370$). A Rietveld refinement assuming a random distribution of potassium atoms in the two K_f (2b) et K_e (6h) prismatic sites leads to a potassium rate x ~ 0.59, which is close to the value measured by ICP-AES (x ~ 0.62).

All the peaks, including those not taken into account with the basic P2 hexagonal cell, can be indexed in the space group $P6_3/m$ using an hexagonal supercell with $a_S = a_0.\sqrt{7}$ ($a_S = 7.517$ Å and $c_S = 12.371$ Å).

The ordered distribution of potassium atoms is shown in figure 7. Both K_c and K_f positions are fully occupied. The first one corresponds to a prism that is only sharing edges with neighboring CoO_6 octahedra whereas the second one is sharing its bases with faces of the two CoO_6 octahedra.

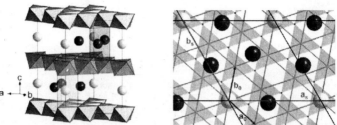

Figure 7. Ordered distribution of potassium atoms in $K_4Co_7O_{14}$. Light spheres represent the atoms occupying K_f sites and dark spheres the atoms in K_c site (see text).

Cobalt atoms are occupying two different positions. For the first one, the six Co1-O distances are identical (1.899 A). The $Co1O_6$ octahedron is sharing, along the c-axis, two faces with K_fO_6 prisms. It shows a D_{3d} symmetry and a large value of the θ angle (60.68°) that denotes a strong compression along the z-axis of the $Co1O_6$ octahedra. The surrounding of the Co2 site is more complex with three different Co2-O distances ranging from 1.876 to 1.919 A. The average value of the θ angle (59.37°) is closer to that found for the disordered phase, $K_{0.6}CoO_2$.

Magnetic properties

The temperature dependence of the magnetic properties strongly differs from one compound to the other.as shown in figure 8.

Excepted for $K_{0.6}CoO_2$, a paramagnetic behavior (in the sense of a magnetization proportional to the field) is observed, at least above 2 K. Two different types of temperature dependence of the magnetic susceptibility were found. One is of Curie-Weiss (CW) type for the lithium and sodium cobaltites whereas the other is of Pauli type for the two potassium compounds.

For the Li and Na cobaltites the magnetic data can be fitted with a CW law including a temperature independent term: $\chi = C/(T-\theta_p)+\chi_0$. The Curie constant (C) and Weiss temperature (θ_p) are given in table I.

Table I : Curie-Weiss parameters and TIP contribution for $Li_{0.6}CoO_2$ and $Na_{0.6}CoO_2$.

Phase	χ_0 (emu/mol)	C (emu.K/mol)	θ_p (K)
$Li_{0.6}CoO_2$	334×10^{-6}	0.017	-3.78
$Na_{0.6}CoO_2$	276×10^{-6}	0.145	-142

For the sodium compound the Curie constant is in fairly good agreement with the spin only value expected for a rate of 0.4 spins S=1/2 (0.152). Most of the authors agree to consider that Co^{3+} is in a low spin (LS) state in this oxide family. Then, the C-value suggests that all the Co^{4+} holes either localized or itinerant are contributing to C. However the large value of the TIP is difficult to explain even after subtracting the second order contribution of LS Co^{3+} (see below). The large negative value of θ_p generally ascribed to strong predominant antiferromagnetic interactions does not well agree with a simple Fermi liquid picture of fully polarized carriers.

The case of lithium is also puzzling. The CW contribution is much weaker with respect to TIP. This suggests a very small concentration of localized moments or spin polarized carriers. Another possibility is the presence of traces of a magnetic impurity. The TIP would then be due to classical paired charge carriers.

Both potassium phases have a magnetic susceptibility almost temperature independent (figure. 8) that can involve the following contributions :

$$\chi = \chi_{dia} + \chi_{VV} + \chi_{Pauli}$$

where χ_{dia} is the diamagnetic contribution of the ion close shells, χ_{VV} is th second order orbital contribution from Co^{3+}(LS) ions and χ_{Pauli} is the classical Pauli type contribution from itinerant spins. Usually χ_{dia} is calculated from tabulated theoretical values[21]. The χ_{VV} contribution can be evaluated to about 150×10^{-6} (emuCGS) per Co^{3+} ion[22, 23]. The resulting Pauli contribution would then evaluates to about 400×10^{-6} cm^3/mol (emuCGS) for both $K_4Co_7O_{14}$ and $K_{0.6}CoO_2$. Such a large enhancement of Pauli paramagnetic susceptibility is generally ascribed to electron-electron correlations and, in the scope of a Fermi liquid behavior, can be modeled by an effective mass increase.

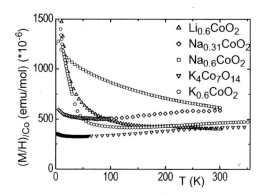

Figure 8. Temperature dependence of the magnetic susceptibility for $A_{0.6}CoO_2$ (A=Li, Na, K). Data for $Na_{0.31}CoO_2$ are also shown.

Specific heat capacity

Sommerfeld coefficient γ, has been determined modeling lattice vibrations with a Debye model, between 1.8 and 10 K : $C_p = \gamma T + \beta T^3$. The coefficient β depends on atom number (n_D) and on Debye temperature (Θ_D): $\beta = (12\pi^4/5)(n_D/\Theta_D^3)$. Variation of C_p/T as a function of T^2 is plotted in figure 9. Values of γ and Θ_D are listed in table II. For calculating Θ_D, n_D has been taken equal to the total atom number in a A_xCoO_2 molecular unit, i. e. $n_D = 3.6$.

For the lithium phase, C_p/T steeply increases below ca. 15K. Such a behavior could result from a Schottky anomaly the origin of which is still unknown [24] and data for T<2K would be useful. In this case Cp/T can be modeled by: $\dfrac{C_p(T)}{T} = \gamma + \beta T^2 + c_0 \dfrac{T_0^2}{T^3} \dfrac{\exp(T_0/T)}{(\exp(T_0/T)+1)^2}$.

Large values of γ (table II) can be interpreted in terms of an effective mass enhancement resulting from electronic correlations.

Table II. Sommerfeld coefficient (γ) and Debye temperature (Θ_D) determined at low temperature for A_xCoO_2 (A=Li, Na, K and x=0.6).

Phase	γ (J/mol.K^2)	Θ_D (K)
$Li_{0.61}CoO_2$	16.6×10^{-3}	501
$Na_{0.62}CoO_2$	26.1×10^{-3}	562
$K_{0.61}CoO_2$	9.0×10^{-3}	401
$K_4Co_7O_{14}$	8.8×10^{-3}	404

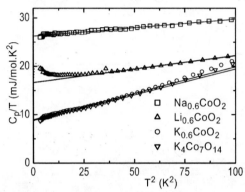

Figure 9. Specific heat capacity for alkali cobaltites with x=0.6. Solid line represents the fitting of the data assuming an usual Debye approximation for lattice vibrations. For the lithium compound doted line corresponds to a model including a Schottky anomaly.

Electrical resistivity

Electrical resistivity (ρ) measurements do not evidence any change of the scattering mechanism upon heating for the Li and Na cobaltites, as the T-dependence of the resistivity is more or less linear in the whole temperature range suggesting that phonon scattering dominates down to 4K or that much more complex effects whose discussion is out of the scope of the present paper are involved (figure 10).

On the contrary both K cobaltites, ordered $K_4Co_7O_{14}$ and random $K_{0.6}CoO_2$, show a change from a temperature independent scattering mechanism to a classical phonon scattering as expected for any metallic phase in which a departure from the periodicity of the potential energy gives rise to a dominant elastic scattering at low temperature. For $K_4Co_7O_{14}$ it could originate from uncontrolled impurities, a residual disorder in K-atom distribution or randomly distributed oxygen vacancies.

However, the change in the dρ/dT slope at low temperature could also originate from electron-electron collisions expected in highly correlated systems according to the process known as Baber scattering[25] and reviewed by Mott[26] that predicts a T^2 dependence of ρ at low temperature. The electrical resistivity of $K_4Co_7O_{14}$ is plotted against squared temperature in figure 11 showing a straight line at least up to 100K.

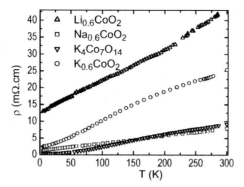

Figure 10. Temperature dependence of the electrical resistivity of alkali cobaltites with x=0.6.

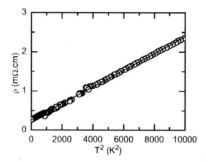

Figure 11. Showing the T^2 dependence of the electrical resistivity of $K_4Co_7O_{14}$ for 4<T<100 K.

Thermoelectric power

Temperature dependence of Seebeck coefficient for Li, Na and K cobaltites with an x-value near 0.6 are compared in figure 12. It tends to become linear above *ca.* 100K with a slope similar for all the samples.

Such a behavior can be compared to the one expected for a metal from Mott formula [27]

$$\alpha = \frac{\pi^2}{3} \frac{k_B^2 T}{e} \left\{ \frac{\partial \log \sigma(E)}{\partial E} \right\}_{E=E_F}.$$

Assuming, in the neighborhood of the Fermi level, $\sigma(E)$=const. E^x, gives:

$$\alpha = \frac{\pi^2}{3} \frac{k_B^2 T}{eE_F} x.$$

For spherical Fermi surfaces (E=const.$\mathbf{k}^2$) and forgetting the wave vector (**k**) dependency of the scattering process, x=3/2. Obviously, the slope $d\alpha/dT$ is strongly enhanced with respect to the free electron gas, which can be attributed to electronic correlations with an effective mass of charge carriers much larger than the electron mass. Such a behavior is in agreement with specific heat measurements and magnetic susceptibility data. It is also in agreement with the usual narrowness of d-bands in transition metal oxides and the 2D character of the crystal structure. While the $d\alpha/dT$ slopes are similar for the four compounds, at least for T>100K, the TEP values at room temperature are very different. It is much larger in the case of sodium, which has been ascribed to the spin entropy of the carriers[28], in agreement with the Curie-Weiss behavior of the oxide. It is much lower for the Li and K cobaltites the thermopower of which becomes negative at low temperature. Such a behavior raises the question of the contribution of two types of carriers. Actually the t_{2g} levels are split into two bands that could overlap at the Fermi level. Calculations of the band structure including accurately correlation effects are required to better understand the behavior. Another qualitative interpretation could be found in the already mentioned spin entropy contribution to the Seebeck coefficient. This contribution is not expected in the case of $K_{0.6}CoO_2$ as a Pauli type magnetic susceptibility is observed, suggesting that most of the electrons are paired. On the contrary, the effect should be strong for the Na compound that exhibits a Curie-Weiss paramagnetic behavior.

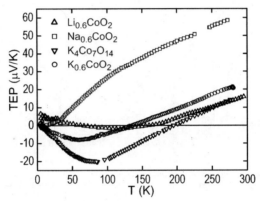

Figure 12. Temperature dependence of the thermoelectric power for selected alkali layered cobaltites with x=0.6.

PHASE TRANSITION IN P'3-Na$_x$CoO$_2$ (x~0.6)

The P'3-Na$_x$CoO$_2$ (x ~ 0.62) cobaltite undergoes a first order phase transition near $T_c = 350$ K as shown by the temperature dependence of X-ray diffraction diagrams (figure 13). Below T_c, the unit cell is monoclinic (Space group: C2/m). Above T_c, the monoclinic cell is reversibly converted into a rhombohedral cell (Space group: R3m). The crystallographic change mainly affects the Na-atom site symmetry and distribution. In both forms, the Na ions are shifted along the c-axis as described above. In the high temperature form the site is more symmetric, with a 3-fold axis parallel to z-axis.

Temperature dependence of the heat capacity, magnetic susceptibility and electrical resistivity measured upon cooling are given in figure 14. Correlation between phase transition as revealed from XRD data and changes in electrical resistivity behavior are still not well understood. Actually, comparing temperature dependence of the electrical resistivity upon cooling and heating reveals an hysteretic behavior and a partial irreversibilty that was not immediately found from the XRD study.

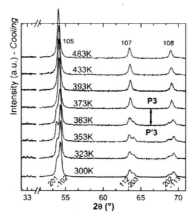

Figure 13. Temperature dependence of XRD patterns for $Na_{0.6}CoO_2$.

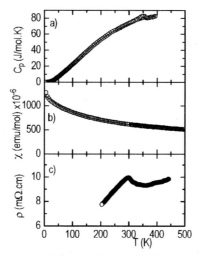

Figure 14. Temperature dependence of the specific heat capacity (a), magnetic susceptibility (b) and electrical resistivity (c) for $Na_{0.6}CoO_2$. Solid lines correspond to fits with the models mentioned in the text.

^{23}Na MAS-NMR spectroscopy has been used to investigate changes in the environment and in the distribution of the sodium cations occurring by raising the temperature. The gradual suppression of the second order quadrupolar interactions has been ascribed to a site exchange mechanism. Neighboring Na-sites are similar, but oriented up and down, which results in opposite V_{zz} components of the qradrupolar field. When the exchange frequency reaches the characteristic time of NMR the

signal is averaged[29]. Whether this effect is directly coupled with the structural transition observed by XRD is still an open question.

DELAFOSSITE OXIDES

In the structure of A'MO$_2$ delafossite type oxide the trivalent element occupy octahedral sites sharing edges and forming layers analogous to those found in the 2D cobaltites (figure 15). The monovalent A'-element is copper or silver. It occupies a 2-fold linear site so that the structure can also be described as a close packing of dumbbells parallel to the z-axis. The twofold linear coordination of the A' d^{10} element is well explained by the hybridization of (n-1)d$_z$ (n=3, 4) with ns orbitals[30].

The sequence of oxygen layers can be *AABB* or *AABBCC* leading to two polytypes denoted 2H and 3R. The size of the M element can vary in a large range from aluminum to lanthanum resulting in a a-parameter of the hexagonal cell varying from 2.8 to 3.8 Å, whereas the c-parameter remains more or less constant. This is explained by the MM electrostatic repulsion through the octahedron common edges[31]. For large enough values of the a-parameter (that corresponds to A'-A' distances) intercalation of oxygen atoms is possible and can reach very large rates even leading to an oxidation number of copper larger than 2+.

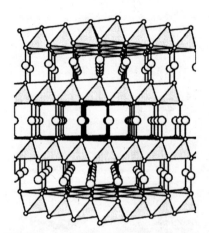

Figure 15. Structure of the 3R-delafossite CuFeO$_2$. Large spheres are copper atoms. small spheres are oxygen atoms. Iron atoms are located at the center of the compressed octahedra.

Most of the delafossite oxides (A'= Cu, Ag) are insulators or semiconductors. However, when A' is Pd or Pt, they are metallic with a very large electrical conductivity in the (a,b) plane[32]. PdCrO$_2$ is a unique example of an oxide where monovalent palladium is found. Actually, the excess electron is itinerant and the chemical stability of the oxide is linked to its highly metallic character. The hybridized (n-1)d$_z$-ns orbital presents a large expansion perpendicularly to the z-axis (that is parallel to the c axis of the hexagonal unit cell) allowing an important overlap in the A' plane. This result suggests that, in semiconducting delafossites, the carriers should move in the A' plane. Investigation of p-type CuFeO$_2$ single crystals supported this idea whereas the opposite seems to hold in the case of n-type CuFeO$_2$[33]. While it appears relatively easy creating holes in the A' layers attempts to dope them with electrons often result in metallic Cu^0 precipitation. For instance, CuFe$_{1-x}$V$_x$O$_2$ solid solutions are no longer stable for x values larger than about 0.67[34]. Beyond this value Cu^0 and VO$_2$ coexist with the border phase. Recently chromium copper delafossites and nickel-doped iron delafossites appeared as interesting candidates for thermoelectric generation[35].

CONCLUSION

Transport properties of transition metal oxides spread over a very broad range from dielectrics to metallic conductors as conductive as the best metallic elements such as copper or silver. In between a large vatiety of semi-conducting behavior can be found. Among them highly correlated metallic or semi-conducting oxides such as layered cobaltites have drawn attention as interesting candidates for thermoelectric generation, for more than ten years. Despite extensive studies during that time some basic questions regarding the origin of a large thermopower value remain open.

Reviewing the electronic properties of transition metal oxides leads to think that it could be rather difficult to reach large ZT values and even to reach large power factor values in **bulk** materials. Therefore devices using bulk oxides would hardly compete with conventional cooling systems for large scale house equipment. However, we believe that **environment compatible** oxides can be found for the recovery of waste heat provided that their relatively low efficiency is balanced by a large availability and a low cost.

ACKNOWLEDGMENTS

Useful and interesting discussions with Professor M. Pouchard and A. Villesuzanne particularly concerning the electronic structure of transition metal oxides are greatly acknowledged.

REFERENCES

[1] I. Terasaki, Y. Sasago, K. Uchinokura, *Phys. Rev. B*, **56**, 12685 (1997).

[2] C. Delmas, J. J. Braconnier, C. Fouassier, P.Hagenmuller, *Solid State Ionics*, **3-4**, 165 (1981).

[3] J. Molenda, C. Delmas, P. Hagenmuller, *Solid State Ionics*, **9-10**, 431 (1983).

[4] J. Molenda, C. Delmas, P. Dordor, A. Stoklosa, *Solid State Ionics*, **12**, 473 (1984).

[5] J.-P. Doumerc, *J. Solid State Chem.*, **110**, 419 (1994).

[6] D.B. Marsh and P. E. Parris, *Phys. Rev. B*, Vol. 54, No 11 (1996), pp. 7720-7728.

[7] K. E. Stitzer, J. Darriet, H.-C. zur Loye, *Curr. Opin. Solid State Mater. Sci.* **5**, 535 (2001).

[8] J. B. Goodenough, *Prog. Solid State Chem.*, Vol. 5 (1971), pp. 145-399.

[9] C.J. Ballhausen, Introduction to Ligand Field Theory, McGraw-Hill (New York and London, 1962).

[10] Drawing Program: Orbital Viewer, http://www.orbitals.com/orb

[11] S. Landron and M.B. Lepetit, *Phys. Rev. B* **74**, 184507 (2006).

[12] S. Landron and M.B. Lepetit, arXiv:0706.1453v2

[13] J.J. Braconnier, C. Delmas, C. Fouassier, and P. Hagenmuller, *Mat. Res. Bull.* **15**, 1797 (1980).

[14] Q. Huang, M. L. Foo, R. A. Pascal, Jr., J. W. Lynn, B. H. Toby, Tao He, H. W. Zandbergen, and R. J. Cava, *Phys. Rev. B* **70**, 184110 (2004).

[15] K. Isawa, Y. Yaegashi, M. Komatsu, M. Nagano, S. Sudo, M. Karppinen, and H. Yamauchi, *Phys. Rev. B*, **56**, 3457 (1997).

[16] J. Tate, M.K. Jayaray, A.D. Draeseke, T. Ulbrich, A.W. Sleight, K.A. Vanaja, R. Nagarajan, J.F. Wager, and R.L. Hoffman, *Thin Solid Films* **411**, 119 (2002).

[17] P. Zhang, *Phys. Rev. B*, **71**, 153102 (2005).

[18] M. Roger, D. J. P. Morris, D. A. Tennant, M. J. Gutmann, J. P. Goff, J.-U. Hoffmann, R. Feyerherm, E. Dudzik, D. Prabhakaran, A. T. Boothroyd, N. Shannon, B. Lake and P. P. Deen, *Nature*, **445**, 631 (2007).

[19] Yanli Wang and Jun Ni, *Phys. Rev. B*, **76**, 94101 (2007).

[20]M. Blangero, R. Decourt, D. Carlier, G. Ceder, M. Pollet, J.-P. Doumerc, J. Darriet, C. Delmas, First experimental evidence of potassium ordering in layered $K_4Co_7O_{14}$, *Inorg. Chem.* **44,** 9299 (2005).

[21]W.R. Myers, *Rev. Mod. Phys.* **24,** 15 (1952).

[22]F.E. Mabbs, and D.J. Machin, Magnetism and Transition metal complexes, Chapman Hall, London (1973).

[23] G. Lang, J. Bobroff, H. Alloul, P. Mendels, N. Blanchard, and G. Collin, *Phys. Rev. B,* **72,** 94404 (2005).

[24]P. Limelette, S. Hébert, V. Hardy, R. Frésard, C. Simon, and A. Maignan, *Phys. Rev. Lett.* **97,** 46601 (2006).

[25]W. G. Baber, *Proc. R. Soc. Lond.,* **A158,** 383 (1937).

[26]N. F. Mott, Metal-Insulator Transitions, Taylor & Francis (London, 1990), p. 72.

[27]N. F. Mott, and H. Jones, The Theory of the Properties of Metals and Alloys, Dover Publications (New York, 1958), p. 311.

[28]Y. Wang, N.S. Rogado, R.J. Cava, and N.P. Ong, *Nature* **432,** 425 (2003).

[29]M. Blangero, D. Carlier, M. Pollet, C. Delmas and J. P. Doumerc, *Phys. Rev. B, submitted.*

[30]L. E. Orgel, *J. Chem. Soc.* 4186 (1958).

[31]J. Tate, M. K. Jayaray, A. D. Draeseke, T. Ulbrich, A. W. Sleight, K. A. Vanaja, R. Nagarajan, J. F. Wager, and R. L. Hoffman, *Thin Solid Films* **411,** 119 (2002).

[32]C. T. Prewitt, R. D. Shannon, and D. B. Rogers, *Inorg. Chem.* **10,** 719 (1971).

[33]P. Dordor, J.-P. Chaminade, A. Wichainchai, E. Marquestaut, J.-P. Doumerc, M. Pouchard, and P. Hagenmuller, *J. Solid State Chem.* **75,** 105 (1988).

[34]K. El Ataoui, J.-P. Doumerc, A. Ammar, P. Gravereau, L. Fournes, A. Wattiaux, M. Pouchard, *Solid State Sciences* **5,** 1239-45 (2003).

[35]K. Hayashi, T. Nozaki, and T. Kajitani, *Jpn. J. Appl. Phys.* **46,** 5226 (2007).

DEFORMATION AND TEXTURE BEHAVIORS OF CO-OXIDES WITH MISFIT STRUCTURE UNDER HIGH TEMPERATURE COMPRESSION

Hiroshi Fukutomi[1], Kazuto Okayasu[1], Yoshimi Konno[2], Eisuke Iguchi[3] and Hiroshi Nakatsugawa[1]

[1] Division of Solid State Materials Engineering, Graduate School of Engineering, Yokohama National University, Tokiwadai, Hodogaya-ku, Yokohama, 240-8501 Japan

[2] Graduate student of Yokohama National University

[3] Professor emeritus, Yokohama National University
Presently at Seimi Chemical Co., Ltd. Chigasaki 3-2-10, Chigasaki, 253-8585 Japan

ABSTRACT

In order to control the orientation distribution of polycrystalline cobaltites with misfit layered structure, thermo-mechanical treatments consisting of high temperature uniaxial compression deformation and heat treatments are examined on $Bi_{1.5}Pb_{0.5}Sr_{1.7}Y_{0.5}Co_2O_{9-\delta}$ and $Ca_3Co_4O_9$. The materials were produced by the usual sintering method. High temperature compression deformation was carried out in air at high temperatures where the activation of slip deformation together with the other complementary deformation mechanisms such as grain boundary sliding and dynamic recrystallization are expected. After the deformation, measurements of density, texture and resistivity, and microstructure observation were performed. The density increases by the deformation up to a true strain of -1.0 in both oxides. No further densification was observed by the deformation above -1.0 in strain. The formation of a (001) texture (compression plane) is found after the deformation. The texture sharpens monotonously with an increase in strain. The sharpening continues above -1.0 in true strain, indicating that the texture formation can be attributed to the plastic deformation of the oxides. In some cases the maximum pole density for (001) becomes more than eleven times as high as that of the as sintered material. It is experimentally confirmed that the resistivity can be reduced below one tenth of that of the as sintered material by the densification and the texture development originating from the high temperature compression deformation.

INTRODUCTION

Some oxides containing two-dimensional CoO_2 conductive layers, so called the cobaltites with misfit layered structure, have attracted attentions as candidates for the high temperature thermoelectric materials. The common feature of this kind of cobaltites is the high in-plane electric conductivity in (001) of CoO_2 layer. Therefore, how to control the orientation of (001) is one of the key issues for the practical use of this kind of thermoelectric materials. The importance of orientation control for the layered cobaltites has been noticed out by many researchers, and hence various methods have been examined; reactive templated grain growth [1], hot pressing [2], magnetic texturation [3], sinter-forging [4] and thermoforging [5] have been proposed. Many of these methods are designed on the basis of the rotation of oxide platelets by the application of pressure before and/or during the sintering.

In the field of metallurgical engineering, various methods have been developed to control textures. Two basic textures are known; deformation and recrystallization textures. Since no texture appears after the recrystallization of materials without texture in the deformed state, it can be said that the essential process for the texture development is the deformation. Plastic deformation of oxide is, however, generally difficult because of brittleness due to high Peierls potential at room temperature. High temperature deformation is attractive to overcome the difficulty, because dislocations can be activated with the help of thermal energy and recovery during deformation might be effective to avoid the inhomogeneous deformation. In addition, activation of deformation mechanisms other than dislocation movement, such as grain boundary sliding and dynamic recrystallization is expected; these mechanisms may work as complementary mechanisms when the number of independent slip systems is insufficient for the deformation of polycrystalline oxides.

Based on this consideration, the behaviors of deformation and texture development in $Bi_{1.5}Pb_{0.5}Sr_{1.7}Y_{0.5}Co_2O_{9-\delta}$ have been investigated [6]. Although one of the authors found that $Bi_{1.5}Pb_{0.5}Sr_{2.5}Y_{0.5}Co_2O_{9-\delta}$ shows a good feature as a thermoelectric material [7], the material investigated in this study is somewhat Sr-deficient. It was expected that the deficiency of Sr ions might create a large amount of ionic vacancies that make high temperature deformation easier. In the previous reports [6], it was confirmed experimentally that $Bi_{1.5}Pb_{0.5}Sr_{1.7}Y_{0.5}Co_2O_{9-\delta}$ could be plastically deformed at high temperatures, resulting in the development of a sharp texture as well as the large reduction in resistivity.

In this study, detailed examination on the deformation behavior, texture development and the change in thermoelectric properties are performed on $Bi_{1.5}Pb_{0.5}Sr_{1.7}Y_{0.5}Co_2O_{9-\delta}$. Effectiveness of high temperature deformation was examined also on $Ca_3Co_4O_9$ which has a misfit layered structure different from $Bi_{1.5}Pb_{0.5}Sr_{1.7}Y_{0.5}Co_2O_{9-\delta}$.

EXPERIMENTAL PROCEDURE

$Bi_{1.5}Pb_{0.5}Sr_{1.7}Y_{0.5}Co_2O_{9-\delta}$ was produced by two methods. One is the conventional sintering method using element oxide powders, Bi_2O_3, PbO_2, Sr_2CO_3, Y_2O_3 and Co_3O_4 (4N). The mixture has been calcined in air at 1063 K for 12 h. After grinding, the calcined powder was pressed into pellets at room temperature (cold press), and finally sintered in air at 1113 K for 24 h. The other method for producing bulk $Bi_{1.5}Pb_{0.5}Sr_{1.7}Y_{0.5}Co_2O_{9-\delta}$ starts with the powder synthesized by the citrate method[*]. The average powder size was approximately 10μm. The powder has been also cold pressed into pellets. The final sintering condition is the same as that of the conventional sintering method. Two sizes of the sintered specimens were used, i.e., $\phi 8$ mm×16.5 mm and ϕ 11 mm×16.5 mm, depending on the amount of final strains. The powder diffractions measured by Cu Kα X-ray (XRD) at room temperature for the two oxides produced by the different methods showed the identical patterns.

High temperature compression deformation was carried out in air at 1113 K using the Instron type testing machine. The deformation temperature was determined referring to the result of the thermal analysis; it is found out that the melting temperature of this oxide is about 1190 K. The strain rates and final true strains are in the range between $1.0×10^{-5}s^{-1}$ and $1.0×10^{-4}s^{-1}$, and from -0.47 to -2.2, respectively. The bulk $Ca_3Co_4O_9$ was produced by sintering the powder synthesized by the citrate method. The compression deformation was conducted at 1153K in the case of $Ca_3Co_4O_9$.

---

[*] The powder was produced by the citrate method in Seimichemical Co., Ltd.

The microstructure was examined by scanning electron microscopy. The specimens were cooled immediately after the compression, in order to prevent the microstructure from the changing after the deformation. Texture measurements have been performed by the Schulz reflection method using Ni filtered Cu Kα radiation. Based on the reflected X-ray intensity, the (001) pole figure is constructed. The measurements of four-probe dc resistivity and Seebeck coefficient were carried out as a function of temperature over a range of 100 K - 1073 K during both heating and cooling runs in air. In order to investigate the effect of the strain on the densification, density is measured at various strains. Strain transient dip tests was conducted on $Bi_{1.5}Pb_{0.5}Sr_{1.7}Y_{0.5}Co_2O_{9-\delta}$ in air by the compression creep testing machine to examine the deformation mechanism at high temperatures [8]. The stresses employed in the tests are 10MPa and 25MPa.

RESULTS and DISCUSSION
Deformation behavior of $Bi_{1.5}Pb_{0.5}Sr_{1.7}Y_{0.5}Co_2O_{9-\delta}$

Figure 1 shows the true stress-true strain curves for the deformation at 1113K. In this case, strain rate is changed from $1.0 \times 10^{-4} s^{-1}$ to $2.0 \times 10^{-5} s^{-1}$ (solid line) and to $1.0 \times 10^{-5} s^{-1}$ (dotted line) at a strain of 0.9 (In this paper, the absolute value of compression true strain is used.). It is seen that the stress increases monotonously up to a strain of 0.9. The change in strain rates results in the sudden decrease in stress; the stress after the strain rate reduction varies depending on the strain rate after the strain rate change. After the sudden decrease due to the strain rate change, the stress increases again with an increase in strain. The stress-strain curve is smooth. This suggests that the deformation proceeds smoothly up to the large strain such as 2.0. In order to understand the stress increase with an increase in strain, density change during the deformation is investigated. Figure 2 shows the result on density measurement, which is given as a function of true strain. It is seen that the density increases monotonously with an increase in true strain up to about one of the absolute value of true strain. No further densification was

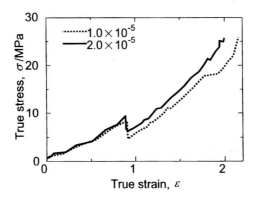

Fig. 1 True stress-true strain curve including the strain rate change at a true strain of 0.9. Initial strain rate is $1.0 \times 10^{-4} s^{-1}$.

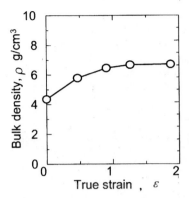

Fig. 2 Density vs strain relationship

observed by the deformation above 1.0 in strain. This indicates that the stress increase in true stress-true strain curve might be attributable to the densification to some extent up to the true strain 1.0. But the further increase in stress should not be attributed to the densification. Since no lubricant was used for the compression test in this study, it is not clear whether the increase in stress with an increase in strain originates from the friction between the specimen and the compression rod and/or work hardening accompanying plastic deformation caused by dislocation motion. However, the sudden decrease in stress by the strain rate change shown in Fig. 1 suggests that the stress increase is the work hardening. In the strain transient dip test, stoppage of

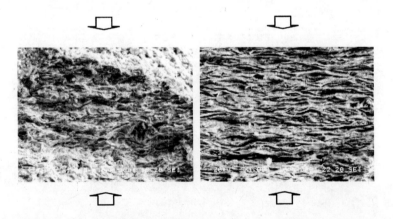

Fig. 3 Microstructure after deformation up to strains of (a) 0.9 and (b) 2.0.

deformation after the reduction in stress from 25MPa to 10MPa was found. This indicates the existence of internal stress, namely contribution of dislocation motion to the deformation. Figure 3 shows the microstructure observed at the cross section after the deformation up to (a) 0.90 and (b) 2.0. The compression was made from the direction indicated by the arrows. Since the observed surfaces are produced by fracture, the shapes of crystal grains are not clearly seen. However, it is suggested that the aspect ratio (width/height ratio) of crystals grains in (b) is larger than that in (a); crystal grains might be plastically deformed. From these various experimental evidences, it was concluded that work hardening contributes to the increase in stress shown in Fig. 1.

Texture of $Bi_{1.5}Pb_{0.5}Sr_{1.7}Y_{0.5}Co_2O_{9-\delta}$

Figure 4 shows an example of pole figure showing the distribution of (001) after the deformation up to a strain of 0.9. Pole densities are projected onto the compression plane. Mean pole density is used as a unit to draw contours. The pole density is distributed concentric circular, indicating a formation of a fiber texture. The area of high pole density exists close to the center of the (001) pole figure, namely (001) aligns parallel to the compression plane. The maximum pole density is 4.7 in this case. In order to study the effect of strain on the texture formation, the maximum pole densities are examined as a function of strain in Fig. 5. It is seen that the maximum pole density increases monotonously with an increase in strain up to a strain of about 2. The texture still sharpens above 1.0 in strain, where densification is complete.

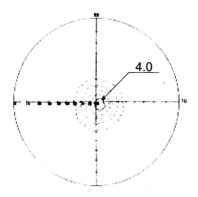

Fig. 4 (001) pole figure of a specimen deformed at 1113K with a strain rate of $1.5 \times 10^{-5} s^{-1}$ up to a strain of 0.9.

As well known, texture develops when translation occurs along the specific crystal plane. The main component of the texture varies depending on the translation plane and direction, and the deformation mode (constraints for the deformation). In the case of compression deformation, the basic process is the rotation of the translation plane parallel to the compression plane. As for the present oxide, two possibilities are considered as the translation process. One is the mutual sliding between oxide powder particles along (001). Another possibility is the plastic deformation by the activation of crystal slip system having (001) as a slip plane. Several researchers reported that some oxide powders are platelet shape and (001) is the surface of the platelet. Application of stress generates the mutual slide of platelets, resulting in the oriented state. In the case of $Bi_{1.5}Pb_{0.5}Sr_{1.7}Y_{0.5}Co_2O_{9-\delta}$ powder, SEM observation showed that powders are not platelet but spherical shape. As already discussed before, results on the effect of strain rate change given in Fig. 2, fracture surface observation given by Fig. 3 suggest that the plastic deformation occurs in the present oxide.

Figure 6 shows the crystal structure of $Bi_{1.5}Pb_{0.5}Sr_{1.7}Y_{0.5}Co_2O_{9-\delta}$. Oxide layers of Sr (Y) and Bi (Pb) exist between CoO_2 conductive layers. As easily found from the figure, the same kind of atoms neighbors in $[1\bar{1}0]$ and $[110]$ directions. This means that the magnitude of Burgers vector of the perfect dislocations is small on this plane. Thus it is expected that $(001)[1\bar{1}0]$ and

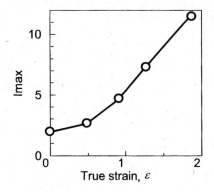

Fig. 5 Effect of strain on the development of (001) texture

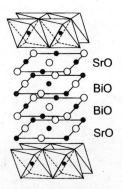

Fig. 6 Crystal structure of $Bi_{1.5}Pb_{0.5}Sr_{1.7}Y_{0.5}Co_2O_{9-\delta}$

(001) [110] might be slip systems of this oxide.

Thermoelectric properties of $Bi_{1.5}Pb_{0.5}Sr_{1.7}Y_{0.5}Co_2O_{9-\delta}$

Figure 7 shows the temperature and texture dependences of resistivity. The resistivity at ε = 0 is close to the values of the (x = 0.5, y = 0.7) specimen in the $Bi_{2-x}Pb_xSr_{3-y}Y_yCo_2O_{9-\delta}$ system [7]. This indicates that Sr-deficiency is mostly compensated by oxygen vacancies, not by hole-doping. It is seen that the resistivity decreases remarkably as the strain increases. The resistivity becomes about 60% of the as sintered state by the deformation up to a strain of 0.47. Up to this strain, drastic increase in density occurs, while texture development is not extensive. Hence the decrease in resistivity can be attributed to the densification. Deformation up to a strain of 1.9 results in the decrease in resistivity to about one tenth of the as sintered state. The decrease in resistivity continues strains above 1.0. Since densification completes at about 1.0 in strain, this means that texture sharpening due to the plastic deformation by the activation of (001) slip system is effective to the improvement in resistivity. The resistivity of the specimen deformed up to a strain of 1.9 is measured at the temperatures from 100K to 973K. It is found that the temperature increase slightly increases the resistivity.

It was found that the thermopower decreased with an increase in strain, namely the texture sharpening. The change in thermopower, however was within 10%. Thermopower of the specimen deformed up to ε =1.9 at 973K was 160μV/K. Since the resistivity at this temperature is 23mΩcm and thermal conductivity is reported as 10mWcm⁻¹K⁻¹ on the oxide with the similar composition, the dimensionless figure of merit ZT at 973K was estimated as 0.11.

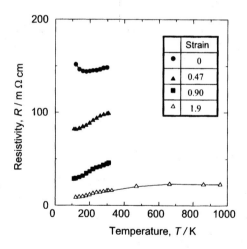

Fig. 7 Effects of texture and temperature on resistivity

Behaviors of $Ca_3Co_4O_9$

In the present study, two kinds of cobaltites with misfit layered structure are investigated. The difference between the two oxides is the structure between CoO_2 conductive layers. In the case of $Bi_{1.5}Pb_{0.5}Sr_{1.7}Y_{0.5}Co_2O_{9-\delta}$, four oxide layers including Bi (Pb) and Sr (Y) exist between CoO_2 conductive layers as shown in Fig. 6, while there are three oxide layers in $Ca_3Co_4O_9$. In both oxides, the distance between the same atoms is long in c direction, and that in (001) plane is short. Therefore, it is considered that (001) slip is also activated and (001) texture is formed in $Ca_3Co_4O_9$.

It was found that $Ca_3Co_4O_9$ could be deformed at 1153K without heavy cracks up to a strain of 1.3. In the case of $Bi_{1.5}Pb_{0.5}Sr_{1.7}Y_{0.5}Co_2O_{9-\delta}$, the compression deformation was performed at 1113K, which is close to its melting temperature. As for the deformation temperature, 1153K is far from the melting temperature of $Ca_3Co_4O_9$. Nevertheless, large amount of deformation was possible in this oxide although small cracks were seen after the deformation.

Density measurements showed that the density increases up to a strain of 1.0 and no further densification was observed. Then, orientation distribution was examined. Figure 8 shows the (001) pole figure after the deformation up to 0.50 in true strain. Pole densities are projected onto the compression plane. Mean pole density is used as a unit. Pole density distribution is similar to Fig. 4; the maximum pole density appears at the center of the pole figure. That is, the (001) texture is constructed. Maximum pole density was 4.8 in this case. It was experimentally found that texture sharpens with an increase in strain similar to the case of $Bi_{1.5}Pb_{0.5}Sr_{1.7}Y_{0.5}Co_2O_{9-\delta}$. Result of resistivity measurement is given in Fig. 9. The measurement was conducted in the temperature ranging from 573K to 1073K. Resistivity decreases with an increase in temperature. The lowest value is about $2m\Omega cm$ at 1073K. This value is close to the reported data on a single crystal. Thus it can be said that the high temperature compression is

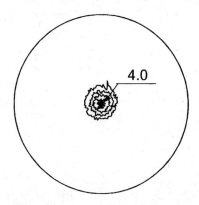

Fig. 8 (001) pole figure for $Ca_3Co_4O_9$ after the
deformation up to 0.50 in true strain

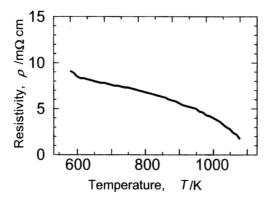

Fig. 9 Temperature dependence of resistivity in
$Ca_3Co_4O_9$ after the deformation up to 1.1.

effective to improve the thermoelectric property of cobaltites with misfit layered structure by developing the sharp (001) texture.

CONCLUSIONS

In order to improve thermoelectric property of cobaltites with misfit layered structure by texture control, possibility of high temperature plastic deformation method is experimentally examined on $Bi_{1.5}Pb_{0.5}Sr_{1.7}Y_{0.5}Co_2O_{9-\delta}$ and $Ca_3Co_4O_9$. As given below, it is shown that the high temperature deformation is promising to reduce resistivity by texture formation. Major results are summarized as follows.

(1) At high temperatures, large amount of compression deformation is possible to the bulk $Bi_{1.5}Pb_{0.5}Sr_{1.7}Y_{0.5}Co_2O_{9-\delta}$ and $Ca_3Co_4O_9$ produced by sintering. The maximum strains at present for $Bi_{1.5}Pb_{0.5}Sr_{1.7}Y_{0.5}Co_2O_{9-\delta}$ and $Ca_3Co_4O_9$ are 2.2 at 1113K and 1.3 at 1153K in true strain, respectively.
(2) The densification and texture development proceed simultaneously by compression deformation. It is experimentally confirmed that densification completes at a true strain of about 1.0 in both oxides, while texture develops up to a strain of 2.2 in $Bi_{1.5}Pb_{0.5}Sr_{1.7}Y_{0.5}Co_2O_{9-\delta}$ and up to a strain of 1.3 in $Ca_3Co_4O_9$.
(3) Strain rate change during compression deformation of $Bi_{1.5}Pb_{0.5}Sr_{1.7}Y_{0.5}Co_2O_{9-\delta}$ results in the change in flow stress. Microstructure observation on the cross section of $Bi_{1.5}Pb_{0.5}Sr_{1.7}Y_{0.5}Co_2O_{9-\delta}$ indicates the shape change of crystal grains. Furthermore, texture develops after the completion of densification in $Bi_{1.5}Pb_{0.5}Sr_{1.7}Y_{0.5}Co_2O_{9-\delta}$ and $Ca_3Co_4O_9$. These experimental evidences suggest that the oxides are plastically deformed and deformation textures are constructed.
(4) Resistivity becomes smaller than one tenth of the as sintered oxide by the deformation up to a true strain of 1.9 in $Bi_{1.5}Pb_{0.5}Sr_{1.7}Y_{0.5}Co_2O_{9-\delta}$. No large change by compression deformation

is seen in thermopower. The thermopower is kept about $160\mu VK^{-1}$ up to 973K. The value of dimensionless figure of merit is 0.11 at 973K. Textured $Ca_3Co_4O_9$ polycrystal shows the resistivity close to the reported value on a single crystal.

ACKNOWLEDGEMENTS
The authors greatly appreciate to Mr. N. Ogawa and K. Shibuya of Yokohama National University for their assistance in the present research. Measurements of thermopower are conducted by ULVAC Co. Ltd.

REFERENCES
[1] T. Tani, H. Itahata, C. Xia, J. Sugiyama: J. Matar. Chem., **13**(2003)1865.
[2] E. Guilmeau E, R. Funahashi, M. Mikami, K. Chong K, D. Chateigner:. Appl. Phys. Lett., **85** (2004)1490.
[3] Y. Zhou, I. Matsubara, S. Horii, T. Takeuchi, R. Funahashi, M. Shikano, J. Shimoyama, Kishio K, W. Shin, N. Izu, N. Murayama: J. Aappl. Phys., **93**(2003)2653.
[4] W. Shin, N. Murayama: J. Mater. Res. 15(2000)382.
[5] M. Prevel, S. Lemonnier, Y. Klein, S. Hebert, D. Chateigner: J. Appl. Phys. **98**(2005)093706.
[6] H. Fukutomi, E. Iguchi, N. Ogawa. Materials Science Forum, **495-497**(2005)1407.
[7] E. Iguchi, T. Itoga, H. Nakatsugawa, F. Munakata, K. Furuya: J. Phys. D. Apply. Phys., 34(2001)1017.
[8] A. A. Solomon:Rev. Sci. Instr. **40**(1969)1025.

FABRICATION OF HIGH-PERFORMANCE THERMOELECTRIC MODULES CONSISTING OF OXIDE MATERIALS

Ryoji Funahashi[1,2], Saori Urata[2], and Atsuko Kosuga[1]
1 National Institute of Advanced Industrial Sci. & Tech., Midorigaoka, Ikeda, Osaka 563-8577, Japan
2 Japan Science and Technology Agency, CREST, Honmachi, Kawaguchi, Saitama 332-0012, Japan

ABSTRACT

Thermoelectric modules composed of eight pairs of p-type $Ca_{2.7}Bi_{0.3}Co_4O_9$ (Co-349) and n-type $CaMn_{0.98}Mo_{0.02}O_3$ (Mn-113) legs were constructed using Ag electrodes and paste including the Mn-113 powder. Dimensions of both oxide legs were 5 mm wide and thick and 4.5 mm high. An alumina plate was used as a substrate, and there was no alumina plate on the other side of the modules. When the substrate side was heated, the module can generate up to 1.9 V and 2.3 W of open circuit voltage (V_o) and maximum power (P_{max}), respectively, at a hot-side temperature at the surface of the substrate of 973 K and a temperature differential of 675 K in air. Because of cracking in the Mn-113 legs, actual internal resistance is about 1.6 times higher than the calculated one. In order to improve the mechanical properties, $Ca_{0.9}Yb_{0.1}MnO_3$/Ag composites were prepared. The sintered composites consisted of two phases of $Ca_{0.9}Yb_{0.1}MnO_3$ and metallic Ag from the X-ray diffraction analysis. The scanning electron microscopic observation indicated that the Ag particles with diameter smaller than $5\mu m$ were homogeneously dispersed in $Ca_{0.9}Yb_{0.1}MnO_3$ matrix for all the composites. Bending strength σ_f of the composite including Ag by 18.8 wt.% was 251 MPa, which was 2 times larger value than that of the monolithic $Ca_{0.9}Yb_{0.1}MnO_3$ bulk.

1. INTRODUCTION

In view of global energy and environmental problems, research and development have been promoted in the field of thermoelectric power generation as a means of recovering vast amounts of waste heat emitted by automobiles, factories, and similar sources. Waste heat from such the sources offers a high-quality energy source equal to about 70 % of total primary energy, but is difficult to reclaim due to its source amounts being small and widely dispersed. Thermoelectric generation systems offer the only viable method of overcoming these problems by converting heat energy directly into electrical energy irrespective of source size and without the use of moving parts or production of environmentally deleterious wastes. The requirements placed on materials needed for this task, however, are not easily satisfied. Not only must they possess high conversion efficiency, but must also be composed of non-toxic and abundantly available elements having high chemical stability in air even at temperatures higher than 800 K. Oxide compounds have attracted attention as promising thermoelectric materials because of their potential to overcome the above-mentioned problems [1-7]. Recently, fabrication and power generation of thermoelectric modules consisting of p-type $Ca_3Co_4O_9$ (Co-349) and n-type $LaNiO_3$ (Ni-113) or $CaMnO_3$ (Mn-113) legs have been reported [8, 9].

Although thermoelectric properties of materials composing the modules should be enhanced, high chemical and mechanical durability of the materials and contact resistance and strength at the junctions are also very important in practical use of the modules. The Mn-113 bulk shows higher dimensionless figure of merit ZT ($=S^2T/\rho\kappa$, S: Seebeck coefficient, T: absolute temperature, ρ: electrical resistivity, and κ: thermal conductivity) than Ni-113 one. Especially, a polycrystalline sample of $Ca_{0.9}Yb_{0.1}MnO_3$ was found to exhibit a moderate ZT of 0.16 at 973 K in air [10]. We found, however, that the Mn-113 legs in the modules broke after the power generation test due to thermal stress, which was attributed to the large difference of thermal expansion coefficient between the Ag electrode and Mn-113 legs, and low mechanical strength of the Mn-113 legs. One possible approach to overcome this problem is to enhance the mechanical properties of the Mn-113 sintered ceramics. There are some material designs to strengthen and toughen ceramics by using composite techniques; incorporating particles, whiskers or platelet reinforcement, and precipitation secondary phases [11, 12]. Among them, ceramic composites having ductile metal dispersion was reported to show excellent mechanical properties such as hardness, Young's modulus, bending strength and toughness [13, 14]. In this paper, after making problems on thermoelectric modules composed of oxide legs clear, mechanical properties of oxide-melt composites will be discussed.

2. THERMOELECTRIC MODULES
2.1. EXPERIMENTAL

Modules fabricated in this study are composed of $Ca_{2.7}Bi_{0.3}Co_4O_9$ (Co-349) and $CaMn_{0.98}Mo_{0.02}O_3$ (Mn-113) for the p- and n-type legs, respectively. The Co-349 powder was prepared by solid-state reaction at 1123 K for 10 h in air. As starting materials, $CaCO_3$, Co_3O_4 and Bi_2O_3 powders were used and mixed thoroughly in the stoichiometric composition. The Co-349 bulks were prepared using a hot-pressing technique. The obtained Co-349 powder was hot-pressed for 20 h in air under a uniaxial pressure of 10 MPa at 1123 K to make density and grain alignment high. Preparation of the Mn-113 was started using $CaCO_3$, Mn_2O_3 and MoO_3 powders. These powders were mixed well and treated at 1273 K for 12 h in air. The powder was densified using a cold isostatic pressing (CIP) technique for 3 h under about 150 MPa. After the CIP process, the precursor pellets were sintered at 1473 K 12 h in air. The Mn-113 bulks can be densified well by CIP and sintering under the atmospheric pressure even without hot-pressing. Both bulks were cut to provide leg elements with a cross-sectional area of 5.0 mm × 5.0 mm and length of 4.5 mm.

Electrodes were formed on one side of surface of an alumina plate (36.0 mm × 34.0 mm × 1.0 mm thick) using Ag paste including Mn-113 powder by 3 wt.% and Ag sheets with a thickness of 50 μm. An alumina plate was used for one side of the module as a substrate. On the other hand, no substrate was used for the other side. This structure is effective to prevent the contact between oxide legs and Ag electrodes from peeling by the deformation of the module. For connection between the oxide legs and electrodes, Ag paste including the Mn-113 powder was used as an adhesive paste because of low contact resistance between oxide legs and Ag electrodes [9]. This oxide/Ag composed paste was applied by screen printing on the Ag electrodes. The eight pairs of p- and n-type oxide legs were put on them alternatively. The precursor module was solidified at 1123 K under a uniaxial pressure of 6.4 MPa for 3 h in air. The perfect Co-349/Mn-113 module is shown in Figure 1.

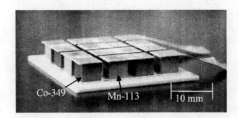

Figure 1   A thermoelectric module composed of eight pairs of Co-349/Mn-113 legs.

2.2. EVALUATION OF POWER GENERATION AND DURABILITY

The module was put between a plate shape furnace and a cooling jacket and heated at 373-1273 K of the furnace temperature and cooled by water circulation of 298 K (Figure 2). Hot-side temperature (T_H) was measured using a thermo-couple put on the surface of the substrate. Cold-side temperature (T_C) corresponds to that of the surface of the cooling jacket. Measurement of the power generation, in which current-voltage lines and current-power curves, was carried out in air by changing load resistance using an electronic load system (E.L.S). Internal resistance R_I of the module corresponded to the slope of the current-voltage lines.

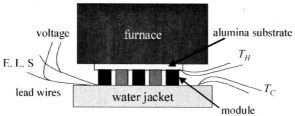

Figure 2   Measurement of power generation from the thermoelectric module

Durability against heating-cooling cycles was investigated for the 15 pieces of modules. The electrical furnace temperature was set at 523 K and on the other side of the modules was cooled by water circulation at 293 K. After heating for 7 h with continuous power generation, the modules were cooled down to room temperature. This heating-cooling cycle was carried out 4-times. R_I of the modules was measured using a standard DC four terminal method before and after the heating-cooling cycles.

2.3. RESULTS AND DISCUSSION

The module can generate up to 1.9 V and 2.3 W of open circuit voltage V_o corresponding to the tangent of the current-voltage line and maximum power P_{max}, respectively at T_H of 973 K and ΔT of 675 K. On the other hand, these values calculated from S and ρ of both p- and n-type bulks reach about 2.0 V and 3.8 W at the same temperature condition (Figure 3).

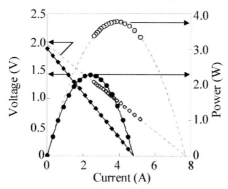

Figure 3   Measured (closed symbol) and calculated (open symbol) power generation characters of eight pairs Co-349/Mn-113 module at T_H and ΔT of 973K and 675 K, respectively. Solid and broken lines are guides for eyes.

The R_i values calculated from resistivity of both p- and n-type bulks are as low as 0.2-0.25 Ω and lower than the •measured one of the virgin modules. Although measured R_i (R_{iM}) is higher than the calculated R_i (R_{iC}) by about 20 % at temperatures lower than 500 K of the heater temperature, the increasing percentage of R_i increases suddenly and reaches 70 % for the Co-349/Mn-113 module (Fig. 4). Meanwhile, in the Co-349/Ni-113 module composed of eight pairs the increasing percentage of R_{iM} from R_{iC} is less than 20 % whole through temperature region. Although no destruction is observed in the Co-349/Ni-113 modules, many cracks are observed in the Mn-113 legs after power generation.

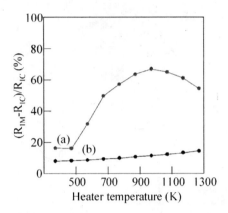

Figure 4   Increasing percentage of R_i of measured one (R_{iM}) from calculated one (R_{iC}). Lines are guides, for eyes.

Figure 5 (a) shows the destruction of the module. All destruction happened in the Mn-113 legs. A reason of this destruction is the differential of thermal expansion coefficient between the alumina substrate. All broken points are not the junctions but in horizontal direction within 1 mm height of the Mn-113 legs from the surface of the alumina substrate. Therefore, the destruction of the Mn-113 legs leads to the remarkable enhancement in R_i by heating the substrate (Fig. 5 (b)). The mechanical strength of the Mn-113 bulks is necessary to be improved.

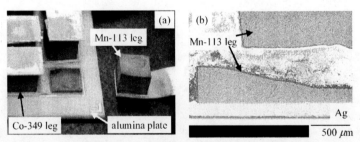

Figure 5   Destruction of the Mn-113 legs in the module after power generation (a) and a scanning electron microscopic (SEM) image of cross-sectional area around the junction between the Mn-113 leg and Ag (b)

Durability against heating-cooling cycles of the 15 pieces of modules was evaluated. R_l was increased after the cycles (Fig. 6). Because 5 modules (module number 4, 7, 8, 14 and 15) were broken after the cycles completely, R_l could not be measured. Considering high R_l even at lower heater temperature than 523 K, the destruction of the n-type legs seems to start before T_H reaching at 523 K and extend with increasing temperature.

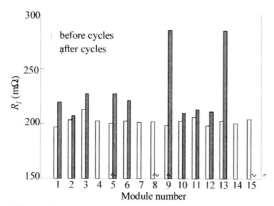

Figure 6 R_l of the modules before and after the heating-cooling cycles. R_l for the module number 4, 7, 8, 14 and 15 could not be measured after the cycles because of complete destruction

3. MECHANICAL PROPERTIES OF MONOLITHIC OXIDE BULKS
3.1. EXPERIMENTAL

The Co-349 and Mn-113 bulk samples for the investigation of mechanical properties were prepared in the same conditions as mentioned above, but without CIP. The bulks were cut into dimensions of 4.0 mm wide, 3.0 mm thick, and 40.0 mm long for three-point bending test and 5.0 mm wide, 5.0 mm thick, and 10.0 mm long for thermal expansion coefficient.

Linear thermal expansion coefficient α was measured using a differential dilatometer (Thermo plus TMA8310, RIGAKU) at 323-1073 K in air. Three-point bending test was carried out at room temperature, the loading speed was 0.5 mm/min and span length was 30.0 mm (Auto graph AG-20kNG, SHIMADZU) as shown in Fig. 7. In the case of Co-349 bulks, measurement was performed in the loading direction perpendicular and parallel to the hot-pressing axis.

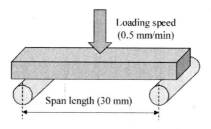

Figure 7   A schematic picture of three-point bending test

Bulk density of the samples was measured by the Archimedean method. Powder density was determined by the picometric method in He gas atmosphere. Relative density for the oxide bulks were calculated using these densities.

3.2. RESULTS AND DISCUSSION

To investigate the cause of the destruction of the Mn-113 legs in the modules, α and three-point bending strength σ_f were investigated. The α values for the Co-349 bulks are lower than those for the Mn-113 and Ni-113 ones and closer to the alumina plates (Fig. 8 (a)). Thermal expansion coefficient increases with temperature for all samples. Differential between thermoelectric oxide bulks and alumina plate is shown in Fig. 8 (b). For the Co-349 and Ni-113bulks, the differential from the alumina plate tend to decrease with increasing temperature. On the other hand, it increases for the Mn-113 bulk. This seems one of the reasons for the destruction of Mn-113 legs. The σ_f values, however, are comparable between Ni-113 and Mn-113 bulks. The difference in α is not an immediate reason for the destruction.

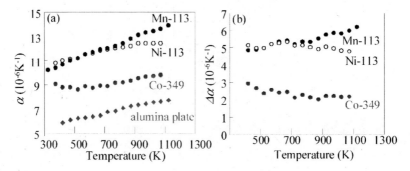

Figure 8   Temperature dependence of α for Co-349, Mn-113, Ni-113 bulks, and alumina plate of the substrate (a). Temperature dependence of differential of α ($\Delta\alpha$) between thermoelectric oxide bulks and alumina plate (b).

σ_f of the Co-349, Mn-113 bulks, and the alumina plates are shown in Fig. 9. This strength corresponds to the maximum load in the load-displacement curve. At this load, the initial cracking happens. The alumina substrates show the highest σ_f. Anisotropy of σ_f is observed in the Co-349 bulks. The strength in the case of loading direction parallel to the hot-pressing axis is higher than perpendicular to the hot-pressing axis. σ_f of the Mn-113 bulk is lower than that of the Co-349 and Ni-113 bulks. This low σ_f is one of the main reasons for the destruction of the Mn-113 legs only.

The relative density of the Mn-113 bulks is lower than the other oxide bulks (Fig. 10). This low density is a reason for the weak σ_f for the Mn-113 bulks. Scanning electron microscopic (SEM) images for the three kinds of oxide bulks are shown in Fig. 11. Many large pores are observed in the Mn-113 bulks clearly. More densification is necessary to enhance σ_f in the Mn-113 bulks. In the SEM images, the Co-349 and Ni-113 bulks are seen as "aggregates of small grains", but Mn-113 grains are grown much more than the other oxide bulks. Such the microstructure in the Mn-113 bulks allows the cracks to run easily. It has been not clear whether cracking happens in the Co-349 and Ni-113 legs in the modules after power generation. Even if it happens, however, the extension of cracks is prevented by the microstructure as seen in Fig. 11. Namely, fracture toughness K_{IC} of these bulks is better than the Mn-113 bulk. σ_f and K_{IC} for the Mn-113 bulk should be improved by densification and microstructure.

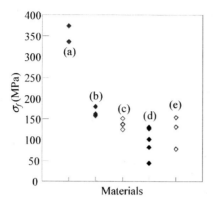

Figure 9   Three-point bending strength σ_f of (a) alumina plates, (b) Co-349 (loading direction parallel to hot-pressing axis), (c) Co-349 (loading direction perpendicular to hop-pressing axis), (d) Mn-113, and (e) Ni-113 bulks.

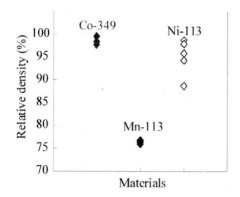

Figure 10   Relative density of Co-349, Mn-113, and Ni-113 bulks

4. MECHANICAL PROPERTIES OF CaMnO₃/Ag COMPOSITES
4.1. EXPERIMENTAL

$Ca_{0.9}Yb_{0.1}MnO_3$ powder was synthesized by a conventional solid-state reaction. The appropriate amounts of $CaCO_3$, Mn_2O_3, and Yb_2O_3 were mixed well and calcined at 1273 K for 15 h and then at 1523 K for 12 h in air with an intermediate grinding. The $Ca_{0.9}Yb_{0.1}MnO_3$/Ag composites were prepared by wet milling various amounts of Ag_2O (0, 5, 10, and 20 wt. %) with the $Ca_{0.9}Yb_{0.1}MnO_3$ powder in an agate pot using ethyl alcohol and agate balls for 24 h. After sintering, the composites included 0, 4.7, 9.4, and 18.8 wt.% of Ag formed by decomposition of Ag_2O. Mixed slurries were dried and milled for 12 h. Green pellets were prepared under a uniaxial pressure of 40 MPa and then fired at 1523 K for 2 h in air. The bulk density was calculated based on the weight and dimension of each sample. The powder density was measured using a pycnometer. The crystallographic structure was analyzed by powder X-ray diffraction (XRD) at room temperature using Cu-K$_\alpha$ radiation. Microscopic structure of the composites was observed by SEM observation. The K_{IC} values were evaluated at room temperature by the indentation fracture technique using a Vickers indenter and the relationship proposed by Niihara [15]. σ_f was measured on the bar-shaped specimens by a three-point bending method at room temperature. The loading speed was 0.1 mm/min and span length was 10.0 mm.

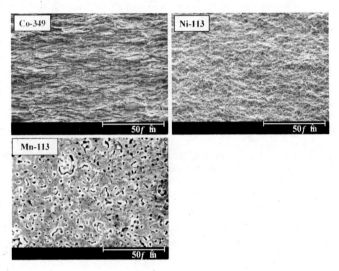

Figure 11   SEM images for Co-349, Mn-113, and Ni-113 bulks

4.2. MECHANICAL PROPERTIES

The Ag content dependence of the relative density for Ca$_{0.9}$Yb$_{0.1}$MnO$_3$/Ag composites is shown in Fig. 12. The powder density increases with increasing Ag content. Relative density of the composites sintered at 1673 K is above 95 %. Their brittleness, however, made the composites difficult to form the shapes for the various measurements. Therefore, although the relative density was around 85 %, the composites sintered at 1523 K were used for measurement of mechanical properties.

The XRD patterns of all the composites are shown in Fig. 13. The starting Ag$_2$O was completely reduced to metallic Ag after sintering at 1523 K. No secondary phases and solid solutions between Ca$_{0.9}$Yb$_{0.1}$MnO$_3$ and Ag were detected. Obvious change in the orthorhombic Mn-113 lattice parameters was not observed in all the composites. Figure 14 shows SEM photographs of the polished surface for the composites including Ag by 0 (a) and 18.8 wt.% (b), respectively. The bright dots in Fig. 14 correspond to the metallic Ag and are homogeneously dispersed in the Ca$_{0.9}$Yb$_{0.1}$MnO$_3$ matrix. The particle size of Ag grew with increasing Ag content. The average particle size of the dispersed Ag was 0.8, 1.6, and 3.9 μm for 4.7, 9.4, 18.8 wt.% composites, respectively.

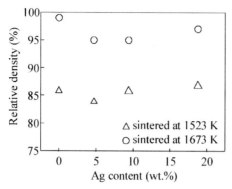

Figure 12   Ag content dependence of relative density for Ca$_{0.9}$Yb$_{0.1}$MnO$_3$/Ag composites

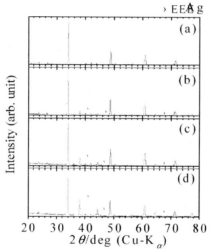

Figure 13   XRD patterns of monolithic Ca$_{0.9}$Yb$_{0.1}$MnO$_3$ (a), 4.7 wt.% (b), 9.4 wt.% (c), and 18.8 wt.% (d) composites

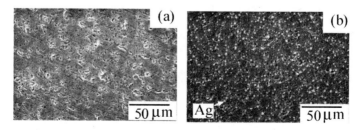

Figure 14   SEM photographs of monolithic Ca$_{0.9}$Yb$_{0.1}$MnO$_3$ (a) and 18.8 wt.% composite (b)

Figure 15 shows the Ag content dependence of K_{IC}. The K_{IC} values increase with increasing Ag content and reaches maximum value of 2.2 MPa m$^{1/2}$ for the 9.4 wt.% composite and then slightly decreases at 18.8 wt.% of the Ag content.

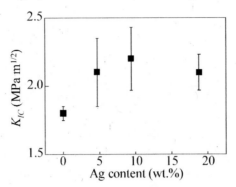

Figure 15   Ag content dependence of the fracture toughness K_{IC} for Ca$_{0.9}$Yb$_{0.1}$MnO$_3$/Ag composites

The improvement mechanism of K_{IC} seems to be the plastic stretching of metallic inclusions by bridging and deflecting the growing crack in ceramic/metal composites [16, 17]. As the crack reached the ceramic/metal interface, the difference in the deformation ability between the ductile particles and the brittle matrix causes the crack to be blunted locally. However, the size of Ag particles becomes larger beyond a critical value, this effect would be small. Thus, the K_{IC} of Ca$_{0.9}$Yb$_{0.1}$MnO$_3$/Ag composites had the maximum value due to the Ag particle exceeding the critical size between 9.4 and 18.8 wt.% of the Ag content.

Figure 16 shows the Ag content dependence of σ_f for the composites. Generally the σ_f values of brittle materials are related K_{IC} and the length of an initial crack c as indicated by the following eq. 1 [18],

$$\sigma_f = \frac{1}{Y} \cdot \frac{K_{IC}}{\sqrt{c}} \qquad (1)$$

where Y is a dimensionless geometrical parameter. σ_f of 6.6 wt.% composite becomes comparable to that of the p-type Ca$_{2.7}$Bi$_{0.3}$Co$_4$O$_9$ bulk and increases by about 25 % than that of the monolithic Ca$_{0.9}$Yb$_{0.1}$MnO$_3$ bulk. Although the relative density is independent of the Ag content, σ_f is increased by incorporation of Ag. The improvement in σ_f seems to be attributed to the enhancement of K_{IC}. Moreover although K_{IC} slightly decreases at higher Ag content than 9.4 wt.%, σ_f still increases up to 18.8 wt.%. The improvement of σ_f is possibly attributed to the change of the parameter c in the eq. 1. In general, c is proportional to the grain size in dense polycrystalline materials. This means that the strength increased when the grain size became fine. Thus, the increased σ_f of the composites is attributed to the refinement of the grains in the region between 9.4 and 18.8 wt.% of the Ag content.

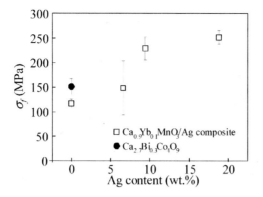

Figure 16   Ag content dependence of σ_f for $Ca_{0.9}Yb_{0.1}MnO_3$/Ag composites

5. CONCLUSION

Thermoelectric modules consisting of eight pairs of p-type Co-349 and n-type Mn-113 legs have been fabricated. The module can generate up to V_0 and P_{max} 1.9 V and 2.3 W, respectively at a hot-side temperature at the surface of the substrate of 973 K and a temperature differential of 675 K in air. R_i of this module is 0.4 Ω which is about 1.6 times higher than the calculated one from resistivity of both p and n-type legs. This is due to the destruction of the Mn-113 legs. This is due to low fracture strength and toughness.

These mechanical properties of the Mn-113 bulk are improved by incorporation of Ag particles. The particle size is within 5 μm. The K_{IC} values increase with increasing Ag content and reach maximum value of 2.2 MPa m$^{1/2}$ at 9.4 wt.% of Ag content and then slightly decrease in higher Ag content. The σ_f value of the 18.8 wt.% composite is higher than that of monolithic $Ca_{0.9}Yb_{0.1}MnO_3$ bulk by 2 times. The increase in σ_f is attributed to the refinement of the matrix grains and improvement of K_{IC} by incorporation of Ag particles.

6. ACKNOWLEDGMENTS

A part of research in this paper is supported by the Japan Society for the Promotion of the Science for finical support under Grand No.196380.

REFERENCES

1. I. Terasaki, Y. Sasago and K. Uchinokura, Large thermoelectric power in $NaCo_2O_4$ single crystals, *Phys. Rev. B*, **56** 12685-12687 (1997).
2. R. Funahashi, I. Matsubara, H. Ikuta, T. Takeuchi, U. Mizutani and S. Sodeoka, An Oxide Single Crystal with High Thermoelectric Performance in Air, *Jpn. J. Appl. Phys.*, **39**, L1127-1129 (2000).
3. Y. Miyazaki, K. Kudo, M. Akoshima, Y. Ono, Y. Koike and T. Kajitani, Low-Temperature Thermoelectric Properties of the Composite Crystal $[Ca_2CoO_{3.34}]_{0.614}[CoO_2]$, *Jpn. J. Appl. Phys.*, **39** L531-L533 (2000).
4. R. Funahashi and I. Matsubara, Thermoelectric properties of Pb- and Ca-doped $(Bi_2Sr_2O_4)_xCoO_2$ whiskers, *Appl. Phys. Lett.*, **79**, 362-364 (2001).
5. M. Ohtaki H. Koga, T. Tokunaga, K. Eguchi and H. Arai, Electrical Transport Properties and High-Temperature Thermoelectric Performance of $(Ca_{0.9}M_{0.1})MnO_3$ (M = Y, La, Ce, Sm, In, Sn, Sb, Pb, Bi), *J. Solid State Chem.*, **120**, 105-111 (1995).

7. W. Shin and N. Murayama, Li-Doped Nickel Oxide as a Thermoelectric Material, *Jpn. J. Appl. Phys.*, **38** L1336-L1338 (1999).

8. R. Funahashi S. Urata, K. Mizuno, T. Kouuchi, and M. Mikami, $Ca_{2.7}Bi_{0.3}Co_4O_9/La_{0.9}Bi_{0.1}NiO_3$ thermoelectric devices with high output power density, *Appl. Phys. Lett.*, **85**, 1036-1038 (2004).

9. S. Urata, R. Funahashi, T. Mihara, Power generation of p -type $Ca_3Co_4O_9$/n-type $CaMnO_3$ module, *Proceedings of 2006 International Conference on Thermoelectrics*, 103 (2006, Wien).

10. D. Flahaut, T. Mihara, F. Funahashi, N. Nabeshima, K. Lee, H. Ohta, K. Koumoto, Thermoelectrical properties of A-site substituted $Ca_{1-x}Re_xMnO_3$ system, *J. Appl. Phys.*, **100**, 084911.1-081911.4 (2006).

11. C. Fan and M. N. Rahaman, Factors Controlling the Sintering of Ceramic Particulate Composites: I, Conventional Processing, *J. American Ceramic Society*, **75**, 2056- 2065 (1992).

12. Y. Nakada and T. Kimura, Effects of Shape and Size of Inclusions on the Sintering of ZnO—ZrO_2 Composites , *J. Amer. Ceram. Soc.*, **80**, 401-406 (1997).

13. T. Sekino, T. Nakajima, S. Ueda, and K. Niihara, Reduction and Sintering of a Nickel–Dispersed-Alumina Composite and Its Properties, *J. Amer. Ceram. Soc.*, **80**, 1139-1148 (1997).

14. M. Nawa, K. Yamazaki, T. Sekino, K. Niihara, Microstructure and mechanical behavior of 3Y-TZP/Mo nanocomposites possessing a novel interpenetrated intragranular microstructure, *J. Mater. Sci.*, **31**, 2849-2858 (1996).

15. K. Niihara, A Fracture Mechanics Analysis of Indentation-induced Palmqvist Crack in Ceramics, *J. Mater. Sci. Lett.*, **2**, 221-223 (1983).

16. A. F. Ashby, F. J. Blunt, and M. Bannister, Flow characteristics of highly constrained metal wires, *Acta Metall.*, **37**, 1847-1857 (1989).

17. B. Flinn, M. Ruehle, and A. G. Evans, Toughening in composites of Al_2O_3 reinforced with Al, *Acta Metall.*, **37**, 3001-3006 (1989).

18. W. H. Tuan, H. H. Wu, and T. J. Yang, The preparation of Al_2O_3/Ni composites by a powder coating technique, *J. Mater. Sci.*, **30**, 855-859 (1995).

INFLUENCE OF GRAIN-BOUNDARY ON TEXTURED Al-ZnO

Yoshiaki Kinemuchi, Hisashi Kaga, Satoshi Tanaka*, Keizo Uematsu*, Hiromi Nakano**, and Koji Watari

National Institute of Advanced Industrial Science and Technology, (AIST)
Nagoya, Aichi, Japan
* Nagaoka University of Technology
Nagaoka, Niigata, Japan
** Ryukoku University
Otsu, Shiga, Japan

ABSTRACT

Thermoelectric properties of ZnO ceramics are largely influenced by the mobility variation because of the formation of a double Schottky barrier at the grain boundary. It was demonstrated that magnetic texturing enabled fabrication of highly c-axis oriented ceramics with orientation degree of 100 MRD (MRD: multiples of random distribution). This effectively enhanced Hall mobility by 80% along the ab-plane. Such a well-conducting grain boundary possessed periodic structures, which was confirmed by edge-on HRTEM images. On the contrary, Hall mobility along the c-axis was identical with that of randomly oriented ceramics. Segregation of Al at the grain boundary is thought to be responsible for the low mobility along the c-axis.

INTRODUCTION

Heat inadvertently emitted from facilities is anindication of ineffectively used energy. Thermoelectric devices enable conversion of such heat into electric energy, which enhances the energy efficiency of the system. Since electric power increases with the square of the temperature difference in thermoelectric devices, high temperature durability is considered an advantage for thermoelectric materials. Oxides have advantages in application to this purpose.

Among n-type thermoelectric oxides, Al-ZnO shows high performance [1] which originates in its high mobility [2-4]. However, the mobility of polycrystalline ZnO is only half that of a single crystal [2,5]. This reduction in mobility is caused by the potential barrier at the grain boundary [6,7] and the anisotropic nature of mobility between a and c axes [4]. The later effect is dominant below 200 K, where mobility of the a-axis is two times higher than that of the c-axis.

It is reported that the mobility of ZnO ceramics strongly depends on doping amount: the higher the doping amount, the higher the mobility [1]. This fact implies two effects of aluminum: as an electron donor and as a suppressor of potential barriers at the grain boundary. Nevertheless, the potential barrier may still exist at near the solubility limit of Al, degrading the mobility of ceramics.

Because of the above influence of the grain boundary on electric transport of ZnO ceramics, the control of the grain boundary is key to improving the mobility. Sato et al. clarified the relation between atomic structure at the grain boundary and electric property across the grain boundary using ZnO bicrystals [8]. They revealed that the atomic structure at the grain boundary strongly influenced the formation of the potential barrier: incoherent boundary structure led to the formation of high potential barrier due to the strong segregation of dopant.

Thus the impact of the grain boundary on the mobility should be understood in order to bring out the potential of ZnO ceramics. This work reports the variation in mobility by magnetic texturing and carrier concentration, and correlates mobility with grain boundary structure.

EXPERIMENT

Magnetic texturing is based on the alignment of particles in a under magnetic field, where particles are suspended in the liquid phase. The driving force of the alignment is the anisotropy in the magnetic susceptibility of the crystals [9]. Disturbances of the particle alignment are Brownian motion, convection of liquid phase, and capillary force during the drying process. The former two issues can be solved by adequate temperature control, and the last is solved by quenching the state of particles by gelation, using the so-called gelcasting process. Detailed procedures are as follows. Starting powders of ZnO (Hakusui Tech Co., Japan) and γ-Al_2O_3 (Sumitomo Chemical, Japan) were weighed to have a composition of $Zn_{0.98}Al_{0.02}O$. A water-based gelcasting [10] was used in this study. The slurry was cast into a Teflon mold at room temperature, and placed in a 10 T magnetic field (TM-10VH10, Toshiba, Japan). In the magnetic field, the mold was rotated with a speed of 30 rpm. The rotational axis was perpendicular to the magnetic flux. Such rotation results in c-axis orientation, while a-axis orientation is achieved without the rotation [11]. Gelation was initiated within 10 min by adjusting the amount of catalyst after casting.

After gelation was completed, the samples were demolded and then dried in a humidity-controlled chamber at room temperature. The dry green bodies were then placed in a furnace to remove organic substances. The green bodies were heated at 848 K for 30 min in air at heating and cooling rates of 0.5 K/min. After cold isostatic pressing at 196 MPa, the green bodies were placed into an Al_2O_3 crucible. To suppress zinc evaporation at high temperatures, the powder bed technique was adopted. Then sintering at 1673 K for 10 h was carried out.

In order to control the carrier concentration precisely, oxygen partial pressure was varied during the sintering procedure. Oxygen partial pressure was monitored by ZrO_2 sensor and controlled by the gas flow of oxygen and nitrogen. Oxygen partial pressure was varied from 10^1 to 10^4 Pa.

Electric conductivity was measured by the DC four-probe method. Seebeck coefficient was measured by the static DC method. Both measurements were simultaneously carried out by a commercially available system (ZEM, ULVAC-Riko, Japan) in He atmosphere of 10 kPa. Carrier concentration and mobility were analyzed by DC Hall measurement at room temperature (Resitest8300, Toyo Corporation, Japan). Next, platinum electrodes were deposited by sputtering for the van der Pauw method. The I-V characteristics of these contacts showed a linear relation. The applied magnetic field was 0.75 T during measurement and reversed polarity measurement was carried out to cancel the voltage offset.

Pole figures were measured in reflection geometry on an X-ray diffractometer (RINT2550, Rigaku, Japan) equipped with a pole figure goniometer. Measurements were performed on polished surfaces of both magnetically processed and randomly oriented specimens for (0002) peak. The measurements were carried out in the range of azimuthal angle of $0^{\circ} < \beta < 360^{\circ}$ (2.5° steps) and polar angle of $0^{\circ} < \alpha < 75^{\circ}$ (2.5° steps). The normal direction of the specimen was set to be parallel to the rotational axis during magnetic texturing.

The microstructure of sintered specimens was observed by transmission electron microscopy (TEM, 3000F, JEOL, Japan). TEM foils were prepared using the standard technique for preparing thin ceramic foils: cutting, grinding, dimpling, and Ar-ion thinning. An energy dispersive X-ray spectroscopy (EDS) was also utilized for elemental analyses of grain boundaries.

RESULTS AND DISCUSSION

Because ZnO has a wurtzite structure (hexagonal structure), the properties have anisotropy between the a- and c-axes, and they are isotropic along the ab-plane. Therefore, a c-axis oriented specimen is identical with a single crystal if the grain boundary effect is negligible. In this sense, c-axis orientation is preferable to enhance the conductivity of ceramics. Figure 1 shows the pole figure of (0002), indicating the c-axis orientation of the specimen. Notic that the specimen possesses an intensively high degree of orientation of 100 MRD. (MRD: multiples of random distribution) Such a high degree of orientation can hardly be obtained by other texturing processes.

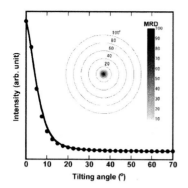

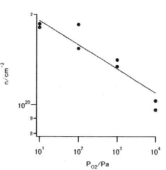

Figure 1 Pole figure of (0002) measured for magnetic textured zinc oxide.

Figure 2 Carrier concentration (n) as a function of oxygen partial pressure (P_{O2}).

Table 1. Electric properties of c-axis oriented and randomly oriented ZnO ceramics.

	n [1/cm^3]	μ [cm^2/Vs]	σ [S/cm]	S [μV/K]
Along a-b plane	9.2×10^{19}	88.4	1331	-96
Along c axis	9.3	48.2	718	-97
Randomly oriented	9.5	49.3	750	-96

The electric properties of c-axis oriented zinc oxide are summarized in Table 1. The results of randomly oriented zinc oxide with the same composition are also shown as a control. The carrier concentration of each specimen was almost the same, which was slightly higher than the reported value of 7×10^{19} cm^{-3} [5]. A remarkable difference can be seen in the mobility. The mobility was enhanced 1.8 times above than that of the control, which resulted in an increase in conductivity. On the contrary, the Seebeck coefficient did not differ much owing to similar carrier concentrations.

Figure 2 indicates the dependence of carrier concentration on the oxygen partial pressure of the sintering procedure. Carrier concentration steadily decreased with an increase in oxygen partial pressure. Since the increase in carrier concentration is owing to the formation of oxygen vacancy, the mean free path of the carrier may be reduced by this defect, leading to a decrease in mobility with an increase in carrier concentration. Such a tendency is significant regarding mobility along the ab-plane. On the contrary, the mobility of randomly oriented ZnO showed the opposite tendency as shown in Fig. 3. Srikant et al. also reported the enhancement of mobility with increase in carrier concentration below the carrier concentration of 3×10^{19} cm^{-3}, and explained the behavior by existence of a double Schottky barrier at the grain boundary [6].

Based on the double Schottky barrier model, the height of potential barrier (Φ) and its effect on mobility (μ) are described as follows.

$$\mu = \frac{v_n L}{V}\exp\left(\frac{eV}{n_g kT}\right)\exp\left(-\frac{e\Phi}{kT}\right) \qquad \Phi = \frac{eN_s^2}{8\varepsilon\varepsilon_0 n}$$

(1)

Here, e is the electric charge, N_s density of trap at grain boundary, ε relative dielectric constant, ε_0 the

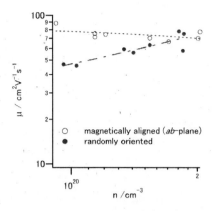

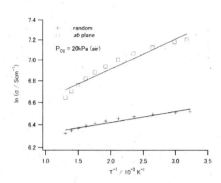

Figure 3 Mobility (μ) vs carrier concentration (n)

Figure 4 Conductivity (σ) vs temperature (T), showing steeper slope for ab-plane than that for randomly oriented ceramics.

dielectric constant in vacuum, n carrier concentration in grain, v_n thermal velocity of the electron, L length of specimen, V applied voltage to the specimen, n_g the number of grain boundaries along L, k the Boltzmann constant, and T absolute temperature [7]. From Eq. (1), the dependence of mobility on carrier concentration is given as:

$$\frac{\partial \ln \mu}{\partial \ln n} = \frac{e\Phi}{kT}$$

(2)

Based on the slope of $\ln(\mu)$ vs $\ln(n)$ (0.79), the potential height, $e\Phi$ is estimated to be 20 meV.

The temperature dependence of conductivity is shown in Fig. 4. A remarkable difference in the slope can be seen between conductivity of the ab-plane and that of a randomly oriented specimen. The difference in activation energy between them was found to be 17 meV, in reasonable agreement with the height of the potential barrier estimated from Eq. 2.

Figure 5 shows edge-on HRTEM images of the grain boundary along the ab-plane. There is no intergranular phase and the two adjacent grains are directly bonded at the atomic level. Tilt angle of the grains, that is, the angle made by the $[1\bar{1}00]$ of both grains, ranges between 13 and 16.5° in this grain boundary. Corresponding to the tilt angle, several periodic structures were observed as shown in Fig.5 (b)-(d). For instance, Fig.5 (d) shows the tilting angle of 16.5° which is close to that of the Σ49 boundary [8]. It is reported that the Σ49 boundary of zinc oxide is composed of three different structural units. These structural units align periodically in order to minimize grain boundary energy. Since such coherent grain boundaries are energetically stable, the segregation amount of impurities is expected to be lower than that of the incoherent grain boundary, which is the normal boundary structure in randomly oriented ceramics. Energy dispersive X-ray spectroscopy analysis of the grain boundary showed the impurity level was lower than a detection limit of 0.5 at%.

On the other hand, segregation of aluminum at grain boundary was observed at an incoherent grain boundary as shown in Fig.6. The arrows in the figure indicate the c-axis direction, thus the grain boundary inclines to the c-axis; moreover ab-planes are slightly twisted at this site. Because of this geometrical configuration, coherency of the grain boundary is much lower than that of the grain

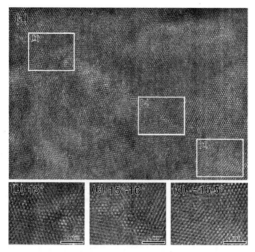

Figure 5 Grain boundary structure of magnetic textured zinc oxide, indicating periodic structures of grain boundary due to high orientation degree.

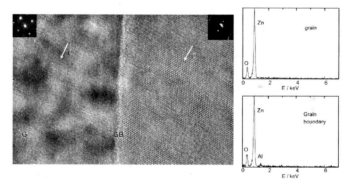

Figure 6 EDS results of incoherent grain boundary, indicating aluminum segregation at the boundary.

boundary shown in Fig.5. This low coherency probably causes the segregation of aluminum. In addition, the detected amount of aluminum at the grain boundary was higher than that of the grain interior, and lower than that of spinel phase ($ZnAl_2O_4$). Sato et al. reported that praseodymium, the well known additive causing the formation of a double Schottky barrier, segregated at the grain boundary, and the amount of segregation was much influenced by the structure of the grain boundary [8]. It is believed that aluminum also behaves similarly. One uncertainty is the origin of the interface state which traps the electron from the grain interior. Since aluminum is a trivalent cation, aluminum itself may not act as acceptor. The valency of praseodymium is also three, and the origin of a double Schottky barrier by this additive is explained by the formation of zinc vacancy [8]. In the case of aluminum, the vacancy of zinc can be created similarly in order to maintain charge neutrality. Localization of carrier near the zinc vacancy may cause the formation of the double Schottky barrier. In fact, addition of aluminum increased the trap density at the grain boundary in ZnO varistors. This phenomenon was observed by deep level transient spectroscopy [12], suggesting an unfavorable effect of Al segregation on the conductivity of ZnO ceramics.

SUMMARY
The present study has been explored the influence of the grain boundary on the conductivity of Al-ZnO ceramics. In the randomly oriented ceramics, the formation of a double Schottky barrier at the grain boundary caused reduction in mobility, which was confirmed by the activation energy of conductivity and dependence of mobility on carrier concentration. Here, the potential height of the Schottky barrier was estimated to be 17-20 meV. Observation of grain boundary structure by HRTEM and compositional analysis by EDS revealed segregation of Al was responsible for the formation of the double Schottky barrier. Magnetic texturing was effective to avoid such segregation, leading to high mobility of 80-90 $cm^2/(Vs)$.

ACKNOWLEDGMENT
The authors would like to thank Dr. W. Shin of AIST for his experimental support.

REFERENCES
[1] M. Ohtaki, T. Tsubota, K. Eguchi and H. Arai: J. Appl. Phys. Vol. 79 (1996), p. 1816
[2] Sol E. Harrison: Phys. Rev. Vol. 93 (1954), p. 52
[3] A. R. Hutson: Phys. Rev. Vol. 108 (1957), p. 222.
[4] P. Wagner and R. Helbig: J. Phys. Chem. Solids Vol. 35 (1974), p. 327
[5] T. Tsubota, M. Ohtaki, K. Eguchi and H. Arai: J. Mater. Chem. Vol. 7 (1997), p. 85
[6] V. Srikant, V. Sergo and D. R. Clarke: J. Am. Ceram. Soc. Vol. 78 (1995), p. 1935
[7] F.M. Hossain, J. Nishii, S. Takagi, A. Ohtomo, T. Fukumura, H. Fujioka, H. Ohno, H. Koinuma and M. Kawasaki: J. Appl. Phys. Vol. 94 (2003), p. 7768
[8] Y. Sato, T. Yamamoto and Y. Ikuhara: J. Am. Ceram. Soc. Vol. 90 (2007), p. 337
[9] Y. Sakka and T.S. Suzuki: J. Ceram. Soc. Jpn. Vol. 113 (2005), p. 26
[10] A.C. Young, O.O. Omatete, M.A. Janney and P.A. Menchofer: J. Am. Ceram. Soc. Vol. 74 (1991), p. 612
[11] S. Tanaka, A. Makiya, Z. Kato, N. Uchida, T. Kimura and K. Uematsu: J. Mater. Res. Vol. 21 (2006), p. 703
[12] J. Fan and R. Freer: J. Am. Ceram. Soc. Vol. 77 (1994), p. 2663

EVALUATION ON THERMO-MECHANICAL INTEGRITY OF THERMOELECTRIC MODULE FOR HEAT RECOVERY AT LOW TEMPERATURE

Yujiro Nakatani[1], Takahiko Shindo[1], Kengo Wakamatsu[1], Takehisa Hino[1], Takashi Ohishi[1], Haruo Matsumuro[2], Yoshiyasu Itoh[1]
1. Power and Industrial Systems R&D Center, Toshiba Corporation, Yokohama, Japan.
2. New Business Promotion Department, Toshiba Corporation, Tokyo, Japan.

ABSTRACT

We are currently developing a low-cost and high-efficiency Bi–Te thermoelectric module for power generation. This module is aimed at heat recovery at temperatures of 423 K or less. Though the total amount of waste heat energy below 423 K is considerably large, it is not utilized because of its low density. In addition, the utilization of such low-density energy is considered economically unviable. Therefore, our objective is to enhance the thermoelectric efficiency and improve the cost performance of the module. One of the methods to improve the cost performance is to achieve long-term operation without maintenance through enhanced module reliability. This necessitates evaluation of the thermomechanical integrity of the thermoelectric module. This paper describes analytical and experimental studies on the mechanical aspects of the integrity of the thermoelectric module against thermal stress and/or heat-cycle damage.

To estimate the reliability of the thermoelectric module, three-dimensional elastic-plastic finite element analysis (FEM) was performed, and heat-cycle tests were conducted with a heat-cycle testing machine for thermoelectric modules, which was developed in this study. The analytical and experimental results indicated the design concept for the mechanical aspects of the structural integrity of the thermoelectric module.

INTRODUCTION

In order to mitigate global warming, which may have serious consequences for the global environment, targets for reductions in emissions of greenhouse gases that cause climate change were established for developed countries in the Kyoto Protocol at COP3 (the Third Session of the Conference of the Parties to the United Nations Framework Convention on Climate Change). Recently, various attempts have been made to reduce greenhouse gas emissions, such as by converting fossil-fuel-based energy into other forms of energy or through effective utilization of waste heat. Meanwhile, although the total amount of waste heat or unutilized energy below 423 K is considerable, it is not recovered because of its low density[1]. In addition, the utilization of such low-density energy is not considered to be economically viable. Several industrial equipments emit large amounts of low-temperature heat below 423 K, but most of this waste heat still remains unused. This research was conducted aiming at the utilization of such unutilized energy from industrial infrastructure.

In this research, one of our objectives was to enhance the thermoelectric efficiency of thermoelectric materials and modules[2]. The thermoelectric materials developed in this study were bismuth telluride (Bi–Te)-based alloys doped with Se and Sb. The materials were cast by the Bridgman method, by which directionally solidified (DS) thermoelectric elements could be prepared. High-efficiency thermoelements could be obtained as a result of close control of the microstructure and improvements in the proportions of doping elements[2]. The developed thermoelectric module is shown in Figure 1. The module is of the skeleton type and consists of Bi–Te thermoelements of the p-type and n-type, electrodes of Al and Mo formed by atmospheric plasma spraying (APS) coating, and mold resin[2,3].

Another objective was to achieve long-term operation without maintenance by achieving enhanced module reliability, because reliability is an important issue for applying the modules to actual industrial equipment. The failure or malfunction of the thermoelectric module should not influence the functioning of the equipment while the equipment is in operation. Additionally, it is also important to

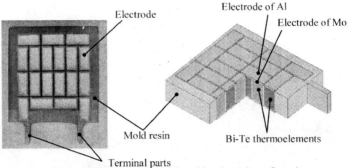

Figure 1. Thermoelectric module and its schematic configuration.

achieve long-term operation without maintenance for a reduced life-cycle cost.

In this paper, in order to discuss the concept of structural design for increased module integrity, numerical analyses were performed to evaluate the behavior of thermal stress and strain on the developed thermoelectric modules, and heat-cycle tests were conducted with a heat-cycle testing machine. In addition, the application of a thermoelectric power generation system for the utilization of unutilized energy is reported.

EXPERIMENTAL DETAILS

Finite Eement Mthod Aalysis

To evaluate the stress and strain state of a Bi–Te thermoelectric module subjected to variable thermal conditions, thermomechanical elastic-plastic finite element method (FEM) analyses were conducted using the MARC 2005 code. As shown in Figure 1, the module consists of Bi–Te elements of the p-type and n-type, a Mo coating on the elements acting as a diffusion barrier, an Al coating as the electrode layer, and mold resin. A total of 18 couples of p-type/n-type thermoelements are placed in the module. In this study, a couple of thermoelements and an entire module were modeled to evaluate the local strain and macroscopic stress behavior, respectively, as shown in Figures 2-(a) and (b). The finite element mesh used for the calculation consisted of 4320 isotropic three-dimensional 8-node hexahedron elements for the local model and 80784 elements for the complete model. The elements at

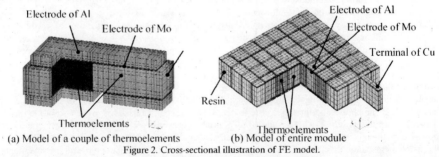

(a) Model of a couple of thermoelements   (b) Model of entire module

Figure 2. Cross-sectional illustration of FE model.

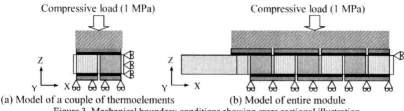

(a) Model of a couple of thermoelements   (b) Model of entire module

Figure 3. Mechanical boundary conditions showing cross-sectional illustration.

the interfaces of each component were 0.05 mm thick. The initial temperature of all the finite elements was set to 298 K, which simulates room temperature. The hot side was then varied to 398 K while the cold side was maintained at 298 K. The mechanical boundary conditions for the thermomechanical analyses are shown in Figure 3. The elastic modulus, Poisson's ratio, heat expansion coefficient, and plasticity data are listed in Tables I and II.

Table I. Material Properties for FEM Analysis

	Elastic modulus (GPa)	Poisson's ratio	Thermal expansion coefficient (10^{-6} /K)
Bi–Te	73.3	0.32	15.1
Al coating	84.0	0.37	23.9
Mo coating	317	0.29	5.1
Resin	1.9	0.40	60.0

Table II. Plasticity Data for FEM Analysis

Al coating		Mo coating	
Plastic strain (%)	Stress (MPa)	Plastic strain (%)	Stress (MPa)
0.0	3.1	0.0	44.4
4.8	15.5	4.8	61.2

Heat-cycle Test

In this study, a heat-cycle testing machine for thermoelectric modules was developed. This testing machine can simulate the actual operating conditions. It has a mechanism to supply heat in cycles to the upper surface of the module while concurrently cooling the lower surface. During the tests, the hot side of the module was loaded with continuous heat cycling from 323 K to 403 K, while the cold side was maintained at 298 K. The modules were also compressed at 1 MPa to represent the actual operating conditions. The testing machine can detect changes in the internal resistance of the thermoelectric module accompanied by crack growth. The thermoelectric module employed for the heat-cycle tests was a 40 × 40 mm module composed of 18 couples of directionally solidified thermoelectric elements. These modules were polished into three types of modules with electrode coating thicknesses of 300 μm, 400 μm, and 480 μm. The target number of cycles was defined as 3,000 cycles. Furthermore, the test was interrupted arbitrarily when a module did not fail after 3,000 cycles. Then, the region of failure was observed by an optical microscope.

RESULTS AND DISCUSSION

Thermal Stress and Strain Estimation

Although the temperature variation in the thermoelectric module is not very extensive, the multilayer structure of the module may induce thermal shear strain owing to the differences in local thermal expansion among the individual materials, as shown in Figure 4-(a). On the other hand, as the module has a temperature difference between the hot and cold sides, thermal stresses caused by macroscopic thermal deformation may also occur, as shown in Figure 4-(b). Therefore, numerical analyses were carried out to estimate the distribution of (a) the thermal strain caused by the differences in local thermal expansion and (b) the thermal stress caused by macroscopic thermal deformation.

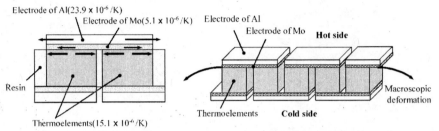

(a) Differences in local thermal expansion (b) Macroscopic thermal deformation

Figure 4. Illustration showing local strain and macroscopic deformation.

Thermal Strain Caused by Differences in Local Thermal Expansion

To estimate the occurrence of thermal strain caused by differences in local thermal expansion, as shown in Figure 4-(a), thermomechanical FEM analysis was performed with the model of a couple of thermoelements. The FEM analysis revealed that shear strain appeared dominantly. Figures 5-(a) and (b) show contour maps of the shear strain obtained from the FEM analysis. The contour map in Figure 5-(a) shows that the shear strain in the ZX-direction distributes in the vicinity of the hot side of the thermoelements. Figure 5-(b) shows that the shear strain in the YZ-direction distributes in the vicinity of the interface between the Al coating and the Mo coating. The numerical values in the figures are the totals of the plastic strain and elastic strain. The maximum values of shear strain at (A) and (B) are 0.57% and 0.90%, respectively. This result suggests the necessity for reduction of the strain at (B). To evaluate the configurational parameters affecting the thermal strain, sensitivity analysis was carried out by varying the thickness of the electrode coating. As shown in Figure 6, decreasing the thickness of the Al coating was found to be effective in reducing the strain at (B).

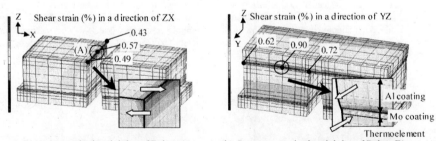

(a) Contour map in the vicinity of Point (A) (b) Contour map in the vicinity of Point (B)

Figure 5. Contour maps of the shear strain obtained from the elastic-plastic FEM analysis.

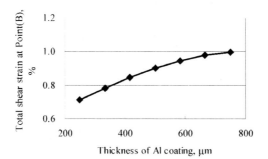

Figure 6. Variation in thermal shear strain accompanied by variation in thickness of Al electrode coating.

Thermal Stress Caused by Macroscopic Thermal Deformation

To estimate the occurrence of thermal stress caused by macroscopic thermal deformation, as shown in Figure 4-(b), thermomechanical FEM analysis was performed with the model of the entire module; the result of heat transfer analysis is shown in Figure 7. This figure shows that the temperatures of the thermoelements change almost linearly from 298 K to 398 K. Subsequently, elastic-plastic thermomechanical analysis was carried out based on the temperature distribution obtained from the heat transfer analysis. In this analysis, we focused on the thermal stress on the electrode coating caused by the macroscopic thermal deformation; the results shown in Figure 8 reveal that the thermal stress appears in the vicinity of the center of the electrode coating. As such stress is caused by the macroscopic bending deformation of the whole module, it is expected that the thermal stress can be reduced by increasing the thickness of the electrode coating.

Figure 7. Result of heat transfer analysis using model of entire module showing temperature distribution.

Figure 8. Result of thermomechanical analysis of model of entire module showing stress distribution.

Direction of Structural Design for Thermoelectric Module

The results described above are summarized in Table III. It shows that (a) the design direction for reducing the thermal strain caused by the differences in local thermal expansion and (b) the design direction for reducing the thermal stress caused by macroscopic thermal deformation are in conflict with each other. Therefore, experimental verification is necessary to identify the area in which the module is damaged and how this damage occurs.

Table III. Summarized Result of Suggested Design for Reduction of Stress and Strain

Cause of damage	Expected morphology of damage	Direction for reduction of strain or stress
(a) Thermal strain caused by differences in local thermal expansion	Delamination or interlayer crack propagation of electrode coating	Decreasing the thickness of the electrode coating
(b) Thermal stress caused by macroscopic thermal deformation	Crack propagarion in the centeral part of electrode coating	Increasing the thickness of the electrode coating

Heat-cycle Estimation for Thermoelectric Module

A heat-cycle testing machine was developed to verify the effect of the thickness of the electrode coating. A photograph of the developed testing machine and its schematic configuration are shown in Figure 9. It has a mechanism to supply a heat cycle to the upper surface of the module while concurrently cooling the lower surface. The modules can also be compressed by a spring load to simulate the actual operating conditions. The compression load is detected and verified with a load cell. The testing machine can also detect changes in the internal resistance of the thermoelectric module accompanied by crack growth. These features are appropriate to estimate the heat-cycle properties of modules for power generation.

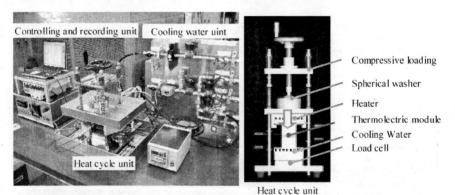

Heat cycle unit

Figure 9. Photograph and schematic illustration of the developed heat-cycle testing machine.

Heat-cycle tests on the thermoelectric modules were carried out with the developed heat-cycle testing machine. Since the thickness of the electrode coating was found to be a significant design parameter for enhancement of reliability as described, three kinds of modules with different electrode coating thickness of 300 µm, 400 µm, and 480 µm were prepared for the heat-cycle test.

Figure 10 shows the result for the module with the 300-µm electrode coating. The internal resistance was elevated at 1200 cycles, following which failure occurred. Microscopic observation revealed that failure was induced by a crack that propagated at the center of the electrode coating, as shown in Figure 11. This corresponds well to the area where the stress appeared, as shown in Figure 8. Therefore, it is suggested that the failure was induced by the cyclic bending of the electrode coating caused by the macroscopic thermal deformation.

As shown in Table III, the thermal stress due to macroscopic thermal deformation could be reduced by increasing the thickness of the electrode coating. This is also convenient from the electrical

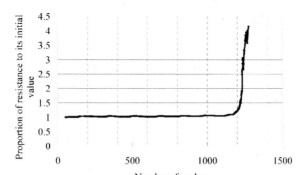

Figure 10. Result of heat-cycle test in the module with a 300-μm electrode coating showing the changes in internal resistance.

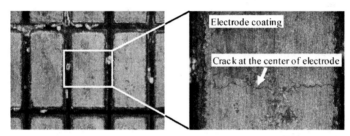

Figure 11. Microscopic observation of crack propagation at the center of the electrode coating.

design standpoint as the increased thickness of the electrode coating will also decrease the electrical resistance of the electrode.

Table IV shows the results of the heat-cycle tests on the modules with electrode thicknesses of 300 μm, 400 μm, and 480 μm; the table also shows that the heat-cycle lives of the modules with 400-μm and 480-μm electrodes were both greater than 3,000 cycles. Thus, it is clarified that the objective of 3,000 cycles in heat-cycle life was achieved due to the effect of increasing the electrode thickness. This result also indicates that such analytical and experimental estimation processes are effective in increasing the structural integrity of thermoelectric module.

Table IV. Results of Heat-cycle Test

Module No.	Thickness of electrode coating μm	Broken or not broken	Number of cycles
1	300	Broken	1200
2	400	Not broken	3800
3	480	Not broken	3900

CONCLUSION

In order to expand the practical applications of thermoelectric systems in ind infrastructure, it is essential for thermoelectric modules to have long-life integrity. In this st structural design method for the development of reliable thermoelectric modules was develope experimental verification of the heat cycle was carried out.

The analytical and experimental results indicated the following design concepts f mechanical aspects of the structural integrity of the skeleton-type thermoelectric module:

(1) In order to reduce the thermal strain caused by the differences in local thermal expansio effective to decrease the thickness of the Al electrode coating. And for the reduction of the tl stress caused by macroscopic thermal deformation, it is effective to increase the thickness electrode coating.

(2) A heat-cycle testing machine was developed in order to simulate the actual operating conc The experimental results revealed that the integrity of the module could be enhanced by the dev design concept obtained from the analytical examinations, and the objective of 3,000 cyc heat-cycle life was achieved.

(3) It was revealed that the analytical and experimental estimation processes discussed in this stu effective for increasing the structural integrity of thermoelectric modules.

ACKNOWLEDGMENTS

This research was conducted as a part of the Japanese National Project on the Developm Advanced Thermoelectric Conversion Systems. The aid received for this research from the Energy and Industrial Technology Development Organization (NEDO) is gratefully acknowledge

REFERENCES
[1]D. M. Rowe, G. Min, S. G. K. Williams, and V. Kuznetsov, An Up-date on the Thermoe Recovery of Low Temperature Waste Heat, Proc. of World Renewable Energy Congre: 1499–1504 (2000).
[2]R. Takaku, T. Hino, T. Shindo, and Y. Itoh, Development of Skeleton-type Thermoelectric Mod Power Generation, Proc. of 23rd Int. Conf. on Thermoelectrics, #038 (2004).
[3]Y. Nakatani, R. Takaku, T. Hino, T. Shindo, and Y. Itoh, Mechanical Aspects of Stru Optimization in a Bi-Te Thermoelectric Module for Power Generation, Mater. Res. Soc. Sympo. **842**, S4.7.1–S4.7.6 (2005).

TRANSPORT PROPERTIES OF $Sn_{24}P_{19.3}Br_8$ and $Sn_{17}Zn_7P_{22}Br_8$

Stevce Stefanoski[1], Andrei V. Shevelkov[2], George S. Nolas[1]
[1]Department of Physics, University of South Florida, Tampa, Florida 33620
[2]Inorganic Synthesis Laboratory, Department of Chemistry, Moscow State University, 119992 Moscow, Russia

ABSTRACT
        Transport properties of two type-I clathrate compounds $Sn_{24}P_{19.3}Br_8$ and $Sn_{17}Zn_7P_{22}Br_8$ are investigated. These are non-conventional clathrate phases formed by cationic framework sequestering anionic guest. The crystal structure as well as resistivity, thermal conductivity, Seebeck coefficient and Hall measurement for both compounds is reported. Their potential for thermoelectric application is also discussed.

INTRODUCTION
        In pursuit of alternative, environmentally friendly energy sources and solid state refrigeration, thermoelectric materials are of interest. Any effective thermoelectric material should have a high electrical conductivity, σ, a large Seebeck coefficient, S, and low thermal conductivity, κ, on the appropriate temperature of operation. The dimensionless figure of merit is defined as

$$ZT = \frac{S^2 \sigma T}{\kappa} \tag{1}$$

and is a measure of the thermoelectric properties of a material. The theoretical value of this figure of merit is not limited, and could go up to infinity, however, the practical upper limit for the past thirty years has been $ZT \sim 1$, and only recently exceeds unity[13] in bulk materials. Equation (1) shows that the value of ZT could be maximized either by increasing the power factor $S^2\sigma$ or decreasing the total thermal conductivity $\kappa = \kappa_l + \kappa_e$, or both. Here κ_l and κ_e are the lattice and electron contributions to the thermal conductivity, respectively.

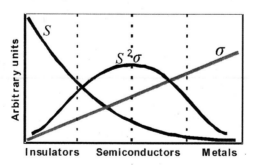

Figure 1. Optimizing a thermoelectric material.

However, these parameters are interrelated and their optimization is not a simple matter. From the Wiedemann-Franz law, $\kappa_e/\sigma = 3/2(k_B/e)^2 T$, we see that the ratio between the κ_e and σ is essentially constant at a given temperature. Therefore, decreasing one property decreases the other. Also, for typical materials an increase in σ will lead to a decrease in S (see Figure 1). From this simple analysis a good thermoelectric material should be a semiconductor due to the fact that the maximum power factor may be established for a particular carrier concentration.

The "phonon glass electron crystal" (PGEC) concept introduced by Slack identifies a range of crystal structures potentially relevant for thermoelectric applications[14]. A cage-like structure sequestering guest atoms is a model structure. The guest atoms can "rattle" around their equilibrium positions inside the atomic "cages" thereby creating disorder. This leads to a suppression of the thermal conductivity from this effective phonon scattering mechanism[15]. A class of materials known as type-I clathrates has such a structure. Typically clathrates contain cationic alkali, alkaline-earth or europium guest atoms. But the two compounds that are considered here have cationic framework sequestering anions as guests.

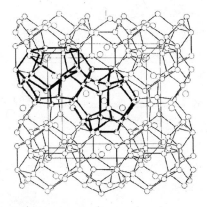

Figure 2. A polyhedral presentation of the crystal structure of a type-I clathrate. Pentagonal dodecahedra are drawn in the center, and tetrakaidecahedra are on the left.
Reprinted from [12]. Copyright 2000, American Physical Society.

The term clathrate was first introduced by Powell[5] in 1948 to identify organic inclusion compounds, and expanded on gas hydrates by Pauling[10]. Shortly after, inter-metallic compounds having the crystal structure of the chlorine hydrates were discovered[10]. They have a cubic unit cell with space group $Pm\bar{3}n$. Two kinds of building blocks participate in the unit cell structure, dodecahedron and tetrakaidecahedron, as shown in Figure 2.

SYNTHESIS AND CRYSTAL STRUCTURE

Two samples of tin-clathrates with stoichiometry $Sn_{24}P_{19.3}Br_8$ and $Sn_{17}Zn_7P_{22}Br_8$ were sent by Dr. Andrei Shevelkov from the Moscow Sate University in Russia. The following materials were used for synthesis[1,2]: Metallic tin (Reakhim 99.99%), Red phosphorus (Reakhim 97%) were purified, Tin (II) and Tin (IV) bromide were synthesized. Samples with overall $Sn_{24}P_{19.3}Br_x I_{8-x}$ (x=0,1,2,3,4,5,6,7,8) composition were prepared by heating the respective stoichiometric mixtures of tin, red phosphorus, tin (IV) iodide and tin (II) bromide in sealed silica tubes under vacuum at 725 K for 5 days. The samples were then reground and heated again at 675 K for another 14 days, and then furnace-cooled to room temperature. The compound $Sn_{17}Zn_7P_{22}Br_8$ was prepared by heating the respective stoichiometric mixtures of tin, red phosphorus and tin (IV) bromide in sealed silica tubes under vacuum at 773 K for 5 to days. The samples were then reground and heated in sealed ampoules at 573 K for another 14 days, followed by slow cooling to room temperature.

Fragments of the crystal structures of both compounds are shown on Figure 3. These are not conventional Zintl phases, formed by cationic framework hosting anions inside the two different polyhedra.

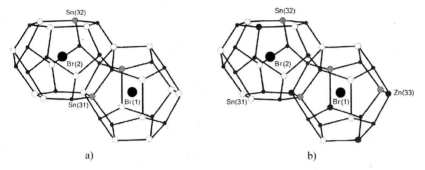

Figure 3. Fragments of the crystal structure of a) $Sn_{24}P_{19.3}Br_8$ and b) $Sn_{17}Zn_7P_{22}Br_8$
Sn(31,) large white spheres; Sn(32), large light gray spheres; P, small gray spheres;
Zn, large dark gray spheres; Br, large black spheres.
Reprinted from [2] with permission from Elsevier Science.

The bromine atoms can occupy two distinct sites, Br(1) and Br(2), both inside the polyhedra that form the crystal structure of $Sn_{24}P_{19.3}Br_8$ (Fig. 3a). The pentagonal dodecahedra sequester the Br(1) atoms, and tetrakaidecahedra sequester Br(2) atoms. The Br(1) atom is surrounded by 20 atoms, 12Sn+8P forming a pentagonal dodecahedron, and the Br(2) atom is surrounded by 24 atoms 12Sn+12P forming a tetrakaidecahedron. The two kinds of phosphorus atoms P(1) and P(2) occupy $6c$ and $16i$ sites, respectively, while the Sn atoms occupy $24k$. This site splits into two positions occupied by Sn(31) and Sn(32). Sn (31) is tetrahedraly bonded with 3 phosphorus atoms and one tin atom Sn(32). Sn(32) has two phosphorus and one tin neighbor in the first coordination sphere plus three more distant Sn(32) atoms in the second coordination sphere.

The framework of the compound $Sn_{17}Zn_7P_{22}Br_8$ is composed of Sn, Zn and P, the Br being encapsulated inside the polyhedra. This compound is vacancy-free on the framework sites since there are 46 atoms that form the framework. It crystallizes in the cubic system, with lattice parameter a=10.7449 Å. Again, there are two bromine sites Br(1) and Br(2) encapsulated in a 20-vertex dodecahedron, and 24-vertex tetrakaidecahedron, respectively (Fig. 3b). The closest distance from the bromine to the framework is 3.55 Å. Br(2) atoms show high and anisotropic Atomic Displacement Parameters[4]. The unique metal site Sn(3) splits into two closely lying atomic positions set at Sn(31) and Sn(32). Each Sn(32) atom forms two bonds with P(2)-atoms and one Sn(31)- Sn(32) bond. The P(2) atom forms one homonuclear P(2)-P(2) bond, further surrounded with three metal atoms Sn(31) or Sn(32) forming a distorted tetrahedron. The P(1) atom is surrounded by 4 metal atoms forming almost regular tetrahedron. The sites at which atoms are located are as follows: Br(1) at $2a$, Br(2) at $6d$, P(1) at $6c$, P(2) at $16i$ and Sn and Zn atoms at $24k$. Some of the inter-atomic distances are: P(1)-

Sn(31) - 2.458 Å; P(1)-Sn(32) - 2.554 Å; P(1)-Zn(33) - 2.422 Å; Sn(31)-Sn(32) - 2.548 Å; Sn(32)-Zn(33) - 2.570 Å.

EXPERIMENTAL PROCEDURE

The two specimens were densified for transport measurements by hot pressing at 450 ^{0}C and 422 MPa for two hours. The polycrystalline pellets were cut into parallelepipeds of size $2 \times 2 \times 5$ mm^3 using a wire saw to reduce the surface damages. Steady state Seebeck coefficient, thermal conductivity as well as four probe resistivity measurements were performed using a radiation-shielded vacuum probe in a custom-designed closed-cycle refrigerator[16]. To measure the temperature gradient, a differential thermocouple was employed. 0.001 in. copper wires were used as voltage probes. Hall measurements were carried out at room temperature employing a 2 T electromagnet, using pellets cut in parallelepipeds of size $2 \times 0.5 \times 5$ mm^3 at successive positive and negative magnetic fields to eliminate voltage probes misalignments.

RESULTS AND DISCUSION

Fig. 4 shows the resistivity, ρ, versus temperature for $Sn_{24}P_{19.3}Br_8$. ρ increases slowly with temperature in the temperature interval 12-55 K, then decreases rapidly with increasing temperature in the temperature interval 55-100 K and continues to decrease up to room temperature, reaching a value of 15 mΩ-cm at room temperature. The maximum value of the resistivity is $5,4 \times 10^3$ mΩ-cm at 55 K.

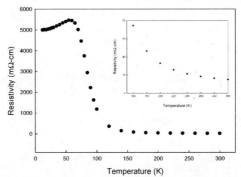

Figure 4. Resistivity versus temperature for $Sn_{24}P_{19.3}Br_8$.

ρ for $Sn_{17}Zn_7P_{22}Br_8$ increases with increasing the temperature in the temperature interval 10-200 K, reaching a peak 6.6×10^5 mΩ-cm at 200K, and then continuously decrease up to 2.6×10^5 mΩ-cm at room temperature.

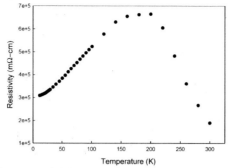

Figure 5. Resistivity versus temperature for $Sn_{17}Zn_7P_{22}Br_8$.

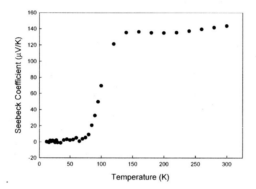

Figure 6. Seebeck coefficient versus temperature for $Sn_{24}P_{19.3}Br_8$.

As shown in Figure 6, S continuously increases with temperature throughout the whole temperature interval and is positive. S has a room temperature value of $40\,\mu V/K$.

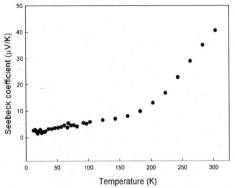

Figure 7. Seebeck coefficient versus temperature for $Sn_{17}Zn_7P_{22}Br_8$

S for $Sn_{17}Zn_7P_{22}Br_8$ increases throughout the entire temperature range (Fig.7) reaching a room temperature value of $40 \,\mu V/K$.

Simultaneous view of the thermal conductivities for both compounds is given on Fig. 8

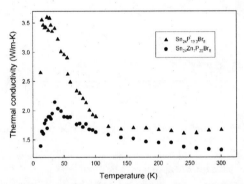

Figure 8. Thermal conductivities for $Sn_{24}P_{19.3}Br_8$ and $Sn_{17}Zn_7P_{22}Br_8$

κ shows typical dielectric behavior. The room temperature values of κ are $1.6 \,Wm^{-1}K^{-1}$ for $Sn_{24}P_{19.3}Br_8$ and $1.3 \,Wm^{-1}K^{-1}$ for $Sn_{17}Zn_7P_{22}Br_8$. This value is similar to that for the commercially used thermoelectric alloys based on Bi_2Te_3. The maximum values for κ are $3.6 \,Wm^{-1}K^{-1}$ for $Sn_{24}P_{19.3}Br_8$ and $2.1 \,Wm^{-1}K^{-1}$ for $Sn_{17}Zn_7P_{22}Br_8$. The thermal conductivity for $Sn_{17}Zn_7P_{22}Br_8$ is lower throughout the whole temperature range, which implies that Zn produces alloy scattering which further suppresses κ.

Room temperature ZT values for both compounds are 0.03 for $Sn_{24}P_{19.3}Br_8$ and 0.0001 for $Sn_{17}Zn_7P_{22}Br_8$. κ for the latter compound is lower than for the first one, but the resistivity is 2 orders of magnitude greater throughout the largest part of the temperature range. This implies that the presence of Zn in $Sn_{17}Zn_7P_{22}Br_8$ also causes scattering of the charge carriers thereby reducing σ greatly.

The Hall measurements yielded carrier concentration of 1.6×10^{22} cm^{-3} for $Sn_{24}P_{19.3}Br_8$, and 3.4×10^{17} cm^{-3} for $Sn_{17}Zn_7P_{22}Br_8$.

CONCLUSION

Transport properties for two non-conventional Zintl compounds $Sn_{24}P_{19.3}Br_8$ and $Sn_{17}Zn_7P_{22}Br_8$ were investigated. These compounds contain cationic framework sequestering anionic atoms. Presence of additional Zn atom in $Sn_{17}Zn_7P_{22}Br_8$ leads to effective phonon scattering and lowering of κ. The low carrier concentration for $Sn_{17}Zn_7P_{22}Br_8$ gives rise to its resistivity compared with the lower resistivity and higher carrier concentration for $Sn_{24}P_{19.3}Br_8$. However, the ZT values for both compounds are insufficient for significant thermoelectric applications.

REFERENCES

[1]Crystal Structure, thermoelectric and magnetic properties of the type-I clathrate solid solutions $Sn_{24}P_{19.3(2)}Br_xI_{8-x}$ (0 ¡Ük ¡Ü8) and $Sn_{24}P_{19.3(2)}Cl_yI_{1-y}$ (y ¡Ü0.8), Julia V. Zaikina et al., Solid State Sci. (2007), doi 10.1016/j.solidstatescience.2007.05.008

[2]Novel compounds $Sn_{20}Zn_4P_{22-v}I_8$ ($v = 1.2$), $Sn_{17}Zn_7P_{22}I_8$, and $Sn_{17}Zn_7P_{22}Br_8$, Kiril A. Kovnir, Mikhail. M Shaturk, Lyudmila N. Reshetova, Igor A. Pesniakov, Evgeny V. Dikarev, Michael Baitinger, Frank Haarmann, Walter Schnelle, Michael Baenitz, Yuri Grin, Andrei Shavelkov, Solid State Sci. (2005), doi 10.1016/j.solidstatescience.2005.04.002

[3]Transport properties of polycristaline type-I Sn clathrates, G.S. Nolas, J.L. Cohn, J.S. Dyck, C. Uher, J. Yang, Physical Review B., Volume 65, 165201, 2002

[4]G.S. Nolas, G.A. Slack, S.B. Schjuman, in: T.M. Tritt (Ed), Recent Trends in Thermoelectric Materials Research, Academic Press, San Diego, CA, 2001

[5]$Sn_{19.3}Cu_{4.7}As_{22}I_8$: a New Clathrate-I Compound with Transition-Metal Atoms in the Cationic Framework, Kiril A. Kovnir, Alexei V. Sobolev, Igor A. Presniakov, Oleg L. Lebedev, Gustaaf Van Tendeloo, Walter Schnelle, Yuri Grin and Andrei V. Shevelkov, Inorganic Chemistry, 2005, 44, 8786-8793

[6]Transport properties of tin clathrates, G.S. Nolas, J.L. Cohn and E. Nelson, 18th International Conference of Thermoelectrics

[7]$Sn_{20.5□.5}As_{22}I_8$: A Largely Disordered Cationic Clathrate with a New Type of Superstructure and Abnormaly Low thermal Conductivity, Julia V. Zaikina, Kiril A. Kovnir, Alexei V. Sobolev, Igor A. Presniakov, Yuri Prots, Michael Baitinger, Gustaaf Tendeloo, Walter Schnelle, Yuri Grin, Oleg I. Lebedev and Andrei V. Shevelkov, Chemistry, A European Journal, 2007, 13, 5090, 5090-5099

[8]Are type-I clathrates Zintl phases and 'phonon glasses and electron single crystals'?, S. Paschen, V. Pacheco, A. Bentien, A. Sanchez, W. Carillo-Cabrera, M. Baenitz, B. B. Iversen, Yu. Grin and F. Steglich, Physica B, 328, 2003, 39

[9]Structural Characterization and Thermal Conductivity of Type-I Tin Clathrates, G. S. Nolas, B.C. Chakoumakos, B.Mahieu, Gary J. Long and T.J.R. Weakley, Chem. Matter. 2000, 12, 1947-1953

[10]Unusually High Chemical Compressibility of Normally Rigid Type-I Clathrate Framework; Synthesis and Structural Study of $Sn_{24}P_{19.3}Br_xI_{8-x}$ Solid Solution, the Prospective Thermoelectric Material, Kiril A. Kovnir, Julia V. Zaikina, Lyudmila N. Reshetova, Andrei V. Olenev, Evgeny V. Dikarev and Andrei V. Shevelkov, Inorganic Chemistry, 2004, 43, 3230-3236

[11]First Tin Pnictide Halides $Sn_{24}P_{19.3}I_8$ and $Sn_{24}As_{19.3}I_8$: Synthesis and the Clathrate-I Type Crystal Structure, Inorganic Chemistry, 1999, 38, 3455-3457

[12]Thermoelectrics, Basic Principles and New Materials Developments, G.S. Nolas, J. Sharp, H.J. Goldsmid, Springer

[13]G. S. Nolas, J. Poon, M. Kanatzidis, Recent developments in bulk thermoelectric materials, MRS Bulletin Vol. 31, No. 3, pp 199-205, March, 2006

[14]A. G. A. Slack, in CRC Handbook of Thermoelectrics, Ed. By D. M. Rowe (CRC Press, Boca Raton, FL, 1995), p. 407

[15]G. S. Nolas, G. A. Slack and S. B. Schujman in Semiconductors and Semimetals, Vol. 69, edited by T. M. Tritt (Academic Press, San Diego, CA, 2000), p.255

[16]Thermoelectric properties of silicon-germanium type I clathrates, J. Martin, G. S. Nolas, H. Wang, J. Yang, Journal of Applied Physics 102, 103719, 2007

TEMPERATURE IMPACT ON ELECTRICAL CONDUCTIVITY AND DIELECTRIC PROPERTIES OF HCl DOPED POLYANILINE

Shuo Chen[1]   Weiping Li[2]   Shunhua Liu[2]   William J. Craft[1]   David Y. Song[3]
[1]Department of Mechanical Engineering, [3]Department of Electrical and Computer Engineering
North Carolina A&T State University, Greensboro, NC 27411
[2] Department of Material Engineering
Dalian University of Technology DALIAN 116024, P.R.CHINA

ABSTRACT

This paper focuses on the electrical conductivity and dielectric properties of HCl doped polyaniline under various heat-treatment temperatures. Using differential scanning calorimetry (DSC), FT-IR spectra, and X-ray diffraction (XRD), it is discovered that PANI-HCl appears an excellent electrical conductivity below 120°C; as the temperature rises to 160°C, its electrical conductivity drops sharply and is almost like a dielectric. Furthermore, PANI-HCl demonstrates an excellent dielectric constant below 100°C and a loss tangent $\tan(\delta_c)$ between 0.35 and 0.49, when the temperature increases to 140°C, the dielectric constant decreases rapidly and the loss tangent falls below 0.1. HCl volatilization and changes in the molecular chain also have a distinct influence on the conductivity and dielectric properties.

INTRODUTION

Polyaniline (PANI) is one of the most widely studied conducting polymers due to its relatively high conductivity and easy of processing. Antistatic materials, electromagnetic shielding, rechargeable battery electrodes, electrochromic devices and electrocatalysis are parts of the potential application list [1-3]. Although the temperature during synthesizing, doping and drying of PANI is usually lower than 100°C, higher temperatures (over 100°C) are inevitable during machining and processing of PANI and other macromolecule materials. For example, some scientists blend PANI with PVC□ABS□nylon□ thermoplastic polyester and so on in order to mitigate the problem of contour machining of PANI. The temperature sometimes exceeds 200°C or even approaches 300°C when these composites are machined in the process of extrusion, injection and compression [4,5]. Therefore, it is important to investigate the impact of high temperature on PANI's structure, composition and conductivity.

In this paper, we present both theoretical and experimental observations on electrical conductivity and dielectric properties of HCl doped polyaniline under different heat-treatment temperatures. Differential scanning calorimetry (DSC), FT-IR spectra and X-ray diffraction (XRD) were used to study structural transformations and phase transition.

BRIEF DESCRIPTION OF THE EXPERIMENT

The synthesis of polyaniline (PANI) was carried out by chemical oxidative polymerization of a 0.1M solution of aniline in an organic protonic acid medium. The acid used was hydrochloric acid (HCl) 0.1M. Ammonium persulphate as the oxidant with 1:1 molar ratio to HCl in an acid solution was used for the present study. The oxidant was added very slowly to the reaction vessel containing aniline and hydrochloric acid solution kept at 0°C. The solution was stirred for 2h. This reaction mixture was then stirred continuously for 24 hours, and then filtered. The dark blue salt form of HCl doped polyaniline was collected and washed in distilled water and acetone four times, respectively, until the filtrate became colorless. This PANI-HCl was first separated into six equal parts, and each portion was dried at a different temperature: $60^\circ C$, $80^\circ C$, $100^\circ C$, $120^\circ C$, $140^\circ C$, $160^\circ C$, under forced air convection for 24 hours. Each purified and dried PANI-HCl was powdered in a mortar and blended with silastic in a mass ratio of 1:1. Once thoroughly homogenized, Fig 1, the mixture was compression molded into plates of 20 mm ×100 mm to study its electrical conductivity and dielectric properties.

Fig 1. Scanning electron microscopic (SEM) images of PANI-HCl

These SEM photos prove that its components are mixed well. PANI-HCl powder covered with silastic is dispersed (the granular component) evenly over the entire area of the matrix. As a result of phase separation, PANI-HCl pathways are created, which, in some cases, is responsible for the conductivity of these composites.

RESULTS AND ANALYSIS

Temperature Impact on Electrical Conductivity of PANI-HCl

The impact of different heat-treatment temperatures on electric conductivity of PANI-HCl is shown in Fig 2. Because there is such a large range of conductivity values, the ordinate has been converted to non-dimensional form and plotted in log format in which the following transformation is used:

$$\log_{10}\{\frac{\sigma(T^\circ C)}{\sigma(60^\circ C)}\}$$

As can be seen from the plot, the electrical conductivity of PANI-HCl is almost constant provided the temperature is below 100° C. However, once the temperature reaches 120° C, the log of the non-dimensionalized electrical conductivity drops to 10^{-1}. As the heat treatment temperature further increases beyond 120° C, a rapid decrease in electrical conductivity is observed. Compared with 60° C, this log plot decreases 5 and 7 fold at 140° C and 160° C, respectively. At 160° C, it exhibits a nondimensional conductivity of almost 1.47×10^{-7}.

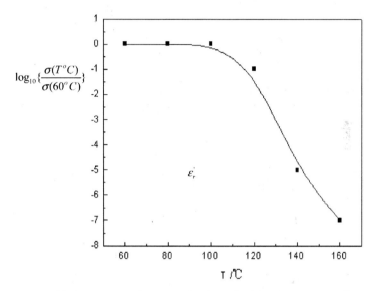

Fig 2. Temperature impact on electric conductivity of PANI-HCl

Heat Treatment Temperature Impact on Dielectric Properties of PANI-HCl

The dielectric constant is a measure of the ability of a polymer to store a charge from an applied electromagnetic field and then transmit that energy. It is also determined by molecular structure, vibration, and polarization. Fig 3 shows the components of dielectric constant ε_r' and ε_r'' as a function of heat treatment temperature and frequency for PANI-HCl. At 60° C and 100° C, Fig 3 also suggests that ε_r' and ε_r'' decrease with frequency. However, both ε_r' and ε_r'' are relatively constant above 140° C. At lower heat-treatment temperatures, the decrease of ε_r' and ε_r'' may be related to charge transport via small mobile polarons involving pair-wise charges of polarons and bipolarons in the "unprotonated" amorphous portions of the bulk emeraldine base polymer [6]. This mechanism is illustrated in Fig 4 in which the HCl is dedoped at higher temperatures leading to the sharp reduction of the number of charges. This causes a decrease in polarization which lowers the variation of ε_r' and ε_r''.

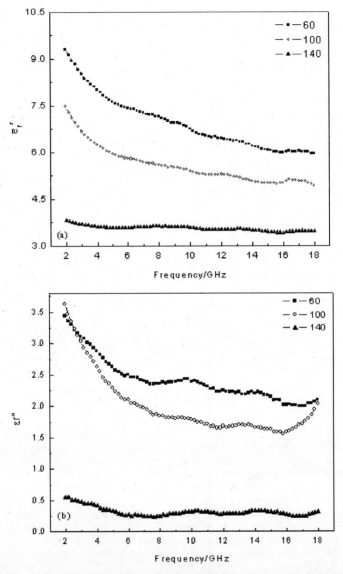

Fig 3. Effect of heat-treatment temperature on the complex dielectric constant of PANI-HCl in the frequency range of 2~18 GHz

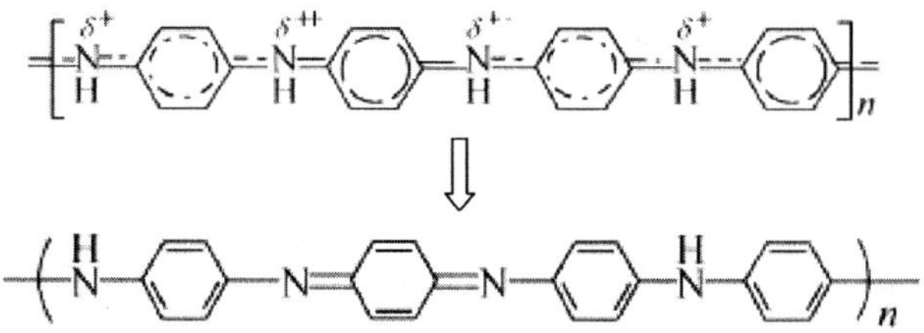

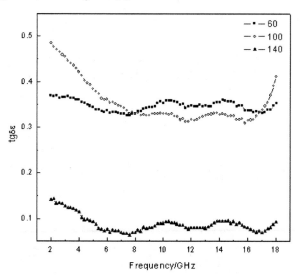

Fig 4. Mechanism of charge-lowering of HCl dedoped from PANI

The frequency dependence of tan(δ_ε) in Fig 5 shows relatively strong dependence on temperature in the range of 2-8GHz, but there is no substantial change in tan(δ_ε) above 8GHz. Changes at higher frequencies could be a result of extreme disorder present that lo Calizes charge leading to severe pinning of polarons, thus restricting their contribution at higher frequencies. Also, Fig 5 suggests that PANI-HCl presents an excellent dielectric property below 100° C and loss angle of tangent tan(δ_ε) is between 0.35 and 0.49. As temperature increases to 140° C, the dielectric property drops so sharply that tan(δ_ε) is below 0.1. For under certain frequency, tan(δ_ε) is tightly related to its structure. The higher the heat-treatment temperature, the less the molecular polarization; and the smaller is the degree of polarization, the lower the loss tangent [7]. Also, when the heat-treatment temperature is much too high, the crystal lattice and molecular chains of polyaniline are destroyed so that the electrical conductivity decreases leading to a rapid lowering of tan(δ_ε).

Fig 5. Temperature impact on loss tangent tan(δ_ε) of the complex dielectric constant of PANI-HCl in the range of 2~18GHz

DSC Analysis

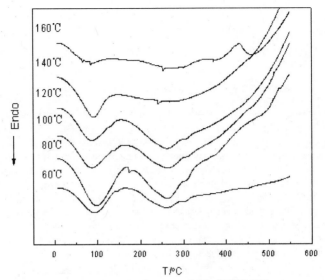

Fig 6. Temperature impact on DSC of PANI-HCl

Differential scanning calorimetry (DSC) is most commonly used to determine thermal transition temperatures such as glass transition, melting, cross-linking reaction, and decomposition. Fig 6 shows the DSC measurement results of PANI-HCl under different heat-treatment temperatures. Two endothermic peaks at 100° C and 260° C were observed. Based on work previously reported, PANI-HCl has discernable moisture content [8]. Therefore, the first endothermic peak is most likely attributed to the moisture evaporation and perhaps outgassing of unknown small molecules, but these changes have negligible influence on electrical conductivity of PANI-HCl. The second endothermic peak at 260° C indicates that the dopant has come out from the polymer backbone. Polyanilines have suitable structures that facilitate solitonic, polaronic and bipolaronic conformation when doped with HCl. Unlike all other conducting polymers, the conductivity of polyaniline depends on two variables: (i) the degree of oxidation of PANI and (ii) the degree of protonation of the material [9]. The bipolarons formed during the doping process are unstable with respect to a spontaneous internal redox reaction, leading to the formation of polarons (radical cations). Also in the protonic acid doping process, the number of electrons associated with the polymer chain remains the same; hence, the conduction is mostly due to polaronic hopping [10-12]. However, the dedopant of HCl greatly reduces the degree of protonation and conjugation. Fig 6 also shows that there is another endothermic peak at 160° C. We consider the endothermic peak to be due to a cross-linking reaction [13, 14] and decomposition. The mechanism of cross-linking reaction of PANI-HCl is shown in Fig 7.

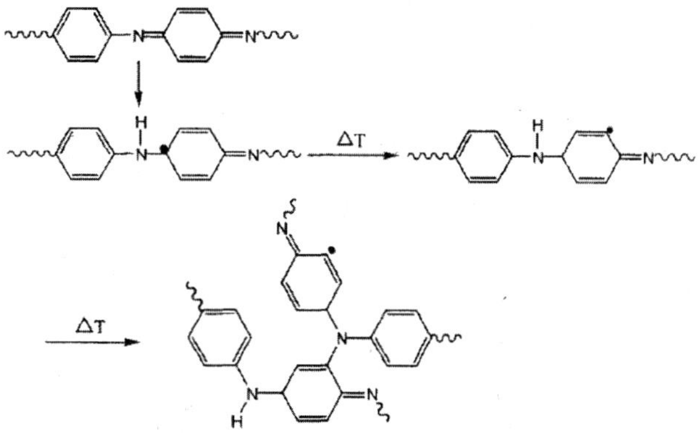

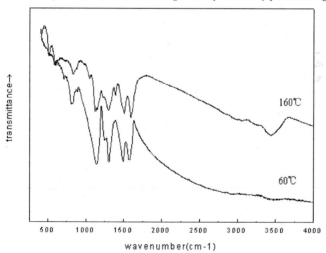

Fig 7. Mechanism of cross-linking reaction of PANI-HCl

FT-IR Analysis

The principal absorption bands observed in the FT-IR spectra of PANI-HCl is shown in Fig 8. In the range 250-4000cm^{-1}, the five absorption bands represent the C=C stretching vibration of quinoid ring☐ C=C stretching vibration of benzenoid ring☐C-N stretching vibration☐N=Q=N deformation and C-N stretching, respectively [15]. Bands of 60°C at 1570cm^{-1} and 1489cm^{-1} are the characteristic bands of nitrogen quinoid and benzenoid ring and are present in the doped polymer matrix. The peak at 1570cm^{-1} and 1489cm^{-1} is shifted to 1589cm^{-1} and 1504cm^{-1} when the heat-treatment temperature rises to 160°C and the intensity of these bands decreases significantly. These imply that at a higher

Fig 8. Temperature impact on the FT-IR spectra of PANI-HCl

heat-treatment temperature, the dopant moiety H⁺ attached to polymer backbone is completely removed leading to the destruction of the conjugation and the metallic polaron band transitions in the polymer matrix. The dedoping of HCl results in the decrease of electrical conductivity. The finding that 1589/1504=1.056>1570/1489=1.054 indicates that the polymer gets converted from quinoid form to benzenoid form, which was not expected. Also, the quinoid form is more stable than benzenoid form at higher heat treat temperatures. We also believe that since our heat-treatment was not conducted under vacuum, there was thermal oxidation leading to an increase of the quinoid form. The mechanism of oxidation of PANI is shown in Fig 9.

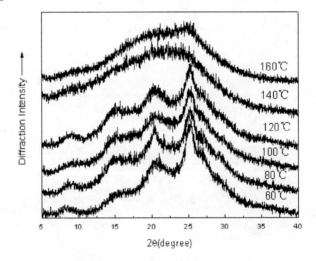

Fig 9. Oxidation of PANI

XRD Analysis

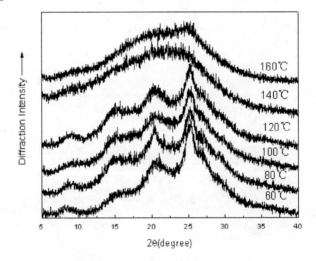

Fig 10. Heat treatment temperature impact on XRD spectra of PANI-HCl

The X-ray diffraction pattern of PANI-HCl film of different heat-treatment temperatures is shown in Fig 10. From $60°C$ to $120°C$, the diffraction peaks which appear at 2θ 8.5°, 15.0°, 20.6°, and 25.3° indicate the partly crystallization of PANI-HCl. The absence of any peak above $120°C$ suggests the amorphous structure of the film. However, the presence of very small peaks in the diffract graph indicates the presence of a mixed state of polycrystalline and amorphous structure of the film. We assume that the decreasing crystallization is mostly due to the removal of the HCl at higher temperatures.

CONCLUSION

Electrical conductivity and dielectric properties under different heat-treatment temperatures of HCl doped polyaniline are studied and the experimental conclusions are listed as follow:

(1) PANI PANI-HCl appears an excellent electrical conductivity below $120°C$; as the temperature rises to $160°C$, its electrical conductivity drops sharply and is almost like a dielectric.

(2) PANI-HCl indicates an excellent dielectric property below $100°C$ and the loss angle is between 0.35 and 0.49. When the temperature increases to $140°C$, the dielectric property is reduced greatly and $\tan(\delta_\varepsilon)$ below 0.1.

(3) HCl volatilization and changes in the molecular chain also have a distinct influence on the conductivity and dielectric properties.

ACKNOWLEDGMENTS

This work was supported in part by a grant from Department of Material Engineering Dalian University of Technology DALIAN 116024, P.R.CHINA.

REFERENCES

[1] F. Trinidal, M.C. Montemayour and E. Fialas, J Electro Chem Sot, 138(1991)3186.

[2] D. C. Trivedi and S.K. Dhawan, Meter J Chem 2 (1992)) 1092.

[3] E. W. Paul, A. J. Ricco and M. S. Wrighton, Phys J Chem, 89(1985)1441.

[4] Renhao Wu et al.Macromolecule Communication,1965,7(2):105.

[5] Yinhan Ju. Latex Industry,1986,(12):35.

[6] F. Zuo, M. Angelopoulos, A.G. MacDiarmid, A.J. Epstein, Phys.Rev. B 39(1989) 3570.

[7] Liu Fengqi , Tang Xinyi. Polymer Physics [M]. Beijing: Higher Education Publishing House, 1994:375□377.

[8] A.J. Conklin, S. Huang, S. Huang, T. Wen, B.R. Kaner, Macromolecules 28(1995)6522.

[9] Chilton, J. A. and Goosy, M. T. (ed.). Special Polymers for Electronics and Oproelectronics. Chapman & Hall.London, 1995.

[10] Cao, Y., Smith, P. and Heeger, A. J., Synth. Met., 1989,32, 263.

[11] Cao. Y., Synch. Met., 1990, 35, 319.

[12] Ohsawa, T., Kumura, O., Onoda, M. and Yoshino, K., Synth. Met., 1991, 41-43, 719.

[13] Y. Wei, K.F. Hsueh, J. Polym. Sci., Part A 27(1989)4351.

[14] S. Chen, L. Lin, Macromolecules 28(1995)1239.

[15] S. Chen, H. Lee, Macromolecules 26(1993)3254.

Geopolymers

PREPARATION OF CERAMIC FOAMS FROM METAKAOLIN-BASED GEOPOLYMER GELS

J. L. Bell and W. M. Kriven
University of Illinois at Urbana Champaign
1304 W. Green St.
Urbana, IL 61801, USA

ABSTRACT

The average pore diameter of hardened $K_2O \cdot Al_2O_3 \cdot 4SiO_2 \cdot 11H_2O$ geopolymer gel was measured by standard porosimetry techniques and determined to be extremely small (6.8 nm in diameter). On heating the geopolymer, significant capillary pressure (> 21 MPa) was expected due to vaporization of water from small pores. These capillary forces were enough to cause cracking and failure of monolithic geopolymer bodies. To avoid this problem, foaming agents, including spherical Al powder and hydrogen peroxide, were added to the geopolymer paste to engineer controlled porosity into the material and to shorten the diffusion distance for entrapped water to leave the samples. In order to control the internal pressure, the mixed pastes were cast into sealed metal dies and cured at elevated temperatures. Armoloy® coating on steel dies was found to be effective at improving the mold life and preventing geopolymer from sticking to the mold after curing. Foamed samples made using 0.5 and 1.5 wt% H_2O_2 and curing at 200°C for 7 h, had good machinability and high compressive strengths (44-77 MPa), but did not produce crack-free ceramics on heating. Mercury intrusion porosimetry and SEM results suggested that these samples did not obtain a percolating network of porosity due to hydrogen peroxide addition. Samples made using 60 wt% spherical Al as a foaming agent had pores of irregular shape with a larger pore size distribution, and were successfully converted to crack-free ceramics on heating. The Al foamed samples appeared to have attained a percolating pore network and exhibited minimal shrinkage on heating.

INTRODUCTION

Upon drying, pressure gradients develop within geopolymers that give rise to capillary forces, which in many cases, are sufficient to induce both microscopic and macroscopic cracking in geopolymers. Perera et al.[1] found that rapid drying, even in the presence of controlled humidly, typically led to cracking due to water loss. The dryout cracking problem can be reduced substantially in fly ash-based geopolymer systems due to the low water demand[2, 3] and reinforcing properties of fly ash. However, in the interest of better compositional control, enhanced reactivity, and higher purity, metakaolin is often used instead of fly ash, despite its higher cost. Unfortunately, metakaolin has a much higher water demand compared to fly ash due to its plate-like particle shape and higher specific surface area. Much of the additional water which is required to mix the metakaolin-based geopolymer exists as "free" water after curing, which can be easily removed on heating below 150°C.[4]

Avoiding this cracking problem is even more important in refractory applications or in situations where the formation of a ceramic on heating is desired. Metakaolin-based geopolymers are typically used in these situations, since refractory phases such as leucite ($KAlSi_2O_6$) or pollucite ($CsAlSi_2O_6$) can be formed on heating. For example, heating of $K_2O \cdot Al_2O_3 \cdot 4SiO_2 \cdot 11H_2O$ composition geopolymer results in high shrinkage[4] and the formation of a leucite glass-ceramic above 1000°C.[5]

One method of preventing cracking during drying and thermal conversion of a geopolymer into a glass-ceramic is to limit the diffusion distance that water must travel from the interior of the geopolymer to the solid/vapor interface. By engineering porosity into a geopolymer (i.e. making a geopolymer foam), the diffusion distance can be sufficiently limited, such that the geopolymers will not fail upon drying and thermal conversion. As shown in Figure 1, by adding porosity, one can reduce d_{max}, the maximum distance that entrapped water must travel from the interior of the geopolymer to the solid/vapor interface for both a solid (Figure 1a) and foamed material (Figure 1b).

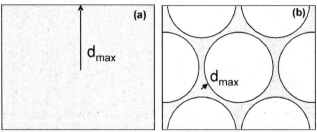

Figure 1. Cartoon showing the minimum distance that entrapped water must diffuse to leave the interior of the geopolymer for a solid sample (a), foam (b).

Conventionally, porous or aerated cements are produced by the addition of zinc or aluminum to a cement paste, which react to evolve hydrogen gas. Careful control of the curing conditions is required to create desirable pore morphologies. Porous cements can also be produced by generating foam separately from the cement paste using foaming agents in a pressurized vessel.[6] The pre-made foam is then mixed with the cement paste prior to casting.

In principle, geopolymer foam could be produced by the addition of hydrogen peroxide (H_2O_2) or Al to a geopolymer paste and allowing the added foaming agent to decompose in a sealed vessel.[7, 8] H_2O_2 decomposes into water and oxygen gas while the Al addition leads to the formation of hydrogen gas according to equations (1) and (2). The decomposition of H_2O_2 is very fast due to the high pH.[9] The reaction of Al in alkali solution is also very fast, and will consume OH⁻ as part of the reaction.[6, 10] As the foaming agents decompose, unfilled volume within the container and air entrapped within the geopolymer paste will become pressurized. The size of the resulting pores will be inversely proportional to the pressure following the Young-Laplace Equation (3):

$$H_2O_2 \rightarrow 2H_2O + O_2 \qquad (1)$$

$$2Al + 6H_2O + 2KOH \rightarrow 2K[Al(OH)_4] + 3H_2 \qquad (2)$$

$$\Delta P = 2\,\gamma\,/\,radius \qquad (3)$$

where γ is the surface energy of the liquid vapor interface and is equal to 7.29×10^{-2} J/m² for water. By adjusting the amount of foaming agent added, the pressure applied, and the free volume within the container, both the pore size and total porosity within the geopolymer foam could be controlled.

In this study, foams were made by adding Al or hydrogen peroxide (H_2O_2) to $K_2O \cdot Al_2O_3 \cdot 4SiO_2 \cdot 11H_2O$ composition geopolymer paste and curing in pressure-sealed containers. The hardened samples were then characterized using scanning electron microscopy (SEM), X-ray analysis, mercury intrusion porosimetry, pycnometry, and compressive testing. Additionally, samples were heated to 1200°C for 3 h in order to test for the possibility of crack-free thermal conversion.

Experimental Procedures

Alkali-silicate solutions were prepared by dissolving Cab-O-Sil® fumed silica (Cabot Corp., Wheaton, IL) into solutions of potassium hydroxide and deionized water. MetaMax® metakaolin powder (Engelhard Corporation, Iselin, NJ) was then mixed with the alkali-silicate solutions using a dispersion mixer to form the geopolymer paste. Given that the addition of porosity is generally expected to reduce strength, a geopolymer mixture which was capable of attaining a high compressive strength was desired. Geopolymers with a Si/Al ratio near 2 are known to have a superior compressive strength and a higher degree of microstructural density compared to lower silica composition (Si/Al < 1.40).[5, 11-14] Therefore, the geopolymer composition was chosen as $K_2O \cdot Al_2O_3 \cdot 4SiO_2 \cdot 11H_2O$. In addition, $K_2O \cdot Al_2O_3 \cdot 4SiO_2 \cdot 11H_2O$ geopolymers are known to convert to refractory leucite ($KAlSi_2O_6$) glass-ceramics on heating above 1000°C.[5]

H_2O_2 geopolymer foamed samples were made by dropping hydrogen peroxide solution (Lab grade H_2O_2, 30 wt%, Fisher Scientific) into the mixed geopolymer paste according to Table I. The addition of the 30 wt% H_2O_2 solution caused the overall water content of the geopolymer to increase as is shown in the Table I. After mixing, samples were vibrated into an 8.89 cm internal diameter cylindrical steel die. Enough paste was added to the die to form a 2.54 cm thick cylinder. Teflon discs, which were 5 mm thick and had a diameter slightly larger than the steel die, were placed above the geopolymer paste to ensure an air-tight seal. Prior to its use, the steel die was coated with Armoloy® (Armoloy Corporation, Dekalb, IL) to improve the mold lifetime and prevent geopolymer from sticking to the mold after curing. Samples were cured at 200°C and an initial pressure of ~1.5 MPa using a hydraulic uniaxial press for 7 h. Heat was applied to the sample by wrapping a Si heating tape (FluidX Equipment, Salt Lake City, UT) around the die. During the course of curing, the pressure increased to as high as ~12.8 MPa for the KGP-1.5HP sample due to gas pressure buildup. The sample was allowed to cool prior to removal from the mold.

Table I. Hydrogen Peroxide (H_2O_2) Geopolymer Foam Sample Composition

Sample name	wt% H_2O_2 added	Resultant $K_2O \cdot xH_2O$
KGP (control sample)	0.0	x = 11.00
KGP-0.5HP	0.5	x = 11.42
KGP-1.0HP	1.0	x = 11.85
KGP-1.5HP	1.5	x = 12.28

All of the H_2O_2 foamed samples tested were adequately removed from the Armoloy® coated steel die. The Cr-based Armoloy® coating was effective at preventing geopolymer adhesion and was easy to clean after use, thus extending mold life. In previous attempts to form H_2O_2 geopolymer foams in uncoated steel vessels or in polypropylene containers, the geopolymer adhered strongly to the container walls, and the force required during extraction caused fracture of the sample. Curing was carried out at 200°C in this work, to ensure that gases created from dissolution of foaming agents were allowed to escape prior to mold removal. In samples cured at lower temperature, pressure pockets created within pores due to H_2O_2 decomposition caused the samples to violently explode after removal from the mold.

Compression testing of H_2O_2 samples was carried out in accordance with ASTM C773-88. Cylindrical samples were core drilled from the larger 8.89 cm diameter x 2.54 cm high cylinders after removal from the mold and lathed to a final dimension of 1.27 cm diameter x 2.54 cm high. Samples

KGP and KGP-1.0HP were not tested as they failed on cutting. In order to test the ability to thermally convert the samples into a ceramic, an additional 1.27 cm diameter x 2.54 cm high cylinder was cut from each sample and was heated to 1200°C for 3 h at a heating and cooling rate of 5°C per minute.

Additional foamed geopolymer samples were prepared by adding 3.0-5.0 µm sized spherical aluminum powder (Alfa Aesar, Ward Hill, MA) to the geopolymer paste. Multiple attempts were made to fabricate samples using the Armoloy® coated, cylindrical steel die, as was done for the H_2O_2 samples. However, the samples failed by delamination cracking after mold removal. The cracks ran perpendicular to the pressing direction and caused the geopolymer body to crack into a series of layers. However, crack-free samples were prepared by mixing 60 wt% Al powder into geopolymer paste and vibrating into a 12.7 x 4.1 x 3.1 cm rectangular steel die and curing at 200°C for 12 h. The initial pressure on the sample was 4 MPa but increased to 12 MPa after curing for 24 hours. The sample was allowed to cool prior to removal from the mold. In order to test the thermal conversion of sample to a ceramic, 10 g sections were cut from the larger bar and were heated to 1200°C for 3 h at 5°C per minute heating and cooling rate.

Microstructure analysis of sample fracture surfaces was performed using a Hitachi S-4700 high resolution SEM. Samples were mounted on Al stubs using carbon tape and were subsequently sputter coated with ~6 nm of a Au/Pd alloy to facilitate imaging. X-ray diffraction patterns were collected using a Rigaku D-Max II X-ray powder diffractometer (Rigaku/USA Inc., Danvers, MA) equipped with a Cu Kα source (λ = 0.1540598 nm) and a single crystal monochromator in the diffracted beam path was used to acquire XRD patterns in Bragg-Brentano geometry over a 2θ range of 5-75° with a step size of 0.02°. Prior to X-ray analysis, all samples were ground to powders and sieved to 325 mesh (<44 µm). X-ray patterns were subsequently examined using Jade 7 software (Minerals Data Inc., Livermore, CA).

MIP results were collected using a Micromeritics Autopore II MIP from 0 - 60,000 psi on single solid fracture pieces. Samples were dried at 100°C for 10 h prior to analysis. A penetrometer sample holder was used to hold the samples and was calibrated prior to measurement. For comparative purposes, the KGP control sample was also sent to Porotech, Inc (Woodbridge, Ontario, Canada) for standard porosimetry analysis using methane as a wetting liquid. This method, the full details of which are given elsewhere,[15-17] allows one to measure smaller pore sizes (~1 nm) compared with MIP (~3 nm).

Pycnometry analysis was done using a Micromeritics Accupyc 1330 pycnometer on both H_2O_2 and Al foamed samples in order to calculate the apparent density and compare with MIP results. Solid fracture pieces (3-4 g) of each sample were selected for analysis. A total of 10 runs were collected for each sample using He gas. Additionally, dilatometer results were collected for the 60 wt% Al foamed sample and the KGP control sample using a Netzsch DIL 402E. Data was collected at a heating rate of 5°C/min to 1205 °C in air on cylindrical samples (6.35 mm diameter x 25.4 mm height).

RESULTS AND DISCUSSION

H_2O_2 Geopolymer Foamed Samples

Similar to conventional cements, geopolymers are expected to contain a variety of pores including air bubbles (mm size range) not removed on vibration, capillary pores from interparticle spaces which are filled by formation of hydrated geopolymer phase on curing, and gel pores within the geopolymer gel structure.[18] As shown in Figure 2a, the unfoamed control sample attained a dense, macropore-free microstructure. It is believed that the application of ~1.5 MPa of initial pressure was sufficient to remove large entrapped air bubbles often seen in cast samples.

The resultant pore size of foamed samples increased with the amount of H_2O_2 added as shown by SEM results in Figures. 2b-d. The largest pores were obtained in the KGP-1.5HP sample and ranged in size between 20-80 μm. Samples KGP-0.5HP and KGP-1.0HP had pore diameters ranging in size from approximately 5-50 μm. Regardless of the amount of H_2O_2 added, it did not appear that a percolating pore network was obtained from SEM analysis.

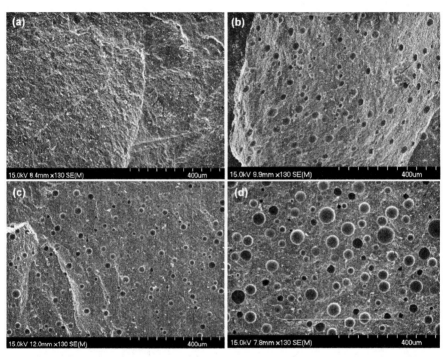

Figure 2. SEM micrographs at 130x magnification for H_2O_2 foamed fracture surfaces for samples (a) KGP (control sample) (b) KGP-0.5HP (c) KGP-1.0HP and (d) KGP-1.5HP.

The addition of 1.5 wt% of the H_2O_2 solution was the upper limit possible for reproducible processing. Above this value, the viscosity of the resultant geopolymer mixture was too low and geopolymer paste was spewed from the mold once pressure was applied. This was due to the fact that a 30 wt% solution of H_2O_2 was added to the geopolymer slurry. In order to add 1 wt% of H_2O_2, a total of 3.33 wt% of H_2O_2 solution had to be added, which caused the water content of the geopolymer paste to increase. The resultant increase in the $K_2O \cdot xH_2O$ ratio with H_2O_2 addition is given in Table I. The pore size is also expected to depend on the amount of water present as bubbles will be more able to expand in a less viscous fluid. Larger pores are more likely to form by agglomeration of smaller pores when the amount of H_2O_2 added is high. This may be why KGP-1.5HP had the largest pore size distribution.

Examination of capillary and gel pores using high magnification SEM is shown in Figures 3a-b for KGP-1.5HP. The geopolymer structure was different within the pores compared to matrix regions surrounding the pores. In the matrix area, the geopolymer consisted of fine precipitates (~10-15 nm) and remnant unreacted metakaolin, typical of what is seen for unfoamed $K_2O \cdot Al_2O_3 \cdot 4SiO_2 \cdot 11H_2O$ composition geopolymers.[5] The area within pores had a similar precipitate structure, but was much more porous. As this sample was cured, the pressure increased from an initial setting of 1.5 MPa to nearly 12.8 MPa due to gas evolution. This increased porosity at pore/gel interfaces could be due to the high internal pressure created within the pores or to the local higher water content. It was unclear from the SEM analysis how deep this increased porous region extended from the pore interface. Because of this, it was believed that a porous percolating network may have been achieved.

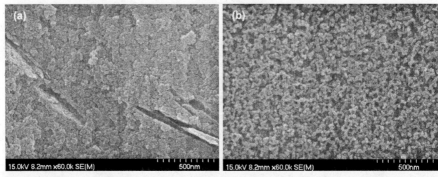

Figure 3. High magnification (60,000x) SEM micrographs for KGP-1.5HP H_2O_2 foamed sample showing (a) matrix area in between pores and (b) area within a pore.

In order to test for the possibility of percolation, mercury intrusion porosimetry (MIP) analysis was conducted on the H_2O_2 foamed samples. The results of this analysis are given in Table II and Figures 4a-b. As shown in Figure 5, the KGP control sample was also submitted for standard contact porosimetry analysis[15-17] using methane as a wetting liquid. This technique afforded higher resolution detection capability (~1 nm) compared with MIP (~3 nm).

MIP results suggested that percolation was not achieved in these systems due to the larger pores created by H_2O_2 addition. However, the critical pore diameter, defined as the pore width corresponding to the highest rate of mercury intrusion per change in pressure,[19] increased from 9.1 to 14 nm with increasing H_2O_2 addition. This is shown by the steeply increasing cumulative intrusion volume between 0.1-0.01 μm pore diameter in Figure 4a as well as the peak of the differential intrusion volume vs. pore diameter in Figure 4b. Sample KGP-1.5 had a sharp increase in cumulative intrusion volume near 30 μm (6 psi of pressure) which was believed to be due to cracking of the sample rather than percolation of 30 μm sized pores. As a consistency check, an additional KGP-1.5 sample was run using MIP. On this second run, the sample had a sharp increase in intrusion volume at even lower pressure (~ 1 psi). This confirmed the suspicion that the sample was cracking rather than being percolated with mercury.

Table II. H_2O_2 Foamed Geopolymer Results for Compressive, MIP, and pycnometry analysis

Sample name	Compressive strength (Mpa)[†]	Total Pore Area $(m^2/g)^{§}$	Total pore volume $(ml/g)^{§}$	Bulk density $(g/cm^3)^{§}$	Apparent density $(g/cm^3)^{§}$	Apparent density $(g/cm^3)^{¥}$
KGP (control)	Cracked (0)	110.56	0.241	1.38	2.07	2.197 ± 0.004
KGP-0.5HP	77.32 ± 7.92	79.81	0.305	1.23	1.97	2.212 ± 0.002
KGP-1.0HP	cracked	50.89	0.173	1.29	1.66	2.204 ± 0.003
KGP-1.5HP	49.51 ± 4.52	42.77	0.358	1.09	1.78	2.214 ± 0.004

[†]For the compression tests a total of 7 samples were tested for KGP-0.5HP and 6 for KGP-1.0HP
[§]As determined from MIP analysis
[¥]As determined from pycnometry analysis, standard deviation given (± for a total of 10 runs)

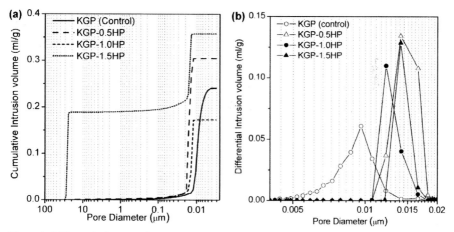

Figure 4. MIP results for H_2O_2 foamed samples showing (a) cumulative intrusion vol vs. pore diameter and (b) differential intrusion volume vs. pore diameter.

It was not clear if percolation was achieved at the critical pore size between 0.1-0.01 μm pore diameter as the observed behavior could be due to fracture or cracking on dryout. Moscou et al.[20] found that MIP results for highly porous silicates were unreliable due to sample damage. In interpreting the critical pore size, it is also important to recognize the "ink-bottle" effect in MIP measurements.[18] In cementitious materials such as geopolymers, there is no guarantee that each pore is connected to the sample surface or to other pores. As the mercury pressure is increased, it will be forced into narrow openings which can appear to be fine pores. Large air pockets can be initially filled with mercury, and will subsequently fracture when the pressure is high enough. In this regard, the pressure at which this occurs may represent the mechanical integrity of the geopolymer pore network. This is consistent with the critical pore size trend observed for the H_2O_2 sample. The geopolymer is expected to be weaker with increasing H_2O_2 due to larger pores and a higher water content in the initial mixture.[14]

As shown in the Table II, the apparent density estimated from the pycnometry data does not change much with increasing H_2O_2 addition, and is collectively higher than the value predicted from

MIP analysis. The He gas used in pycnometry is capable of infiltrating much smaller pores and therefore gives a better representation of apparent density compared to MIP. There is no guarantee that all of the pores will be infiltrated by Hg in MIP analysis, which in this case, resulted in inaccurate density estimates. For example, with increasing H_2O_2, there will be more inaccessible isolated pores, which will cause the MIP to predict a lower apparent density.

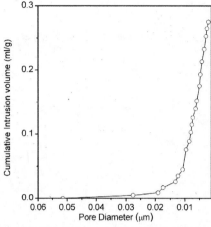

Table III. MIP and Standard Porosimetry Results for KGP Control Sample

Measured Value	Standard Porosimetry Analysis (SPA)	MIP
Average Pore Diameter (nm)	6.74	8.70
Total Porosity (%)	41.06	33.23
Total Pore Area (cm^2/g)	274.69	110.56
Skeletal Density (g/cm^3)	2.05	2.07

Figure 5. Standard porosimetry analysis (SPA) results for sample KGP (control sample) using the standard contact porosimetry technique with methane as a wetting liquid.[15-17] Calculated values for MIP and SPA are also shown within Table III on the right for the KGP control sample.

As a consistency check, samples were also run using standard porosimetry analysis (SPA) as shown in Figure 5. Using SPA, a smaller average pore size, higher total porosity, and higher pore area was obtained due to the improved ability to examine pores down to 1 nm. An average pore size of 6.74 nm was determined from SPA and was representative of the pore structure of the geopolymer gel. Given the high value for total porosity of 41.06 %, it is possible that unfoamed geopolymers have a percolating network of gel pores. However, in water permeability tests on Ca-based cement pastes, the porosity percolation threshold was found to be more related to the pore size distribution rather than the overall porosity.[21] In samples with finer pores, the permeability was lower. Regardless of whether or not percolation is achieved in geopolymer, both the control and H_2O_2 foamed samples were unable to withstand thermal conversion and cracked on heating. It is believed that this is primarily due to the very small intrinsic pore size within the gel, which leads to extremely high capillary pressure $> 2x10^7$ Pa according to equation 1.

Although the H_2O_2 samples could not be converted to ceramics, samples KGP-0.5HP and KGP-1.5HP had excellent machinability and high compressive strength (Table II). KGP-0.5HP, in particular, had a compressive strength of 77.3 ± 7.9 MPa, which is comparable to that found for non-porous geopolymer of a similar composition.[11-13] Therefore, the H_2O_2 based geopolymer foams still show potential as lightweight, high-strength materials which can be utilized at standard temperatures. Al Geopolymer Foamed Samples

A similar procedure was used to fabricate geopolymer foams using Al as a foaming agent. A variety of samples were fabricated using different Al loadings. However, only samples loaded with 60 wt% Al were successfully converted to ceramics on heating. As shown in Figure 6a, 60 wt% polydisperse Al powder was added to the geopolymer. It was confirmed from SEM (Figure 6b) and XRD (Figure 7a), that not all of this powder was consumed during curing. The partially dissolved Al particles were no longer spherical in shape after being partially consumed as shown in Figure 6b.

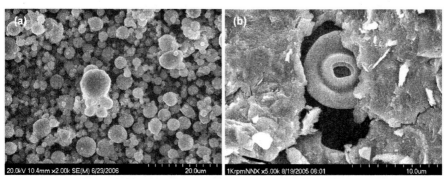

Figure 6. SEM micrographs for Al powder (a) before and (b) after addition to the geopolymer. The Al particles were partially dissolved after being exposed to the alkaline conditions of the geopolymer paste.

It may be possible to optimize the amount of Al added, Al particle size as well as curing conditions, to ensure that it is completely dissolved prior to geopolymer setting. However, given that samples were cured at 200°C, this led to fast geopolymerization kinetics, and rapid setting. It was therefore unlikely that there would be adequate time for complete dissolution, diffusion and homogenization under these conditions. As was stated previously, in samples cured at lower temperatures, entrapped, pressurized gas remained in pores after mold removal, and caused catastrophic failure of the samples.

Given that Al dissolution consumes OH⁻ as shown in equation 2, there was concern that this may hinder geopolymer formation. However, as shown in Fig. 6b, the geopolymer appears to have formed the typical structure expected for this composition, and consisted of a fine precipitate structure and remnant metakaolin. It is expected that the fast reaction kinetics at 200°C caused the geopolymer to set before OH⁻ consumption could lead to a significant drop in the pH.

SEM micrographs for the as-set sample are shown in Figures 8a-d. The resulting geopolymer appeared to be dense to the naked eye; no pores were visible. SEM examination revealed a highly porous microstructure, with pores of various sizes and shapes. High resolution examination of porous regions (Figures 8c-d) revealed that "flower-shaped" crystallites formed. In areas where Al dissolved, the local content of Al was likely to be higher compared to the bulk composition. Combined with the high curing temperature, this may have favored the formation of these crystallites, which are most likely hydrated zeolites. Similar results have been found for hydrothermal treatment of fly ash in KOH.[22-24] Trace crystalline peaks for these phase(s) were observed near 20° 2θ in XRD as shown in Figure 7a. The three large peaks in Figure 7a were due to remnant Al.

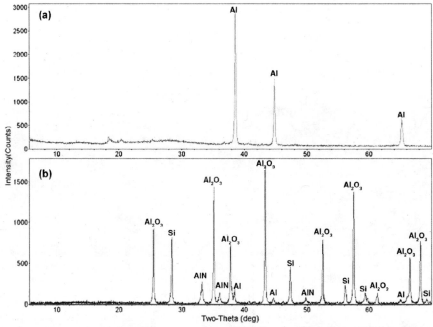

Figure 7. XRD results for Al-foamed geopolymer (a) after hardening and (b) after heating to 1200°C (5°C per minute heating and cooling rate). In the as-set state, Al metal (PDF #98-000-0062) was present along with trace crystalline phase(s). After heating, Al_2O_3 (PDF #00-010-0173), Si metal (PDF #04-004-5099), AlN (PDF #00-025-1133), and remnant Al metal (PDF #98-000-0062) were observed.

To test the possibility of percolation, MIP results were also collected for the Al foamed samples and are shown in Table IV and Fig. 9. Although the overall intrusion volume was low, the Hg was able to infiltrate the Al foamed sample at a lower pressure than for the H_2O_2 samples. As shown in Fig 9b, there does not appear to be a characteristic pore size in which infiltration occurred. Rather, infiltration occurred over a range of pressures corresponding to pores over the 0.01-1.00 μm diameter size range. Given that only part of the sample was infiltrated, the apparent density estimated from MIP is less that that determined from pycnometry analysis.

The Al foams did not crack after heating to 1200°C for 3 h. XRD results, shown in Figure 9, showed the presence of Al_2O_3, Si metal, AlN, and a small amount of remnant Al after being heated. It is expected that on heating, the Al reacted with SiO_2 to form Al_2O_3 and Si metal. Dilatometry results suggest that both the Al-foamed and geopolymer control sample have a fair amount of shrinkage over the 100-300°C due to loss of free and chemically bound water,[25] although it is less for the foam (Fig. 10). Over the 300-800°C range, there is little shrinkage in either material. However, above 800°C, the geopolymer has significant shrinkage while the foam actually expands a little. The shrinkage in the geopolymer is expected to be related to viscous sintering, densification, and leucite crystallization above 1000°C.[4, 5] The expansion in the foam may be due to the variety of new phases formed on heating (Al_2O_3, Si metal, and AlN). The microstructure of the heated ceramic retained its porous structure (Figure 10a), but contained crystallites of varying size and shape which were visible at higher magnifications (Figures 10b-d).

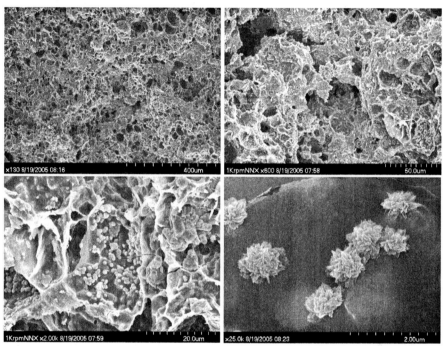

Figure 8. SEM micrographs of an Al-foamed geopolymer fabricated under pressure at (a) 130x, (b) 600x, (c) 2000x, and (d) 25,000x magnification showing zeolitic crystallites.

Table IV. Al Foamed Geopolymer Results for MIP and pycnometry analysis

Sample name	Total Pore Area $(m^2/g)^\S$	Total pore volume $(ml/g)^\S$	Bulk density $(g/cm^3)^\S$	Apparent density $(g/cm^3)^\S$	Apparent density $(g/cm^3)^¥$
Al foamed geopolymer	0.785	0.018	1.5574	1.6023	2.291 ± 0.003

$\S$As determined from MIP analysis
$¥$As determined from pycnometry analysis, standard deviation given (± for a total of 10 runs)

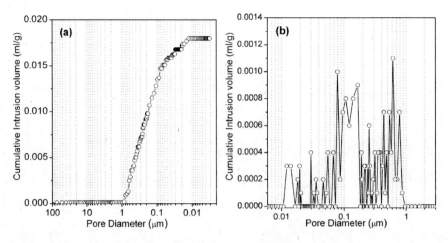

Figure 9. MIP results for Al foamed sample showing (a) cumulative intrusion vol vs. pore diameter and (b) differential intrusion volume vs. pore diameter.

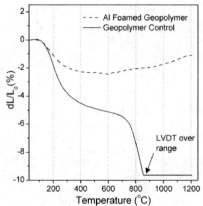

Figure 10. Dilatometer results for unfomed geopolymer (control sample) compared to the 60 wt% Al foamed samples. Samples were heated at 5°C/min up to 1205°C in air.

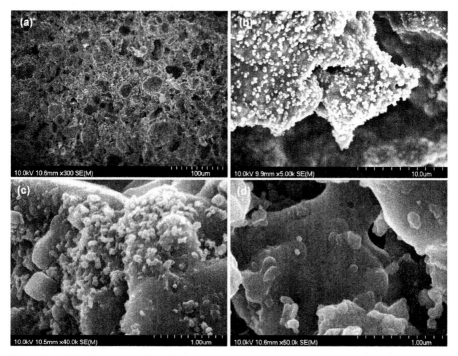

Figure 11. SEM micrographs of an Al-reinforced geopolymer after heating to 1200°C for 3 h at (a) 300x, (b) 5000x, (c) 40,000x, and (d) 50,000x.

CONCLUSIONS

High strength geopolymer foams were produced by adding H_2O_2 or Al foaming agents to geopolymer paste and curing at high-temperature (T = 200°C) in sealed containers. The use of H_2O_2 produced foams with spherical, non-percolating pores, which could not be converted to a ceramics on heating without cracking. However, unheated H_2O_2 foams had high compressive strengths of 77.3 ± 7.9 MPa and 49.5 ± 4.5 MPa for 0.5 wt% and 1.5 wt% H_2O_2 added systems respectively. These materials show potential as lightweight, high-strength materials which can be used in non-refractory applications. Foams made from polydisperse (3.0-5.0 μm) spherical Al powder had pores of varying sizes and shapes. The use of 60 wt% Al powder produced foam of sufficient porosity and was subsequently converted to a monolithic ceramic on heating. Remnant Al, which was not completely consumed during curing, reacted with SiO_2 on heating to form Al_2O_3 and Si.

ACKNOWLEDGEMENTS

This work was supported by Air Force Office of Scientific Research (AFOSR), USAF, under Nanoinitiative Grant No. FA9550-06-1-0221. The authors also acknowledge the use of facilities at the

Center for Microanalysis of Materials, in the Frederick Seitz Research Laboratory (at the University of Illinois at Urbana-Champaign), which is partially supported by the U. S. Department of Energy under grant No. DEFG02-91-ER45439. The authors would also like to thank Mr. Matthew Gordon for his help with this work.

REFERENCES

[1]D. Perera, O. Uchida, E. Vance, and K. Finnie, Influence of Curing Schedule on the Integrity of Geopolymers, *J. Mater. Sci.*, **42**, 3099-106 (2007).

[2]L. Jiang, and V. Malhotra, Reduction in Water Demand of Non-Air-Entrained Concrete Incorporating Large Volumes of Fly Ash, *Cem. Concr. Res.*, **30**, 1785-89 (2000).

[3]M. Rafalowski, Fly Ash Facts for Highway Engineers, American Coal Ash Association, (2003).

[4]P. Duxson, G. Lukey, and J. S. J. van Deventer, Thermal Evolution of Metakaolin Geopolymers: Part 1 - Physical Evolution, *J. Non-Cryst. Solids*, **352**, 5541-55 (2006).

[5]W. Kriven, J. Bell, S. Mallicoat, and M. Gordon, In *Intrinsic Microstructure and Properties of Metakaolin-Based Geopolymers*, Int. Workshop on Geopolymer Binders – Interdependence of Composition, Structure and Properties, Weimar, Germany, 71-86, (2006).

[6]T. Tonyan, and L. Gibson, Structure and Mechanics of Cement Foams, *J. Mater. Sci.*, **27**, 6371-78 (1992).

[7]J. Davidovits, and J. Legrand, Expanded Minerals Based on Potassium Poly(sialates) and/or Sodium, Potassium Poly(sialate-siloxo). French Patent FR2512805A1, (1981).

[8]V. Li, and S. Wang, Lightweight Strain Hardening Brittle Matrix Composites. U.S. Patent 6,969,423, (2005).

[9]N. Greenwood, and A. Earnshaw, *Chemistry of the Elements*. 2nd ed.; Butterworth-Heinemann: Oxford, UK, 1997.

[10]J. Head, Cellular Cement Composition. U.S. Patent 3925090, (1975).

[11]P. Duxson, S. W. Mallicoat, G. C. Lukey, W. M. Kriven, and J. S. J. van Deventer, The Effect of Alkali and Si/Al Ratio on the Development of Mechanical Properties of Metakaolin-based Geopolymers, *Colloids Surf., A.*, **292**, 8-20 (2007).

[12]P. Duxson, J. L. Provis, G. C. Lukey, S. W. Mallicoat, W. M. Kriven, and J. S. J. van Deventer, Understanding the Relationship between Geopolymer Composition, Microstructure and Mechanical Properties, *Colloids Surf., A*, **269**, 47-58 (2005).

[13]M. Rowles, and B. O'Connor, Chemical Optimisation of the Compressive Strength of Aluminosilicate Geopolymers Synthesised by Sodium Silicate Activation of Metakaolinite, *J. Mater. Chem.y*, **13**, 1161-65 (2003).

[14]M. Steveson, and K. Sagoe-Crentsil, Relationships between Composition, Structure and Strength of Inorganic Polymers - Part 2 - Flyash-derived Inorganic Polymers, *J. of Mater. Sci.*, **40**, 4247-59 (2005).

[15]Y. M. Volfkovich, and V. E. Sosenkin, Standard Noncontact Porosimetry, *Soviet Electrochemistry*, **14**, 67-69 (1978).

[16]Y. M. Volfkovich, V. E. Sosenkin, and E. I. Shkolnikov, Measurement of Pore-Size Distribution by Standard Contact Porosimetry, *Soviet Electrochemistry*, **13**, 1583-86 (1977).

[17]Y. M. Volfkovich, and E. I. Shkolnikov, Application of Standard Porosimetry for Determination of Contact Angles of Wettability, *Zh. Fiz. Khim.*, **52**, 210-11 (1978).

[18]F. Moro, and H. Bohni, Ink-bottle Effect in Mercury Intrusion Porosimetry of Cement-based Materials, *J. Colloid Interface Sci.*, **246**, 135-49 (2002).

[19]D. N. Winsolw, and S. Diamond, A Mercury Study of the Evolution of Porosity in Cement, *ASTM J. Mater.*, 564-85 (1970).

[20]L. Moscou, and S. Lub, Practical Use of Mercury Porosimetry in the Study of Porous Solids, *Powder Technol.*, **29**, 45-52 (1981).

[21]G. Ye, Percolation of Capillary Pores in Hardening Cement Pastes, *Cem. Concr. Res.*, **35**, 167-76 (2005).

[22]X. Querol, N. Moreno, J. C. Umana, A. Alastuey, E. Hernandez, A. Lopez-Soler, and F. Plana, Synthesis of Zeolites from Coal Fly Ash: an Overview, *Int. J. Coal Geol.*, **50**, 413-23 (2002).

[23]H. Mimura, K. Yokota, K. Akiba, and Y. Onodera, Alkali Hydrothermal Synthesis of Zeolites from Coal Fly Ash and Their Uptake Properties of Cesium Ion, *J. Nucl. Sci. Technol*, **38**, 766-72 (2001).

[24]C. Amrhein, G. H. Haghnia, T. S. Kim, P. A. Mosher, T. Amanios, and L. DelaTorre, Synthesis and Properties of Zeolites from Coal Fly Ash, *Environ. Sci. Technol.*, **30**, 735-42 (1996).

[25]P. Duxson, G. C. Lukey, and J. S. J. van Deventer, Physical Evolution of Na-geopolymer Derived from Metakaolin up to 1000 degrees C, *J. of Mater. Sci.*, **42**, 3044-54 (2007).

PREPARATION OF PHOTOCATALYTIC LAYERS BASED ON GEOPOLYMER

Z. Černý, I. Jakubec, P. Bezdička and V. Štengl,
Institute of Inorganic Chemistry, v.v.i., Academy of Science of the Czech Republic, 250 68 Řež.
P. Roubíček,
České lupkové závody a.s., Nové Strašecí č. p. 1171, 271 11 Nové Strašecí.
Czech Republic.

ABSTRACT

Metakaolinite-based geopolymer composites containing about 10 wt% of TiO_2 powders were prepared directly by a mixing of metakaolinite, sodium silicate solution (water glass) and TiO_2 powders. Prepared materials in form of layers on steel plates were investigated by XRD, SEM, BET and BJH (Barrett-Joiner-Halenda) methods. In all cases the TiO_2 powders are to different extents encapsulated in the geopolymer matrix. All studied materials, including the initial geopolymer matrix free of synthetic TiO_2, exhibited photocatalytic properties.

INTRODUCTION

Surfaces that exhibit photocatalytic character are able to oxidize or decompose substances which come from the outside into contact with them. These photocatalytic layers are usually based on the common semiconductor - titanium dioxide.[1-3] There are general methods for preparing these photocatalytic layers:

a) The active component of a photocatalyst is applied or coated onto the surface of a substrate using organic binders. These binders are usually organic polymers, such as polyviny alcohol (PvOH), copolyamides (PA6/12), polyvinyl acetates (PvAc), polyvinyl pyrrolidone (PvPy) etc. The photocatalysts react with the binder itself to oxidize or decompose it. Moreover, coated films are formed even on the surface of the photocatalyst, thus reducing the active surface area of the catalyst with respect to its surroundings. In this context the processes taking place at the catalyst, governed by diffusion, are decelerated.

b) More efficient methods involve the deposition of photocatalytic layers directly on the surface of a substrate. These methods are based on plasma or flame spray coating processes. However, the resulting layers are usually significantly reduced in their surface area and sophisticated techniques or equipment are needed. Other methods exist based on the preparation of photocatalytic layers "in situ of the surface" using either spraying or coat dipping of the substrates directly into $TiCl_3$ or $TiOSO_4$ solutions. However, the durability of these prepared layers is doubtful due to limited mechanical stickiness of the coating on the substrate.

c) Frequently techniques are used in which nanoparticles of the active component are immobilized on support materials in the form of films prepared by sol-gel methods. However, in this case the resulting thin layers are relatively expensive due to the price of initial compounds.

The present work describes a comparatively simple and cheap method for the preparation of photocatalytic layers based on TiO_2 pigments in a geopolymer matrix.[4,5]

anions (sodium silicate solution) and metakaolinite mixed "in situ" with TiO_2 powders at room temperature.

Chemicals and raw materials

Sodium silicate solution - water glass, supplied by Koma ltd, Praha, Czech Republic, of analytical composition: 27 wt% SiO_2, 9.2 wt% Na_2O, density 1.38 g/cm^3, content of Fe_2O_3 < 1 wt%.

Metakaolinite, product MEFISTO KO5, supplied by České Lupkové Závody Corp., Nové Strašecí, Czech Republic, analytical composition: 40.7 wt% Al_2O_3, 57.7 wt% SiO_2, 0.53 wt% TiO_2 and 0.58 wt% Fe_2O_3.

TiO_2 powders: i) commercially available standard TiO_2 product P 25, supplied by Degussa, GmbH, Essen, Germany, ii) samples 1710 and TiT 144, represent laboratory products of Institute of Inorganic Chemistry, AV ČR, v.v.i., Czech Republic. TiT 144 represents a pure anatase TiO_2, 1710 represents kaolin modified by anatase TiO_2.

All the above products were used without any other treatment.

Characterization methods

Surface of the samples was determined by BET and the pore volume was determined by BJH (Barrett-Joiner-Halenda), for both methods based on gas adsorption instrument Coulter SA 3100 was used.

X-ray powder diffraction patterns (XRD) were obtained by Siemens D5005 and Philips RW instruments using $CuK\alpha$ radiation (40kV, 30mA) and a diffracted beam monochromator.

SEM (scanning electron microscopy) studies were obtained by Philips XL30 CP microscope at 80 kV, HRTEM data (transmission electron microscopy) were obtained by JEOL JEM 3010 at 300 kV.

Preparation procedure

An initial geopolymer suspension was prepared by mixing of 100 g of water glass, 50 g of metakaolinite in well stirred 0.5 liter vessel for 1 hour at ambient temperature.

Photocatalytic materials were prepared by mixing of 20 g of above geopolymer suspension with 2 g of the corresponding TiO_2 powder and 10 g of water. The mixtures were stirred for next 1 h at ambient temperature. For a reason of comparison, a "blind" pure geopolymer layer was prepared by diluting of 20 g of the geopolymer suspension with 10 g of water.

Prepared suspensions were deposited on steel plates 150x100 mm by a common painting technique. After drying of the plates for 12 hours at ambient temperature, the layers were cured at 180°C for 4 hour. A typical loading of dried photocatalytic layers on the steel plates was 10 ± 2 mg of per cm^2. A final calculated composition: i) initial geopolymer material – 64.5 wt% SiO_2, 23.4 wt% Al_2O_3, 10.6 wt% Na_2O and 1.5 wt% of Fe_2O_3, molar ratios: Si/Al = 2.3, Al/Na = 1.4, ii) photocatalytic materials – 59 wt% SiO_2, 21 wt% Al_2O_3, 9.5 wt% Na_2O, 1.45 wt% of Fe_2O_3, molar ratios: Si/Al = 2.3, Al/Na = 1.4, Si/Ti = 4.3 (calculated on 100% TiO_2).

Photocatalytic characterization

Photocatalytic properties of the prepared layers were investigated in an original gas phase, flat plate photoreactor, using butane as a model volatile organic compound for the photodegradation process. The photoreactor was coupled on line with a detector of butane and CO_2. A UV light source, low-pressure 7 W lamp operating at $\lambda = 254$ nm, was placed along the longer axis of the plate. The calculated UV intensity was 3.10^{-6} Einstein s^{-1}.

RESULTS AND DISCUSSION

The morphology and surface characteristics of initial TiO$_2$ powders are shown in Figure 1.

P 25   TiT 144

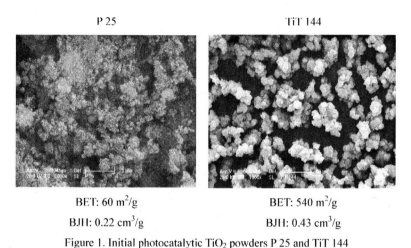

BET: 60 m^2/g   BET: 540 m^2/g

BJH: 0.22 cm^3/g   BJH: 0.43 cm^3/g

Figure 1. Initial photocatalytic TiO$_2$ powders P 25 and TiT 144

While P 25 and TiT 144 samples represent more or less aggregated spherical nanoparticles, the sample 1710 exhibited non regular particles derived from the tabular particles of the kaolin substrate, as seen in Figure 2.

Initial kaolin   Kaolin particle treated by anatase, TiO$_2$

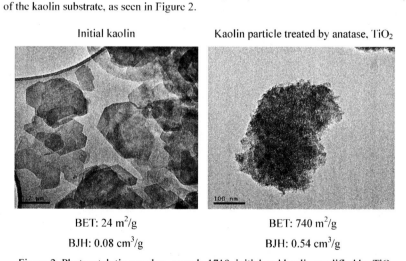

BET: 24 m^2/g   BET: 740 m^2/g

BJH: 0.08 cm^3/g   BJH: 0.54 cm^3/g

Figure 2. Photocatalytic powder - sample 1710, initial and kaolin modified by TiO$_2$

The morphology and XRD characterization of the initial geopolymer matrix are shown in Figures 3 and 4. SEM figures, Figure 3, show expected glass - like structure obviously containing metakaolinite relicts. Homogeneity of the layer is defected by impurities of quartz grains. Beside of the quartz RXD pattern, Figure 4, reveals traces of illite and confirmed amorphous character of geopolymer matrix. Layer exhibits the lowest value of surface area, about $1 m^2/g$, compared with the all studied materials. (This value of the surface area holds for a given dilution of the initial geopolymer suspension, see experimental part. With increasing dilution of the geopolymer suspension significantly increase the values of the surface area of the resulting layers, however, significantly decrease durability of these layers on given substrates.)

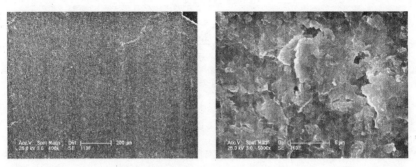

Figure 3. SEM micrographs of the initial geopolymer matrix having a specific surface area 1 m^2/g (as measured by BET) and pore volume 0.006 cm^3/g (as measured by BJH).

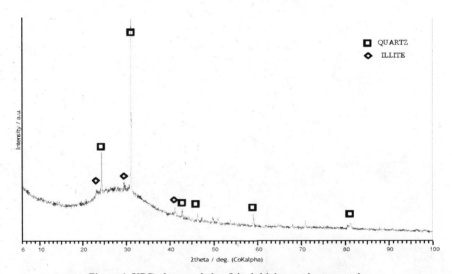

Figure 4. XRD characteristic of the initial geopolymer matrix

Photocatalytic material with TiO_2 - standard P 25- is characterized in Figures 5 and 6. Compared with the initial matrix the layer apparently does not exhibit any TiO_2 particles at a given level of magnification, Figure 5. Introduction of about 10 wt% of TiO_2 powder P 25 with a relatively high value of surface area 60 m^2 and porosity 0.22 cm^3/g, increased the surface area of the resulting layer only to a value of 1.4 m^2/g, (compare it with the value of the surface area of the initial matrix of 1 m^2/g). This indicates that the TiO_2 powder was encapsulated in the geopolymer matrix. In addition to metakailonite impurities, (quartz and illite), the XRD pattern,

Figure 6, indicates anatase and rutille in the approximate ratio 4/1 which is a typical for the Degusa product.

Figure 5. SEM micrographs of the geopolymer matrix with TiO_2 - standard P 25 - having a specific surface area 1.4 m^2/g (as measured by BET) and pore volume 0.008 cm^3/g (as measured by BJH).

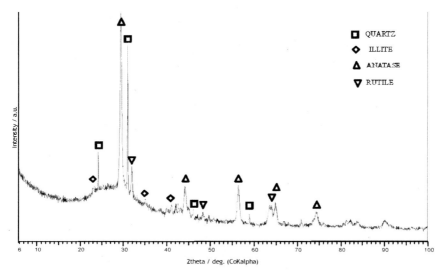

Figure 6. XRD characteristic of the geopolymer matrix with TiO_2 - standard P 25

Photocatalytic material based on TiT 144 is shown in Figure 7. The layer exhibited the highest values of surface area and porosity of the all studies materials. The XRD pattern, Figure 8, confirmed the presence of pure anatase.

Figure 7. SEM micrographs of the geopolymer matrix with TiO_2 - TiT 144 - having a specific surface area 5.9 m^2/g (as measured by BET) and pore volume 0.03 cm^3/g (as measured by BJH).

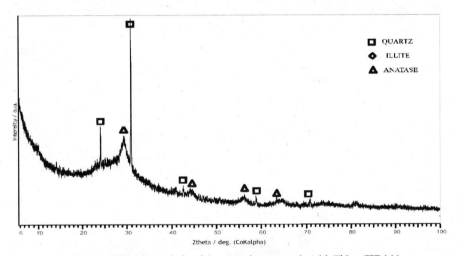

Figure 8. XRD characteristic of the geopolymer matrix with TiO_2 - TiT 144

Characteristics of photocatalytic material based on 1710 are shown in Figure 9. The XRD pattern, Figure 10, indicates the presence of pure anatase and kaolinite that were used for preparation of the powder. In spite of the fact that the initial powder exhibited the highest value of surface area of 740 m^2/g, the value of the surface area of the resulting layer did not exceed 1.5 m^2/g, (for a given dilution of the geopolymer suspension used). This confirmed a high efficiency of geopolymer matrix to wet the above combined photocatalytic powder.

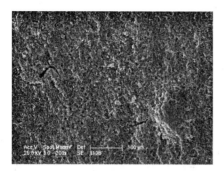

Figure 9. SEM micrographs of the geopolymer matrix with TiO_2 – 1710 - having a specific surface area 1.5 m^2/g (as measured by BET) and pore volume 0.007 cm^3/g (as measured by BJH).

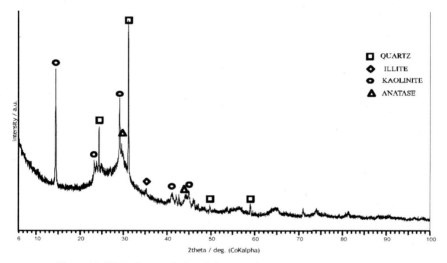

Figure 10. XRD characteristics of the geopolymer matrix with TiO_2 - 1710

Photocatalytic characterization

Plates with photocatalytic layers were characterized in a gas phase flat reactor. A typical experimental run of the decomposition of butane is shown on Figure 11. A decrease in the concentration of butane represents a reaction of pseudo-first order with respect to the initial concentration of butane (0.3 mol%), which is characterized by a constant of the velocity k at constant temperature and relative humidity. Tab. I. shows average values of k that were calculated from 2-3 experimental runs. The photocatalytic activities of the layers correlate with the values of their surface areas. The observed activity of the initial geopolymer matrix could be explained by a relatively high content of natural metal oxides like TiO_2 and Fe_2O_3 in the metakaolinite used.

In all cases the values of k decreased with the number of photocatalytic experiments, typically 15-20% decrease in the value of k between the first and the second runs was observed. Similarly 5-10% decrease between the second and the third runs occurred. This deactivation can be explained rather by deposition of carbon residual impurities originating from the decomposition of butane than by any changes in structure of the photocatalysts resulting from UV irradiation. This was concluded after a long term experiment in which the layers were almost 80 hours UV exposed without the presence of organic compound, and only negligible changes in their photocatalytic activities were found. On the other hand, a decrease of the photocatalytic activities in the prepared materials after two months of free ageing at ambient temperature and humidity in the laboratory was found, see Tab. I. This ageing effect could be explained probably by a contamination of the TiO_2 nanoparticles by impurities originating from the matrix. After the two month period the ageing was apparently stopped.

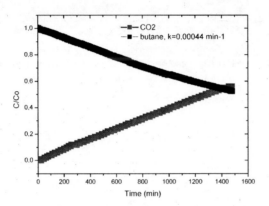

Figure 11. Kinetics of the degradation of butane on photocatalytic material-

1710

Tab. I.

	Geopolymer matrix	P 25	TiT 144	1710
BET [m^2/g]*	1	1.4	5.9	1.5
k 10^5 [min^{-1}]**	0.2	14	69	47
k 10^5 [min^{-1}]***	0.2	12	63	35

FOOTNOTES
*Values of surface area of the corresponding layers.
**"Freshly" prepared photocatalytic materials characterized within one week.
***Material aged for two month at ambient temperature and humidity.

CONCLUSION
In the present work the preparation, characterization and photocatalytic properties of geopolymer layers with photocatalytic TiO_2 powders have been studied. Results show that all studied materials, including initial geopolymer matrix, exhibited photocatalytic activities that correlate with the values of their surface area. All prepared photocatalytic materials however exhibited significantly enhanced photocatalytic activity compared with that of geopolymer

matrix free of TiO_2. In all cases the photocatalytic activities decreased to some extent with the number of experimental runs and with ageing of the photocatalytic material.

ACKNOWLEDGMENTS

The authors thank the Academy of Science of the Czech Republic for project AV0Z40320502 and Ministry of Industry and Trade of the Czech Republic for project 2A-1TP1/063.

REFERENCES

[1]K. Hashimoto, H. Irie, and A. Fujishima, TiO_2 photocatalysis: A Historical Overview and Future Prospects, *Jap. J. Appl. Phys.*, **44**, 8269–8285 (2005).

[2]M. Macounova, H. Krysova, J. Ludvik, and J. Jirkovsky, Kinetics of Photocatalytic Degradation of Diuron in Aqueous Colloidal Solutions of Q-TiO_2 particles, *J. Photochem. Photobiol. A-Chem.*, **156**, 273-282 (2003).

[3]S. Bakardjieva, J. Subrt, and V. Stengl, Photoactivity of Anatase-rutile TiO_2 Nanocrystalline Mixtures Obtained by Heat Treatment of Homogeneously Precipitated Anatase, *Appl. Catal.*, **58**, 193-202 (2005).

[4]J. Davidovits, and J. Davidovits, Geopolymers: Inorganic Polymeric New Materials, *J. Therm. Anal.* 37, 1633–1656 (1991).

[5]A. Buchwald, H. Hilbig, and C.Kaps, Alkali-activated Metakaolin-slag Blends - Performance and Structure in Dependence of their Composition, *J. Mater. Sci.*, **42**, 3024–3032 (2007).

CHARACTERIZATION OF RAW CLAY MATERIALS IN SERBIA 0.063mm SIEVED RESIDUES

Snežana Dević, Milica Arsenović and Branko Živančević
Institut IMS, Bulevar vojvode Mišića 43, 11000 Beograd
11000 Beograd, Serbia
Snezana.devic@institutims.co.yu

ABSTRACT

Particle size is the effective diameter of a particle as measured by sedimentation, sieving, or micrometric methods. Particle sizes are expressed as classes with specific, effective diameter class limits. The broad classes are clay ($< 2\mu m$), silt ($2 - 20\ \mu m$ dust fragments), and sand (over $20\ \mu m$). The physical behavior of a soil is influenced by the size and percentage composition of the size classes. Particle size is important for most soil interpretations, for determination of soil hydrologic qualities, and for soil classification. Physical properties of the soil are influenced by the amounts of total sand and of the various sand fractions present in the soil. Sand particles, because of their size, have a direct impact on the porosity of the clay. This influences other properties, such as saturated hydraulic conductivity, available water capacity, water intake rates, aeration, and compressibility related to plant growth and engineering uses. Soil properties and application depend on many factors such as: raw clay row, mineralogical constitution, quantity and category of accessory minerals, purity and chemical composition, as well as physical and thermal properties. The aim of this study is to determine mineral and chemical composition of 0.063mm sieve residues at a few localities in Serbia. Depending on a quality, clay is used as a raw material in brick or ceramic industry. Sand content is measured in the laboratory by the wet sieving method. Mineralogical composition of samples is determined using optical and XRF methods. Certain fragments are deeper examined with methods used. Results obtained enabled raw clay samples characterization at 4 different locations in Serbia. This way more soil profound evaluation is gained.

INTRODUCTION

Clay deposits are abundant in Serbia. Some of those deposits provide high quality raw material for the ceramic industry, production of facade bricks, roof tiles and blocks[1]. 0.063 mm sieve residue samples of clays presented in this paper originate from clay deposits adequate for the brick industry. These deposits are "A", "B", "C" and "D". Sieved 0.063 mm residues of clays from deposits "A", "B", "C" and "D" clays have been completely examined and are being used for production of various bricks.

EXPERIMENTAL PROCEDURES

Experimental procedures includes chemical and mineralogical (from macroscopic to microscopic scale) examination of 0.063 sieve residue samples from clay deposits marked as A, B, C and D.

RESULT AND DISCUSSION

The results of examination of 0.063 sieve residue samples show certain differences. Results of chemical and mineralogical examinations are presented in sample descriptions in this section. Samples A1 and A2 from deposit "A" are characterized by high SiO_2 and Al_2O_3 contents and low CaO content, as shown in Table 1. Table 1 also shows high contents of the same oxides in samples D1 and D2. In samples C1 and C2, the content of SiO_2 is a bit lower than in previous samples, but CaO content is slightly raised. Samples B1 and B2 have significantly different main oxide contents. The contents of SiO_2 and CaO in them are almost equal. Carbonate fragments in these samples are more frequent, especially loess nodules (Calcium – carbonate type residues contain mostly calcite, limestone, loess nodules, bivalves and gastropods fragments)(Figure 3.). Figure 4.shows increased presence of silicate, quartz fragments, and also of carbonate fragments. DEPOSIT A - Sample A1 - White-yellow sample size up to 18 mm contains coarse sand, sand and dust fractions. Sand fraction is present in less quantity, and it is rich in quartz with little carbonates. Dust fraction present is rich in quartz, mica and accessory minerals. Particle dimensions vary from 0.063 to 18 mm. Reaction with 5 % HCl is moderate. DEPOSIT A - Sample A2 - Yellow colored sample consists of coarse fraction with up to 4 mm size. Sand fraction is also present, and contains quartz, little carbonates, mice and accessory minerals. Dust fraction consists of quartz, mica and accessory minerals. Particles dimension is from 0.063 to 4 mm. Reaction with 5 % HCl is weak. DEPOSIT B - Sample B1- Sample is a brown-like soil. Coarse and sand fractions are present, made of loess nodules up to 10 mm of magnitude. Sand fraction consists also of loess nodules, quartz and some accessory minerals. The dust fraction contains loess (this is type connection for sedimentary brick- petrological) nodules, quartz, a little bit of mica. Particle dimensions is from 0.063 to 10 mm. Reaction with 5 % HCl is turbulent. DEPOSIT B - Sample B2 - White-yellow sample contains coarse sand, sand and dust fractions. The coarse sand fraction consists of alevrolitic particles of size up to 22 mm. The sand fraction is made of loess nodules, macrofloral fragments, quartz and accessory (accessory minerals are following minerals, as rutile, turmaline, magnetite, etc.) minerals. Dust fraction consists of loess nodules, quartz, mica and accessory minerals. Particle dimensions vary from 0.063 to 22 mm. Reaction with 5 % HCl is turbulent. DEPOSIT C – Sample C1 -Yellow-gray colored sample has fragments up to 7 mm.The sand fraction consists of alevrolitic-like particles, loess nodules, quartz, ferrous-alevrolitic fragments, limonite grains, mica and accessory minerals. Particle dimensions vary from 0.063 to 7 mm. Reaction with 5 % HCl is turbulent. DEPOSIT C - Sample C2 - Sample is yellow-gray and consist of a coarse fraction with particle sizes up to 9 mm. Sand fraction contains alevrolitic-like, ferrous - alevrolitic fragments, quartz, little limestone, limonite grains, macrofloral particles and mica. The dust fraction contains alevrolitic-like fragments, mica and accessory minerals. Particle dimensions vary from 0.063 to 9 mm. Reaction with 5 % HCl is moderate. DEPOSIT D - Sample D1 - Brown-gray sample has a small quantity of coarse fraction up to 3 mm. This fraction consists of metallic minerals, limonite grains and quartz. Sand fraction contains quartz, alevrolitic fragments, macrofloral fragments, limonite grains and accessory minerals (with a high content of metallic minerals). The dust fraction contains quartz, alevrolitic fragments, limonite grains, mica and accessory metallic minerals. Particle dimensions vary from 0.063 to 3 mm. The sample does not react with addition of 5 % HCl. DEPOSIT D - Sample D2 - Brown-gray sample contains mostly 22 mm grains, with other grains from 2 to 6 mm. The sand fraction consists of quartz, alevrolitic fragments, macrofloral segments, limonite fragments and accessory minerals with metallic minerals included. The sample does not react with addition of 5 % HCl.

Table I. Chemical compositions sample of Deposits A, B, C and D (wt-%)

	Deposit A		Deposit B		Deposit C		Deposit D	
	A1	A2	B1	B2	C1	C2	D1	D2
SiO_2	87.46	85.89	38.76	47.74	69.93	61.47	76.53	71.12
CaO	1.33	1.33	30.35	23.34	3.00	5.33	1.17	1.67
Al_2O_3	5.15	7.57	2.42	2.72	12.65	15.53	11.29	11.22
MgO	0.36	0.24	0.96	1.44	1.20	1.92	0.48	1.56
MnO	0.020	0.022	0.080	0.152	0.060	0.040	0.042	0.215
Fe_2O_3	0.59	0.83	1.54	0.95	3.08	4.51	1.30	5.22
LOI	1.41	1.27	25.39	20.81	2.27	6.41	1.43	2.86

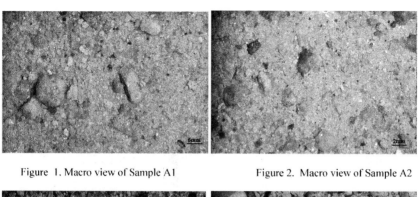

Figure 1. Macro view of Sample A1   Figure 2. Macro view of Sample A2

Figure 3. Macro view of Sample B1    Figure 4. Macro view of Sample B2

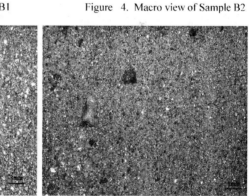

Figure 5. Macro view of Sample C1   Figure 6. Macro view of Sample C2

Figure 7. Macro view of Sample D1   Figure 8. Macro view of Sample D2

Macro views of different sieve residues are presented on Figs. 1 to 8. Pictures are obtained by digital camera. Original size of photographed components is 0.063mm - 20 mm.

Aside from the mentioned minerals, in some samples alevrolitic fragments, metallic minerals, macrofloral fragments and other accessory minerals also occurred. Figure 9 shows microscopic views of siliceous component – mineral quartz. Figure10 shows microscopic view of loess nodules. Figure 11 shows microscopic view of metallic minerals.

Figure 9. Micro view of quartz –
Microscopic analysis in transmitted light[2]

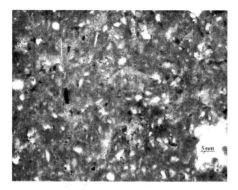

Figure 10. Micro view of loess nodules
Microscopic analysis in transmitted light[2]

Figure 11. Micro view of metallic minerals
Microscopic analysis in transmitted light [2]

CONCLUSION

The results obtained by macroscopic and microscopic identification of 0.063 mm sieved residues have shown significant differences in their mineral composition and content, visual appearance and mass percent of the residue in the sample. The samples have generally shown two composition types: a) silicate and b) calcium-carbonate. Silicate type residues most often contain quartz, micas and other silicate mineral fragments. Calcium-carbonate type residues contain mostly loess nodules, calcite and limestone.

REFERENCES
[1]S.Dević, Z.Radojević, D.Urošević: Macroscopy and microscopy identification of the residue on sieve 0.063mm of the clay deposits in Serbia ,Mines Engineering , vol. 1, 19 – 24 (2007)
[2]S.Dević, L.Marčeta, A.Mitrović: Mineral phases of 0.063 mm sieve residue of Serbian clay deposits identified microscopic method, 39th IOC on Mining and Metallurgy, 2007 Sokobanja, Serbia, Proceedings , TF Bor 409 – 414 (2007)

FIREPROOF COATINGS ON THE BASIS OF ALKALINE ALUMINUM SILICATE SYSTEMS

P. V. Krivenko, Ye.K. Pushkareva, M. V. Sukhanevich, and S.G. Guziy
State Scientific Research Institute for Binders and Materials, Kiev, Ukraine

ABSTRACT

The paper covers the results of theoretical studies and practical experience of manufacturing intumescing inorganic materials in a system: $Na_2O - Al_2O_3 - SiO_2 - H_2O$ due to a target synthesis of zeolite-like hydration products of the heulandites group capable to swell at low temperatures (below 300 celcium degrees). The principles of compositional build-up of reactive mixes made with alkaline cementations materials and fillers to provide the production of intumescing fire resistant coating with required service properties that are environmentally friendly are discussed. The processing parameters have been selected and the results are brought on commercial scale. The results of commercial scale production are discussed.

INTRODUCTION

Securing the safety conditions for the use of civil and industrial buildings envisages taking action to increase the fire resistance of the buildings. Recently, along with traditionally used inorganic fireproof materials – the brickwork, plaster - also the thin-layer coatings and paints have been widely applied. They increase the fire resistance of the structure and do not increase its weight. The most effective fireproof coatings are the bloating coatings capable of enlarging their volume by 15-20 times and increasing the fire resistance durability of the metal constructions for more than one hour.

Nowadays the bloating materials on the basis of organic components – carbamide pitch, phenol-formaldehyde pitch and others are widely used. They are characterized with fireproof parameters that correspond to normative documents and have high bloating ability. An essential drawback of the defined materials is their polymeric base that creates certain environmental problems at the moment of fire, thus, leading to creating a lot of smoke and exhausting toxic gases that inflict irreparable harm to peoples' health.

Eliminating of the said drawbacks is possible by way of utilization of mineral coatings that contain chemically bound water which can be discharged under relatively low heating temperatures. Owing to the fact that bloating of the inorganic substances due to water discharge in the most environmentally safe process. It is reasonable to use materials such as pearlite, vermiculite and hydromica that contain chemically bound water, structurally bound water and zeolitically bound water [1].

The use of the minerals such as pearlite, vermiculite and hydromica is also known in the context of obtaining organic bloating fireproof coatings; at that the said materials either perform the function of filling materials [2-4] without influencing the bloating process, or take part in the bloating process through the discharge of chemically bound water together with organic gasifesers (carbamide pitch, phenol-formaldehyde pitch) [5-6].

Experience was gained in using sodium water glass to obtain the bloating materials including the fireproof materials [2, 7]. By the bloating nature the liquid glass materials are divided into thermal bloating materials and materials that are bloated as result of chemical interaction between the glass and the special additives. However, these materials have a number defects i.e. limited water resistance and insufficient durability. The elimination of such defects is possible by injecting hardeners into the liquid glass that would facilitate formation of the hydro silicate and hydro alumino silicate phases that leads to the formation of durable composite material having special properties.

Numerous works by the Scientific Research Institute for Binding Materials (SRIBM) [8-18] have proved the opportunity to perform direct synthesis of the whole range of alkaline hydro alumino silicates that are substitutes to the natural materials - vermiculite and hydromica – contained in the products of artificial stone hydration. Research into the hydration of such phases has demonstrated that at the higher temperatures the hydrate substances are crystallized into waterless substances [8]. Their composition depending on the temperature may be presented by the whole range of the synthetic minerals having the following structural formula: $R_2O \cdot Al_2O_3 \cdot (2-6)SiO_2$ [11, 13].

Therefore, the development of inorganic bloating materials is based on the purposeful formation inside the alkaline binding hardeners of the zeolite-like new formations having certain composition and structure and capable of dehydration at relatively low temperatures (up to 300°C) and increase the volume by dozens times.

The purpose of this work is to study properties of the bloating alkaline aluminum silicate materials in the system $Na_2O-Al_2O_3-nSiO_2-mH_2O$, and to investigate the opportunity to control their parameters depending on composition of the substance and the technology for obtaining it.

THEORETICAL PREMISES OF THE RESEARCH

It is clear from published work [19, 20] that zeolitization occurs as a result of reactinge silica and alumino silicate constituents in the solutions of alkaline or alkaline earth metals at a temperature of 100-300°C and pressure of 0.1- 0.4MPa. The type of the zeolite being formed will depend on natural component compositions and the steam liquid. Figure 1 shows the diagram of the cryotallization field of natural zeolites, depending on the conditions of the hydrothermal processing as well as upon the gel composition.

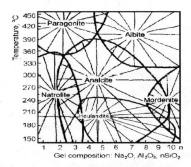

Figure 1. Diagram of the zeolite crystallization fields

From the Diagram of the zeolite crystallization fields one can see that with increase in temperature and gel compositions ($SiO_2/Al_2O_3=2-4$) one can observe the formation of the low temperature Na–zeolites (analcite, gmelenite, nytrolite) which are being crystallized through the formation of the zeolite nuclei in the liquid phase. With the increase in the ratio of SiO_2/Al_2O_3 to 4 and higher, the higher silica content zeolite of modernite group is created.

According to the data of reference [21] zeolites may be heated above the dehydration temperature without destroying their crystalline structure. As the SiO_2/Al_2O_3 ratio in zeolite is higher the destroying of crystalline structure occurs at higher temperatures. The thermal stability of the zeolites is displayed with the parameters as follows: (a) the higher the SiO_2/Al_2O_3 ratio in zeolite, the higher is the temperature when the crystalline structure is destroyed and its thermal resistance raises; (b) availability of Na^+ and K^+ exchange cations in zeolites raises their stability as compared with their calcium varieties;(c) the skeletal topology – the dimensions of T-O-T angles in the skeleton – isothermal rings are steadier than the rings having equal number of tetrahedural. From the above referenced it is clear that under the influence of the temperature factor, destructive changes occur in the oxygen alumino silicate framework without destroying the integrity of the crystalline structure that brings the dehydration product amorphous state, thus creating waterless alumino silicates. These hydro aluminum silicates include zeolite-like new formations of heulandite-types (desmine, epistilbite, laumontite etc.) of the general formula $Me_{x/n}[Al_xSi_yO_{2(x+y)}]zH_2O$, the bloating properties of which are provided through availability inside of their wide porous channels of 10 and more water molecules. Water is kept inside the zeolite structure thanks to Van-der-Vaals' forces and is being removed from the channels at relatively low temperatures.

RESULTS
Physical and chemical research to the aluminum silicate compositions

It was defined by the research in references [13, 16, 17] that in the process of interactions in the $Na_2O-Al_2O_3-SiO_2-H_2O$ system under certain conditions, the content of zeolite-like dehydration products, new formations of heulandite types are formed. They are capable of bloating at low temperatures (up 300°C). The binding compositions within the system $Na_2O-Al_2O_3-(2-10)SiO_2-25H_2O$ were studied. They were made with the use of metakaolin, the sodium soluble glass (Ms=2.8; ρ=1400 kg/M^3), micro silica and alkali Na_2O. Analysis of the former research [12, 14-16] has shown that in the system being studied it is possible to perform the synthesis of zeolite content hydro alumino silicates in the ratio SiO_2/Al_2O_3 =4.7-11.4, which may increase in volume by 7-13 times when influenced by relatively low temperatures from 150 to 350°C.

The content of new formations of the alumino silicate compositions were studied with the methods of physical and chemical analysis including X-ray, DTA (different thermal analysis), IRS (infrared spectroscopy) and electron microscopy.

Thus, at hydration of the composition $Na_2O \cdot Al_2O_3 \cdot 6SiO_2 \cdot 25H_2O$, the hardening products are represented by: - heulandite $Na_2 \cdot Al_2 \cdot Si_7O_{18} \cdot 6H_2O$ (d/n=0.489:0.356:0.307:0.280:0.266:0.243:0.227:0.201:0.166 nm) and the residue of metakaolin that has not reacted (Figure 2, curve 1).

Based on infrared spectroscopy (Figure 3, curve 1) the occurrence of heulandite shall be proved by the existence of absorption peaks in the area 420-500 cm^{-1} (428; 441; 528), that related to deformation fluctuations TO_4, the absorption stripes in 650-820 cm^{-1} (652; 665; 687; 706; 725; 822) correspond to the valent fluctuation tetrahedra TO_4. The position of these peaks is influenced by the ratio Si/Al in the zeolite framework: as the content of tetrahedral coordinated atoms Al which are shifted into the low frequency range. On the DTA curve (Figure 4, curve 1) the appearance of the heulandite is proved by the endothermic effect at the temperature 300-350°C. The therma gravimetric curve (Figure 2, curve 3) points to the stepped decomposition of the heulandite at the temperature range 250-300°C. In the range of temperatures 110-150°C water is discharged slowly. However, at the temperature 200°C the dehydration speed increases sharply, when the mass loss makes 16.5 wt %. After burning the composition at 500°C (Figure 1, curve 2), the amorphization of the structure is observed. The heulandite peaks can hardly be fixed; it is the diffractional reflections of metakaolin that remain. The mass loss after the second burn makes 3.6 wt % (Figure 3, curve 1).

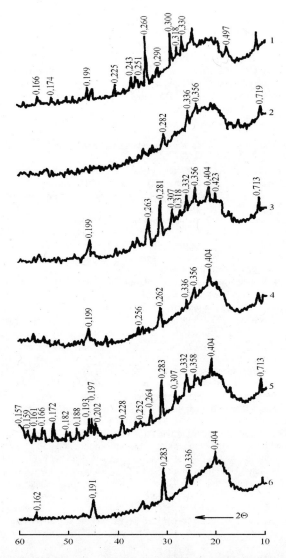

Figure 2. X-ray diffractometry of the hydration products (curves 1, 3, 5) and dehydration products (curves 2, 4, 6) of binding compositions $Na_2O \cdot Al_2O_3 \cdot 6SiO_2 \cdot 25H_2O$ (curves 1, 2); $Na_2O \cdot Al_2O_3 \cdot 8SiO_2 \cdot 25H_2O$ (curves 3, 4); $Na_2O \cdot Al_2O_3 \cdot 10SiO_2 \cdot 25H_2O$ (curves 5, 6)

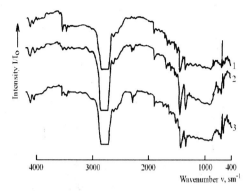

Figure 3. Infra red (IR) spectroscopy of dehydration products of the binding compositions $Na_2O \cdot Al_2O_3 \cdot 6SiO_2 \cdot 25H_2O$ (curve 1); $Na_2O \cdot Al_2O_3 \cdot 8SiO_2 \cdot 25H_2O$ (curve 2); $Na_2O \cdot Al_2O_3 \cdot 10SiO_2 \cdot 25H_2O$ (curve 3)

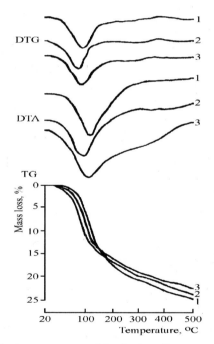

Figure 4. Derivative spectroscopy of the alumino silicates compositions: $Na_2O \cdot Al_2O_3 \cdot 6SiO_2 \cdot 25H_2O$ (1); $Na_2O \cdot Al_2O_3 \cdot 8SiO_2 \cdot 25H_2O$ (2); $Na_2O \cdot Al_2O_3 \cdot 10SiO_2 \cdot 25H_2O$ (3)

The curve (Figure 3, curve 1) points to the stepped decomposition of the heulandite at the temperature range of 250-300°C. In the temperature range of 110-150°C the water is discharged slowly, however, at the temperature of 200°C the dehydration speed increases sharply; at that the mass loss makes 16.5 wt %.

According to scanning electron microscopy (Fig. 5) at the relatively insufficient enlargement of 100x, the structure of the bloated material on the basis of alumino silicate composition $Na_2O \cdot Al_2O_3 \cdot 6SiO_2 \cdot 25H_2O$ may be defined as cellular, with evenly occurring porosity.

With the picture enlargement by x1000-x5000 inside the material structure, one can observe regularly repeated framework elements including also zeolite microregions.

a)   b)

c)   d)

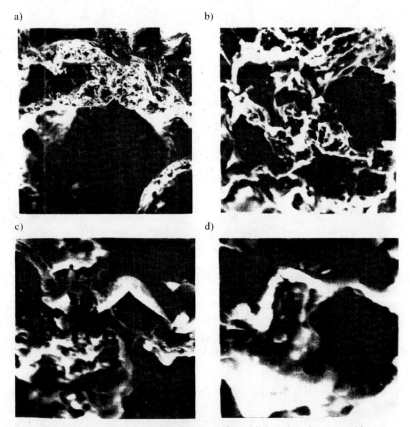

Figure 5. SEM Electron micrographs of stone surface chip based on the composition $Na_2O \cdot Al_2O_3 \cdot 6SiO_2 \cdot 25H_2O$ after the bloating at the temperature of 500°C with enlargement: (a) -x100; (b) -x1000; (c) -x2500; (d) -x5000

The resulting structure obtained on the basis of synthetic zeolite of heulandite content ($Na_2O \cdot Al_2O_3 \cdot 6SiO_2 \cdot 25H_2O$), is by its characteristic similar to the structure obtained by the natural zeolite of the monthmorillonite and the liquid glass (Fig. 6). The difference in the degree of the crystallization of the structures obtained on the basis of natural and synthetic zeolites can clearly be seen when comparing the electronic photos (Fig. 5, 6) at the enlargement of x5000. In so doing, the structure formed in the process of dehydration of the composition based on natural zeolite, which is represented by widely brimmed fiber-like elements. However, the structure obtained on the basis of synthetic zeolite $Na_2O \cdot Al_2O_3 \cdot 6SiO_2 \cdot 25H_2O$, is composed of mainly amorphized elements, set up the same manner as the in the structures formed by natural zeolites and this is due to low degree of crystallization. However, despite the fact that the indicated differences in the structures remain similar, in the composition based on the dehydration of the synthetic zeolite, one can observe the presence of zeolite microregions evenly concentrated over the skeleton of the bloated material.

a)   b)

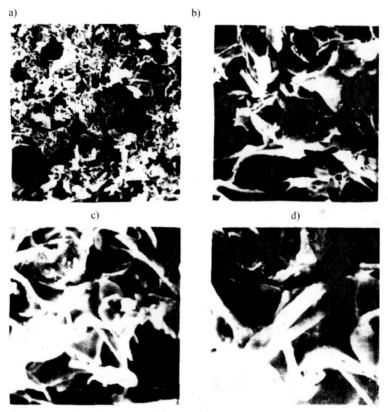

c)   d)

Figure 6. SEM Electron micrographs of stone surface chip based on the composition based on natural zeolite (clinoptilolite) and sodium water glass after the bloating at the temperature 500°C with enlargement: (a) -x100; (b) -x1000; (c) -x2500; (d) -x5000

At the hydration of the composition $Na_2O \cdot Al_2O_3 \cdot 8SiO_2 \cdot 25H_2O$, the hydrate new formations are presented by (Figure 2, curve 3): - Na-modernite (d/n=0.421:0.407:0.356:0.332:0.315:0.283:0.199 nm), and the residual metakaolin (d/n=0.407:0.356:0.349:0.338 nm).

The DTA data prove the availability of Na-modernite residual metakaolin in the content of the composition product hydration. Based on the data of thermal gravimetrical analysis TGA (Figure 4, curve 2) the mass loss observed during the burning at 500°C, makes 16.7 wt % - and corresponds to data published in references [20, 21] concerning changing the mordenite mass at dehydration.

After burning the composition at 500°C one can observe amorphization of the metal structure. According to the X-ray data (Figure 2, curve 4) in the content of burned products the mordenite residue was fixed (d/n=0.407:0.356:0.338:0.282:0.199 nm). The mass loss of the samples does not exceed 2.5 wt % (Figure 4, curve 2), which indicates almost complete discharge of the zeolite water when bloating t the composition $Na_2O \cdot Al_2O_3 \cdot 8SiO_2 \cdot 25H_2O$.

While hydrating composition $Na_2O \cdot Al_2O_3 \cdot 10SiO_2 \cdot 25H_2O$, the content of the new formations is represented by mainly modernite, moreover, its quantity is much higher than in hydrating of the composition $Na_2O \cdot Al_2O_3 \cdot 8SiO_2 \cdot 25H_2O$ (Figure 1, curve 5). According to DTA data (Figure 4, curve 3) dehydration occurs steadily in the range of temperature range 20-400°C, the mass loss based on the thermal gravimetrical analysis, makes 13 wt %. After burning the composition at a temperature of 500°C (Figure 1, curve 6), as in the previous case (Figure 5, curve 4), the content of the dehydration products is represented by amorphized substance that contains an insufficient quantity of modernite (d/n=0.404:0.336:0.28:0.191 nm). Under TGA data (Figure 4, curve 3) the mass loss after the repeated burn makes 4.8 wt %. Therefore, the set of physical and chemical analysis methods have confirmed the assumption of the possible synthesis in the content of the hydration products within the alkali alumino silicate system $Na_2O-Al_2O_3-(2-10)SiO_2-25H_2O$ of the zeolite-like heulandite group. The latter is capable of dehydration at relatively low temperatures with partial amorphization of the structure.

Influence of technological factors on bloating ratio of the alumino silicate compositions

The binding mixtures in the system $Na_2O-Al_2O_3-(6-8)SiO_2-25H_2O$ can be used to prepare the fireproof bloating compositions that are applied to various surfaces such as metal, concrete, brick, wood. The research of the bloating degree of the compositions was performed based on the change of surface bloating factor that is calculated as correlation of the thickness (height) of the coating bloated in a furnace at a temperature of 500°C, and the thickness of the initial coating (that was hardening under normal conditions or drying at the temperature 50-80°C). The fireproof bloating coating was obtained by mixing alumino silicate binding composition over 5 minutes in the "Hobort" type mixer (two composition was kept this composition previously in the closed vessel for no less than one hour and then the filling material – silica- and other additives were added to have the composition defined for this particular experiment. It was better to use the prepared mixture over 15-30 min. Previous testing of the coatings for bloating ability which is an indirect index of fire resistance was done according to the Bartel methodology [22] (Figure 7).

The temperature of the flame was measured by a platinum and platinum rhodium thermocouple, and the temperature of the metal plate was measured by a chrome aluminum thermocouple. Heating of the surface was performed by the open fire (spot influence) from the coating side fixing the temperature on the reverse side of the metal plate within 1 hour. For the purpose of studying the influence of temperature and interval of bloating upon the factor of bloating by the alumino silica composition the studied reacting mixtures were optimized with the use of the bifactorial three level method of experiment planning. The selected variable factors were the bloating temperature (T, °C) and interval (τ, min) of bloating the coating applies to the metal plate [18]. Analysis of the research results shows that the composition $Na_2O \cdot Al_2O_3 \cdot 7SiO_2 \cdot 23H_2O$ had the highest bloating factor (17,83) (Figure 8 (b)).

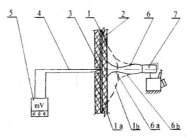

Figure 7. Diagram of Bartel unit: 1) the fireproof bloating coating; 2) metal plate of 3 mm thickness; 3) heat isolating plate; 4) thermocouple; 5) millivoltmeter; 6) torch flame (temperature 1000-1200°C); 7) source of flame (petrol torch). 1a and 1b –thickness and form of flame distribution prior to bloating; 6a and 6b – after bloating of the surface

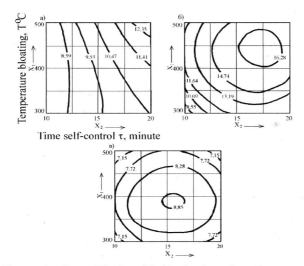

Figure 8. Diagram showing modification of the bloating factor depending on
The temperature and time of exposure of the composition:
a) $Na_2O \cdot Al_2O_3 \cdot 6SiO_2 \cdot 20H_2O$;
b) $Na_2O \cdot Al_2O_3 \cdot 7SiO_2 \cdot 23H_2O$;
c) $Na_2O \cdot Al_2O_3 \cdot 8SiO_2 \cdot 26,5H_2O$.

Isolines of bloating factor are shifted towards the right corner of the factor space which is limited on the X_1 axis by bloating temperature ranging from 410 to 475°C, and on the X_2 – axis is limited by the time of exposure at the given temperature from 15 to 18 min. The composition $Na_2O \cdot Al_2O_3 \cdot 8SiO_2 \cdot 26,5H_2O$ had a somewhat lesser bloating factor – 10.54 (Figure 8, c). Isolines of the bloating factor are concentrated in the central part of the figure which is limited on the X_1 axis by temperature from 360 to 410°C, and on the X_2 – is limited with time of exposure from 14 to 17 min. The different picture of isoline concentration for the bloating ratio has the composition $Na_2O \cdot Al_2O_3 \cdot 6SiO_2 \cdot 20H_2O$ (Figure 8, a). In this case no clear optimal range for the bloating factor is observed. As the temperature and time of exposure increases, the value of the bloating factor increases. The maximum value of the bloating factor – 13.29 is observed after the burning at a temperature of 500°C and time of exposure – 20 min.

This composition will be the most applicable to be used as the basis for obtaining a fire-resistant bloating material, while $Na_2O \cdot Al_2O_3 \cdot (7-8)SiO_2 \cdot 20H_2O$ composition would be used to obtain a fire-resistant bloating filling material.

During the work performance, the research was undertaken to improve the quality of the bloated alumino silicate coating that would lead to increase in the bloating ratio, strengthening of the bloated layer, increase of the adhesion of coating to metal etc. It was determined that the number of binding materials applied was a dominating parameter influencing the factor of composition bloating and the quality of the coating [23].

On a metal plate previously cleaned from dirt and grease, the alkaline alumino silicate binding composition was applied - $Na_2O \cdot Al_2O_3 \cdot (6-7)SiO_2 \cdot (20-23)H_2O$. Each subsequent layer was applied after natural drying of the previous layer under ambient temperature not higher than 20±2°C. The plate protected by the coating was bloated inside the laboratory furnace at a temperature of 500°C for 15-20 min. The value of the bloating factor was calculated according to the above referenced methodology. The resulting influence on the number of the layers applied is given in Table I.

Table I. Modification of the Surface Bloating Factor Depending on the Number of Layers Applied

Content of the reactive mixture of the binding material	Number of coating layers				
	1	2	3	4	5
$Na_2O \cdot Al_2O_3 \cdot 6SiO_2 \cdot 20H_2O$	35	24	18,8	17,7	10,2
$Na_2O \cdot Al_2O_3 \cdot 7SiO_2 \cdot 23H_2O$	39	25,4	20,6	19,4	10,9

The data obtained indicated that, on one hand, the bloating degree decreased in proportion to the number of the layers applied (due to increasing their mass). On the other hand, the more layers of the coating were applied, the higher was the durability of holding it onto the substrate, with the structure of the bloated coating becoming cellular (i.e effect of layer-by-layer bloating wais observed).

As the number of the applied layers increased the internal space was being filled because each layer bloated independently and the layers adhered to each other with a number of interlayer partitions. This strengthened the bloated layer, prevented the tunnel effect at fire and fast heating of the protected surface. While bloating the 4-5- layer coating the cut of the bloated material had a stratified structure of quite low thermal conductivity, as compared to the single-layer coating. The ratio of oxides SiO_2/Al_2O_3 did not influence sufficiently the value of the bloating factor that differs by 7-8% from that value of composition having the ratio of $SiO_2/Al_2O_3=6$ and 7.

Therefore the results of determining of the bloating degree for alkaline alumino silicate content, in combination with the created cellular structure of the material give the grounds to say that it is reasonable to apply 3 to 4 layers of the protected coating onto the metal substrate that would make it possible to obtain a required coating structure having a bloating factor of 17-20.

The authors suggested applying three layers of the alumino silicate coatings onto the metal plate mechanically – using the pneumatic spray, such that the thickness of the coating decreased significantly – up to 2 mm and also adhesive capacity the coating is increased.

Apart for the described factors influencing the bloating degree for the silica-alumina composition of the optimized composition of $Na_2O \cdot Al_2O_3 \cdot 7SiO_2 \cdot 23H_2O$ it should be mentioned about the additive-modifiers significantly influence the bloating degree. Each of those additives controls certain property of the basic composition. It was noticed that the nature of the modifying additive – micro silica SiO_2, added at the stage of basic composition design that facilitates control over the oxygen ratio SiO_2/Al_2O_3 may also influence the value of the bloating factor. The research was done on the single – layer coatings applied to metal plate and bloated at temperature of 500^oC within 15-20 min. The results of the coatings tests are shown in Table II.

Table II. Modification of the Surface Bloating Factor Depending on the Type of Microsilica Spheres in the Composition of the Binding Substance $Na_2O \cdot Al_2O_3 \cdot 7SiO_2 \cdot 23H_2O$

Type of Microsilica Spheres	Bloating factor
Model 1	39.0
Model 2	33.0
Model 3	112.5
Model 4	13.5

The obtained data make it possible to set forth that the application of the microsilica spheres (model 3) would be preferable, as it enables one to obtain coatings with the highest bloating factor.

The variety of the experimental data testifies to different activity of the studied substances as well as to various degrees of their dispersion. The larger is the specific surface area of the micro silica the more actively it is interacting with the other components of the alkaline silica alumina binding material creating micro nuclei of zeolite-like phases of the heulandite group. On the other hand, microsilica spheres obtained from the industrial waste has certain characteristics, in other words, it contains certain additives, vitrified parts etc. that may influence their reactive capacity. It is a known fact that by introducing additive-modifiers of different composition it is possible to control material properties and adjust them as required. This method was used for the study of the alkaline alumino silicate materials' ability to be used in the conditions of fire on metal surfaces. It was noticed that in the process of applying the binding material onto the vertical metal surface the problem occurs of coating sliding from the protected surface under its own weight. To eliminate this shortcoming the experiments were undertaken by modifying the alkaline alumino silicate binding material with inorganic additives that would increase the hardening ability of the coating and its adhesion without lowering the value of the bloating factor. The additives were introduced into the binding composition $Na_2O \cdot Al_2O_3 \cdot 7SiO_2 \cdot 23H_2O$ in the amount of 10 wt % of the composition mass.

The coating was applied to the vertically placed metal surface with a palette knife to be hardened naturally (at a temperature of 20 ± 2^oC and humidity of 60-70%). The bloating was carried out in a laboratory furnace at a temperature of 500^oC within 15-20 min. The results of coating tests are given in Table III.

Table III. Characteristics of the Bloated Coatings Depending on the Type of Additive

Type of additive	Hardening time, min	Value of the bloating factor
NaHCO$_3$ (baking soda)	10-15 sec	11-12
Sodium tetraborate	30	18-20
5% soda + 5% sodium tetraborate	15-20	12-14
Portland cement M400	3-5 sec	1-1,5
Blast furnace slag (Mariupol)	35-45 sec	6-7
CaCO$_3$	30-40	23-24
No additive	55-60	24-25

The analysis of the data obtained allows one to set out a number of additives that can be used to accelerate hardening in the alkaline alumino silicate binding material. They are also used to support the surface of the vertically placed metal plate at the same time they do not cause a decrease the degree of bloating significantly. These additives include calcium carbonate and the mixture of sodium tetraborate and sodium bicarbonate. The other additives (Portland cement, slag and sodium bicarbonate) cause accelerated setting up of the compositions (makes it impossible to apply it to large area) and lower the value of the bloating factor (due to the chemical interaction of the additive with the unbound water glass of the alkaline alumino silicate binding material. The completed research made it possible to set up the principles for compositional building of environmentally safe, bloating coatings based on alkaline alumino silicate binding material, and identify optimal technological parameters for their preparation and application.

CONCLUSIONS
1. The main patterns were set up to obtain the bloated compositions in the system Na$_2$O·Al$_2$O$_3$·(6-10)SiO$_2$·25H$_2$O on the basis of natural and anthropogenic raw materials, using physical and chemical research methods of X-ray, DTA (different thermal analysis), IRS (infrared spectroscopy) and scanning electron microscopy SEM. The hypothesis was confirmed concern the synthesis in the content of the hydration products of the alumino silicate compositions and the zeolite-like new formations of heulandite group. It was demonstrated that the maximum value of the bloating factor (15-20) for the compositions is achieved at molar oxides ratios in the mixture SiO$_2$/Al$_2$O$_3$ equal to 6-7, Na$_2$O/Al$_2$O$_3$=1-1.3 and H$_2$O/ Al$_2$O$_3$=22-25.
2. The fundamentals of the technology were developed to obtain fireproof bloating coatings based on optimal reacting mixtures. The properties of the developed bloating coatings were studied with application of the mathematical methods of experiment planning. Research was done on the influence of technological parameters on the value of the bloating factor. It was determined that the most meaningful technological factors to obtain materials having the highest value of the bloating factor (F$_b$=17) shall be the time for endurance of the composition prior to applying the coating (15-18 minutes), the number of the layers of the coating applied onto the substrate (3-4 layers), as well as the method of applying the coating onto the substrate (e.g.by pneumatic spray).
3. The study was undertaken to research the influence of different microsilica sphere additives on the degree of bloating of the alumino silicate composition Na$_2$O·Al$_2$O$_3$·7SiO$_2$·23H$_2$O, (optimal content). It was determined that the size of the specific surface area and availability of the glassy particles

influence its reacting capacity and facilitate obtaining materials having high bloating ability (F_b ≥30...35).

4. Additives or modifiers were selected (calcium carbonate and a mixture of sodium tetraborate sodium bicarbonate) to improve the technological properties of the fireproof swelling coatings, on the basis of alkaline alumino silicate systems of optimal composition $Na_2O \cdot Al_2O_3 \cdot 7SiO_2 \cdot 23H_2O$, which increased the speed of hardening and adhesion of the coating, without lowering the value of the bloating factor.

REFERENCES

1. Sobolev V.A. Introduction into Mineralogy of Silicates. - Lvov: Publishing House of the Lvov University. - 1949 - 375 p.
2. Romanenkov I.G., Levites F.A. Fireproof Design of Buildings. - Moscow: Stroyizdat, 1991. - 320 p.
3. Ovcharenko F.D., Suyunova Z.E., Teodorovich Ju.N. Disperse Minerals in Fireproof Compositions. - Kiev.: Naukova Dumka, 1984. - 160 p.
4. Gedeonov P.P. Fireproof Coatings Based on Vermiculite / Building Materials. - №6. - 1992. - Pp. 12-17.
5. A.S. the USSR № 883119 Duleba M.T., Trush L.E. Fireproof Coatings // Discoveries and Inventions. - № 43. - 1981. - Pp. 5-7.
6. Zhartovskiy V.M., Ragomov S.Ju. Developments of the Fireproof Coatings // Theses and Reports from the First Interstate Seminar "Issues on the Fireproof Building Materials and Structures", Lvov, 1984. - Pp. 7-12.
7. Matveev V.D., Smirnova K.A. About Hardening Porous Products on a Alkaline-silicate Sheaf // Works Scientific Research Institute of Buildings Ceramics. - Moscow: NIIB, 1950. - Part 3. - 387 p.
8. Gluhovskiy V.D. The Soil Silicates. - Kiev.: Gosstroyizdat, 1959. - 127 p.
9. Gluhovskiy V.D., Petrenko I.Ju. Complex Thermographic Analysis of the Kinetics Process of Formation of Alkaline Alumino Silicates // News of High Schools. A Series Chemistry and Chemical Technologies. – Vol. 8. - Part 6. - 1969. - Pp. 899-992.
10. Gluhovskiy V.D., Starchevskaja E.A., Krivenko P.V. Research to the Silicate Formation in the Mixtures Based on Clays, Quartz Sand and Soda // the Ukrainian Chemical Magazine. - 1985. – Vol. 35. - Part 4. - Pp. 433-435.
11. Krivenko P.V. Synthesis of Special Properties of Binding Substances in System Me_2O - MeO - Me_2O_3 - SiO_2 - H_2O // Cement. - 1990. - №6. - Pp. 10-14.
12. Krivenko P. V., Pushkareva E. K., Sukhanevich M. V. Fireproof Bloating Alumino Silicate Coatings // The Collection "Problems Fireproof Building Materials and Designs" – Lviv, 1994. - Pp. 31-34.
13. Sukhanevich M. V. Inorganic Bloating Materials on the Basis of Alkaline Alumino Silicate Systems. // Abstract of Dissertation Ph.D. (Cand.Tech.Sci.)– Kiev, KTUCA, 1997. - 15 p.
14. Krivenko P. V., Pushkareva E. K., Sukhanevich M. V. Development of Physical and Chemical Basis of the Directed Synthesis Inorganic Binders in System Na_2O-Al_2O_3-SiO_2-H_2O for Obtaining Ecologically Safe Fireproof Bloating Materials // Magazine Construction of Ukraine. - №2. – 1997. – Pp. 46-49.
15. Krivenko P. V., Pushkareva E. K., Sukhanevich M. V. Bloating Concrete Coatings to Improve Fire Resistance of Building Structures // Proc. of the Intern. Conference held at the Dundee, Scotland, UK, 8-10 September, 1999 (Concrete Durability and Repair Technology). - Pp. 415-422.

16. Sukhanevich M.V. Physical and Chemical Regularities of Manufacturing the Bloating Special Purpose Materials Based on Alkaline Alumino Silicate Systems // Proceed. of the Second Internation. Conf. "Alkaline Cement and Concrete". – Kiev: ORANTA Ltd. – 1999. – P. 220-236.
17. Pushkareva E.K., Guziy S.G., Sukhanevich M.V., Borisova A.I. Studying of Influence of Inorganic Modifiers on Structure, Properties and Durability of the Bloating Geocement Compositions // 3rd International Conference "Alkali Activated Materials Research, Production and Utilization, June 21-22, 2007, Prague. Priceeding. – Praha 10, Ceska rozvojova agentura, o.p.s. – Pp. 581-592.
18. Krivenko P. V., Pushkareva E. K., Sukhanevich M. V. Optimization of the Composition of the Fireproof Bloating Coatings in System Na_2O-Al_2O_3-(6-8) $SiO_2 \cdot nH_2O$ // Theses of the International Scientific Conference "Computer Materiology and Maintenance of Quality", MOK"36. – Odessa, 1997. - P. 107.
19. Barrer P., White E. Synthetic Crystal Alumina-silicate // Physical Chemistry of Silicates. - Moscow: Mir, 1965. - Pp. 132-138.
20. Senderov E.E., Hitarov N.I. Zeolites: Synthesis and Formation in the Nature. - Moscow: the Science, 1970. - 128 p.
21. Ovcharenko G.I., Sviridov V.L .Zeolites in Building Materials. - Barnaul, Publishing House AltGtU. - Part 1. - 1995. - 102 p.
22. Sukhanevich M. V., Guziy S.G. Studying of Influence of Technology Factors on Properties Alkaline Alumino Silicate Systems to Obtain the Fireproof Coatings // New Refractory Products - Scientific Technical and Industrial Magazine.: Moscow, "Intermet Engineering" Ltd., 2004. - №3. - Pp. 47-50.
23. Pushkareva E.K., Guziy S.G., Sukhanevich M.V. Ecologically Safe Fireproof Bloating Coatings: Structure, Properties, Technology of Obtaining and Application Peculiarities// the Collection Building Materials, Products and Sanitary Technique – Kiev.- 2007. - №25 - Pp. 95-103.

DETERMINING THE ELASTIC PROPERTIES OF GEOPOLYMERS USING NONDESTRUCTIVE ULTRASONIC TECHNIQUES

Joseph Lawson, Benjamin Varela, Raj S. Pai Panandiker, and Maria Helguera
Mechanical Engineering Department
Rochester Institute of Technology
76 Lomb Memorial Drive
Rochester, NY 14623
United States of America

ABSTRACT

A non-destructive process, using ultrasonic techniques, for evaluating the elastic properties of a material is applied to a series of metakaolin based geopolymers with Si:Al ratios ranging from 1.49 to 6.4. This study evaluates the speed of sound of both a shear and longitudinal ultrasonic wave for each sample as well as the density's relationship to the Si:Al ratio. The trends observed for the Poison's ratio, elastic modulus, and speeds of sound showed a discontinuity around a Si:Al ratio of 3.1:1 supposedly due to the chemical changes between the poly(sialate-siloxo) (PSS) and poly(sialate-disiloxo) (PSDS) regions. Within the PSS region, the elastic modulus and Poison's ratio decreased linearly as the Si:Al ratio was increased. The trends within the PSDS region were discontinuous from the previous trends but not enough data was collected to sufficiently describe them.

INTRODUCTION

Developing non-destructive testing methods for geopolymers has become necessary to effectively identify the mechanical properties without destroying the samples to allow for cheaper and repeatable studies to be performed. Determining a definitive value for such properties can be very difficult because of the numerous variations that can go into the synthesis of a geopolymer such as chemical ratios, curing regiment, particle size, and impurities [1,2]. Previous research has shown that studying the compressive strength is not an effective method of material characterization because the presence of porosities will lead to large deviations due to the destructive nature of the tests [3].

The use of ultrasound for a wide range of mechanical testing has already been deeply explored and theories about the relationship of the speed of sound and the elastic properties have been proposed and tested. The aim of this work is to use this ultrasonic technique to evaluate a method of determining the elastic modulus and Poisson's ratio for a geopolymeric material. This procedure is completely non-destructive, allowing for samples to be tested and retested. A range of metakaolin based geopolymer samples will be evaluated to determine both elastic properties over a range of Si:Al ratios.

THEORY

As an ultrasonic wave propagates through its medium, it causes small sinusoidal displacements of molecules which induce strains. If the displacements could be measured accurately, determining the elastic properties would be a straightforward approach of applying Hooke's Law. With ultrasound however, it is just as convenient to know the speed of sound through a material to determine these same elastic properties. The rate at which a material can deform will determine the rate at which sound can propagate. To calculate the Poisson's ratio

143

and elastic modulus, the longitudinal and shear speeds of sound are measured which represent the propagation velocity of the longitudinal and shear waves through the elastic medium. Assuming that the pores are randomly oriented, uniformly distributed, and that the binder material behaves isotropically, the relationships between the elastic properties and the speed of sound have previously been shown as [4,5,6] :

$$c_l = \sqrt{\frac{E(1-\upsilon)}{(1+\upsilon)(1-2\upsilon)}} \tag{1}$$

$$c_s = \sqrt{\frac{E}{2\rho(1-\upsilon)}} \tag{2}$$

Where ρ is the density of the propagating material, C_l is the speed of sound of the longitudinal wave and C_s is the speed of sound of the shear wave. The elastic properties are given by the elastic modulus, E, and Poisson's ratio, υ. Solving these equations to isolate the elastic properties will yield:

$$E = \rho C_s^2 \left(\frac{3C_l^2 - 4C_s^2}{C_l^2 - C_s^2} \right) \tag{3}$$

$$\upsilon = \frac{1 - 2\left(\dfrac{C_s}{C_l}\right)^2}{2 - 2\left(\dfrac{C_s}{C_l}\right)^2} \tag{4}$$

The speed of sound of a material is simply determined by the thickness of a material divided by the time necessary for a sound wave to propagate through it. Therefore, determining the speed of sound is a matter of measuring the time of flight of an ultrasonic beam through a material as well as the thickness of the propagating material. Measuring the thickness of a solid material is a straightforward procedure. However, measuring the time of flight can sometimes be difficult due to frequency dependant variations of the ultrasonic properties of a material. Mechanical waves traveling through a viscoelastic medium will lose energy due to attenuation which is a frequency dependant phenomenon [6, 7]. Additionally, most materials experience some degree of velocity dispersion where the speed of sound also changes as a function of frequency [6,7]. Since ultrasonic transducers generate signals with a distribution of different frequencies centered around some central frequency, the frequency dependant phenomena of dispersion and attenuation can lead to a distorting of the signal as is demonstrated in Figure 1. The signal shown in Figure 1(a) was transmitted through only water which has negligible effects of attenuation and dispersion. The signal shown in Figure 1(b) was imaged through a geopolymer sample and a deformation of the signal is visable along with the addition of noise from the signal. As can be seen, the profile of the waveform has been altered between the two cases. This alteration can make choosing the location for the time of flight difficult, especially if additional noise is present in the signal.

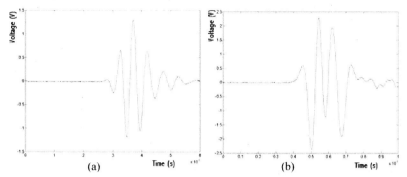

Figure 1: Ultrasonic signals (a) as sent by the transducer through water and (b) as received after attenuation through a geopolymer sample.

The frequency dependant nature of the material properties can be used to determine the speed of sound in spite of the distortion. A Wigner-Ville transform was used to transform the signal in the time domain into a time frequency distribution [8]. This method was chosen over a short time Fourier transform because it is capable of determining the time and frequency nature of a rapidly changing signal, without the use of a windowing function which can be difficult to use do to tradeoffs between resolution of the time and frequency. Instead the Wigner-Ville transform was used to determine the time of flight for discrete frequencies within the bandwidth of the transducer used. These times of flight were then used to evaluate separate phase velocities, or in other words, determine the speed of sound related to each of the chosen frequencies.

The phase velocities are insufficient for determining the elastic properties using the technique described earlier. In order to use the speed of sound to determine the elastic properties, the group velocity, which is the velocity with which the wave packet actually travels, must first be evaluated. The group velocity is commonly defined as:

$$c_\phi = \frac{\partial \omega}{\partial k} \tag{5}$$

where c_ϕ represents the phase velocity, ω is the angular frequency, and k is the wavenumber. This equation is also commonly written in the form:

$$c_g = \frac{c_\phi}{1 - \frac{\omega}{c_\phi} \frac{\partial c_\phi}{\partial \omega}} \tag{6}$$

where c_g represents the group velocity. Once the average group velocity is determined for each sample, the elastic properties can be evaluated.

SAMPLE PREPERATION

Seven groups of samples consisting of at least three specimens each were manufactured for the purpose of this study. In order to reduce the variables of the experiments, the curing regiments, manufacturing processes, and chemical compositions were kept consistent between the different samples so that the only distinguishing characteristic between each sample would be the Si:Al ratio. Each sample was distinguished using the corresponding Si:Al ratio, which were 1.49, 1.51, 1.9, 2.2, 3.1, 4.1, and 6.4.

The geopolymer samples were prepared by reacting commercially available metakaolin purchased by the BASF Corporation under the trade name MetaMax and reagent grade silica purchased from Fischer Scientific with a sodium based alkali solution. The chemical composition of the metakaolin and silica, provided by their respective distributors, is given in Table I. The activating solution consisted of one part of 15 molal NaOH solution in water and two parts sodium silicate. The source materials were then stirred into this solution to create the geopolymeric gel. Different Si:Al ratios were obtained by altering the concentration of metakaolin and reagent grade silica added to the activating solution. After mixing each set of samples, the geopolymer gel was cast into 50 mm cubic molds and vibrated for 10 minutes to remove entrapped air bubbles. After vibrating, each mold was sealed with wax paper and allowed to oven cure for 24 hours at 65°C and then sit at room temperature for an additional 24 hours before being removed from the mold.

Table I: Concentration of compounds found in Metamax used in this study

Compound	SiO_2	Al_2O_3	Na_2O	K_2O	TiO_2	Fe_2O_3	CaO	MgO	P_2O_5	SO_3	LOI*
Metamax % Concentration	53.0	43.8	0.23	0.19	1.70	0.43	0.02	0.03	0.03	0.03	0.46

When using ultrasonic testing methods, it is important that the surface faces transmitting and/or reflecting the ultrasonic signal be parallel with one another. ASTM standard E 494-95 prescribes a minimum included angle of 2° between these sides [9]. The original casted samples were well within this tolerance from retaining the shape of their mold. These original samples were used for analysis of the longitudinal speed of sound; however the shear mode had a much higher rate of attenuation and required that the samples be cut thinner to approximately 10 mm. After cutting each sample they were ground flat and CMM equipment was used to verify that all samples were within the geometric tolerances.

EXPERIMENTAL SETUP

Two methods of determining the speed of sound were used for this procedure. The shear waves were generated and recorded using a 2.25MHz transducer in a pulse echo setup, where a single transducer is used to both generate and record a signal. This requires the signal to propagate through a material reflect off of the back wall and then propagate back to the same transducer where the signal is recorded. The type of shear transducer used requires that the transducer be placed in contact with one side of the sample. In order to transmit the wave into the material, a coupling agent must be used. For this procedure a shear wave coupling gel was applied to each sample and spread out so as to create a thin film. Due to the geometries of this setup, the speed of sound of a shear wave is determined by:

$$C_S = \frac{2d}{t} \tag{5}$$

where d is the thickness of the material and t is the time of flight recorded by the transducer.

The longitudinal speed of sound was measured using a through transmission setup. In this set up two transducers, both with a bandwidth centered around 2.5MHz, were aligned with one another in a water tank. A sample is measured by placing it in between the two transducers as in Figure 2 so that a signal could be sent through the sample using the water in the tank as a coupling medium between the transducers and the sample. Due to the geometries of this setup, the speed of sound of a longitudinal wave is determined by:

$$C_l = \frac{d}{t + t_{H_2O} + \dfrac{d}{C_{H_2O}}} \tag{6}$$

where t_{H2O} is the time of flight without the sample and C_{H2O} is the longitudinal speed of sound through water. The speed of sound through water is calculated as function of the water temperature as was determined by [10].

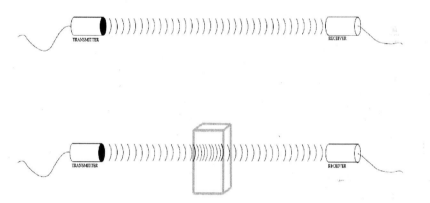

Figure 2: Ultrasonic transducer setup for through transmission tests of geopolymer samples

RESULTS

Speed of Sound Measurements:
After each sample was measured for thickness and scanned in both shear and longitudinal modes, the group speed of sound was determined for each Si:Al sample. The values for the average speed of sound for each sample are provided in Table II and Figure 3. In each case, the maximum speed of sound occurred at the lowest Si:Al ratio of 1.49 and decreased with an inverse relationship to the Si:Al ratio.

Table II: Average speeds of sound for each geopolymer sample.

Si:Al Ratio	1.49	1.52	1.9	2.2	3.1	4.1	6.4
Longitudinal Velocity [m/s]	2632.3	2579.5	2469.2	2393.6	2350.0	1884.1	1923.1
Shear Velocity [m/s]	1575.4	1566.7	1513.1	1494.3	1554.2	1156.6	1321

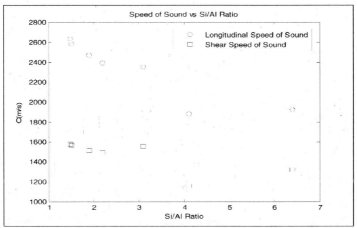

Figure 3: Average speed of sound for each geopolymer sample.

Porosity and Density Measurements:

The density of each sample was measured using the Archimedean principle. However, because the voids caused by porosities cannot transmit ultrasound, the pore contribution to the density was discounted to determine the binder density for each sample. Using microscopy techniques, the percent pore volume, by area, and the average pore diameter were determined for each Si:Al sample batch. Assuming that the porosities occur homogenously, this percent pore volume can be applied to the entire volume. Therefore, the binder density is given by:

$$\rho_{binder} = \frac{\rho_{sample}}{1 - \% PoreVolume} \tag{7}$$

The average pore diameter and percent pore volume measurements collected are presented in Figures 4 and 5. It was observed that the average pore diameter had a tendency to decrease as the Si:Al ratio increased. However, the percent pore volume appeared to behave sporadically as a function of the Si:Al ratio. The likely cause of this apparently random behavior is due to entrapped air bubbles within the geopolymer binder from the casting process. The workability of the geopolymers were observed to have a decreased workability during casting at lower Si:Al ratios. This could cause more air to become trapped within the binder leading to the high standard deviations shown in the average pore diameter plot, Figure 6, as well as the apparently sporadic high values in the percent pore volume plot, Figure7.

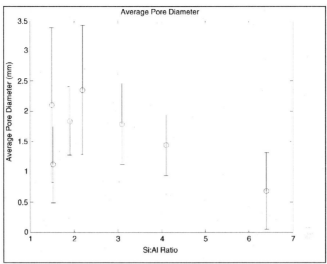

Figure 4: Average pore diameter variations with respect to the sample Si:Al ratio.

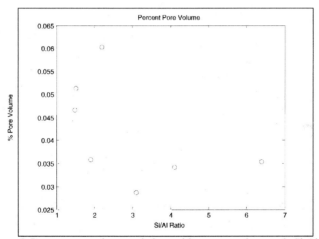

Figure 5: Percent pore volume variations with respect to the sample Si:Al ratio.

The density for the entire sample as well as just the binder were also determined and presented here in Figure 6. It can clearly be seen that the density increases as the Si:Al ratio is increased

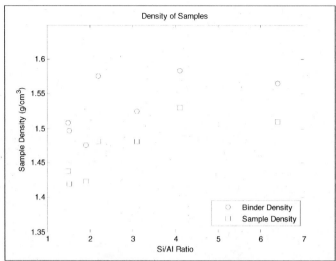

Figure 6: Sample density, determined using the Archimedean principle, and binder density, the density of the geopolymer discounting the porosity, for each Si:Al ratio.

Elastic Properties:

The elastic modulus and Poisson's ratio shown in Figures 7 and 8 and provided in Table III were determined for each Si:Al ratio using the speeds of sound and binder densities found above. The elastic modulus of the first five generally decreases as the Si:Al ratio is increased. This decreasing trend has been demonstrated previously by Duxson et al [11] to be an effect of additional silica molecules actually hindering the reaction at Si:Al ratios above 1.65 and leading to a degradation of the mechanical properties. However, there is a discontinuity between these samples and the two highest Si:Al samples. This discontinuity happens to occur around the transition between the poly(sialate-siloxo) (PSS) geopolymers and the poly(sialate-disiloxo) (PSDS) geopolymers marked with a dotted line in Figures 7 and 8. The terms PSS and PSDS were coined by Davidovits, [12] to describe the chemical formation of geopolymers, where PSS geopolymers contain Si:Al ratio approximately between 2.0 and 3.0 and the PSDS geopolymers contain Si:Al ratios above 3.0.

The values measured for the Poisson's ratio ranged between 0.22 and 0.05 and also tend to decreases as the Si:Al ratio increases. Furthermore, similar to the elastic properties, the Poisson's ratio also experienced a discontinuity in its trend between the PSS and PSDS regions.

Table III: Elastic properties for each set of samples by Si:Al.

Si:Al Ratio:	1.49	1.52	1.9	2.2	3.1	4.1	6.4
Elastic Modulus (GPa)	9.14	8.87	8.11	8.31	8.19	5.08	5.75
Poisson's Ratio	0.221	0.208	0.199	0.181	0.111	0.198	0.053

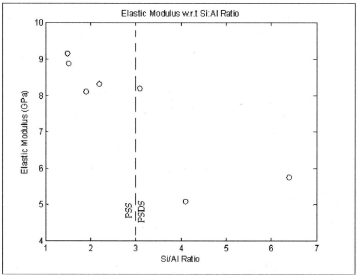

Figure 7: The elastic modulus as a function of Si:Al ratio.

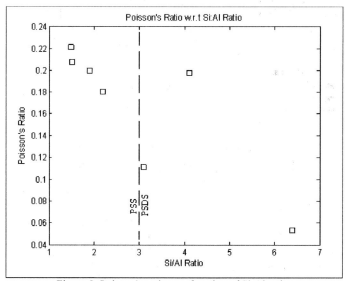

Figure 8: Poisson's ratio as a function of Si:Al ratio.

CONCLUSIONS
The use of ultrasound to measure the elasticity of a material is a technique that has been well documented in the past and can be adequately applied to geopolymer materials. However, because of attenuation concerns, geopolymer samples meant for ultrasonic testing are required to be manufactured very thin, at or under 10 mm, to ensure that a quality signal can propagate through the material.

Using these techniques, this study has corroborated the previous work of Duxson et al, [11] which showed that a maximum of the elastic modulus was observed near a Si:Al ratio of 1.6, after which the elastic modulus tends to gradually decrease if not remain at that value. However, after a PSDS geopolymer matrix is created, the elastic modulus of the material begins to have rapid increases as the Si:Al ratio is increased. This increase in the elasticity of these geopolymers is coupled with a rapidly decaying Poisson's ratio indicating that the geopolymer binder becomes extremely brittle with increasing Si:Al ratio.

REFERENCES
[1] Sindhunata, Effect of Curing Temperature and Silicate Concentration on Fly-Ash-Based Geopolymerization, *Industrial Engineering Chemistry Research*, **45**(10) pp. 3559, (2006).

[2] L. Weng, K. Sagoe-Crentsil, T. Brown, S. Song. Effects of Aluminates on the Formation of Geopolymers, *Materials Science and Engineering B*. 117. pp.163-168, (2004).

[3] P. Duxson, G.C. Lukey, S.W. Mallicoat, W.M. Kriven, J.S.J. van Deventer, Understanding the Relationship between Geopolymer Composition, Microstructure and Mechanical Properties, *Colloids and Surfaces. A, Physicochemical and Engineering Aspects*, **269**(1) pp. 47, (2005).

[4] L. Chang, Characterization of Alumina Ceramics by Ultrasonic Testing, *Materials Characterization*, **45**(3) pp. 221. (2000)

[5] J.G.S. van Jaarsveld, J.S.J. van Deventer, L. Lorenzen., The Potential Use of Geopolymeric Materials to Immobilise Toxic Metals: Part I. Theory and Applications, *Minerals Engineering*, **10**(7), 659-669, (1996).

[6] T.L. Szabo, *Diagnostic Ultrasound Imaging: Inside Out*. Elsevier Academic Press. Burlington, MA, (2004)

[7] C.M. Sayers, Ultrasonic Velocity Dispersion in Porus Materials, *Journal of Physics D: Applied Physics*, 14(3) pp. 413-420, (1981)

[8] L. Cohen, Time-Frequency Distributions-A Review, *Proceedings of the IEEE*, **77**, 941-81 (1989).

[9] Standard Practice for Measuring Ultrasonic Velocity in Materials. ASTM standard E 494-95

[10] W. Marczak, Water as a Standard in the Measurements of Speed of Sound in Liquids, *Journal of the Acoustical Society of America*, **102**(5) , pp. 2667-2779, (1997).

[11] P. Duxson, S.W. Mallicoat, G.C. Lukey, W.M. Kriven, J.S.J. van Deventer, The Effect of Alkali and Si/Al ratio on the Development of Mechanical Properties of Metakaolin-Based Geopolymers, *Colloids and Surfaces A: Physicochem. Eng. Aspects*, **292**(1), 8-20, (2006).

[12] J. Davidovits, "Poly(sialate-disiloxo)-based geopolymeric cement and production method thereof" US Patent No. 0172860A1, August 11, 2005

BI-AXIAL FOUR POINTS FLEXURAL AND COMPRESSIVE STRENGTH OF GEOPOLYMER MATERIALS BASED Na$_2$O-K$_2$O-Al$_2$O$_3$-SiO$_2$ SYSTEMS

C. Leonelli[1], E. Kamseu[1], V.M. Sglavo[2]

[1] Department of Materials and Environmental Engineering, University of Modena and Reggio Emilia Via Vignolese 905, 41100 Modena, Italy

[2] Deparment of Materials Engineering and Industrial Technology, Via Messiano 77, 38050 Trento, Italy

ABSTRACT

Bi-axial four point flexural and compressive strength tests were used for the assessment of mechanical properties of geopolymer material-based, calcined kaolin and kaolinitic clays. Various activating solutions which consist in the mixture of potassium and sodium hydroxide, water and sodium silicate were designed and tested. Six specimens, over a wide number of investigated compositions, were selected with SiO$_2$/Al$_2$O$_3$ varying from 1:1 to 3:1. The compositions 1:1 and 2:1 were obtained by using two different grades of kaolin as raw materials (standard and sand-rich), while the 3:1 was obtained by adding required amounts of silica to either kaolin or kaolinitic clay. All the samples were prepared by slip casting, using density ($\approx$1.5 g/cm^3) as indicator of the optimum viscosity for shaping geopolymer pastes, and cured at room temperature for different periods.

The bi-axial four points flexural strength values vary from 13 to 21 MPa while the compressive strength vary from 45 to 67 MPa, being essentially influenced by curing time, SiO$_2$/Al$_2$O$_3$ and K$_2$O/Na$_2$O/H$_2$O ratios. Increasing the SiO$_2$/Al$_2$O$_3$ ratio from 1:1 to 3:1, the mechanical properties increase but longer setting and curing times were required. Low SiO$_2$/Al$_2$O$_3$ ratio results the appearance of micro cracks and deformations during curing. The flexural and compressive behaviour of the specimens studied were directly correlated to the porosity, density and the final product microstructure. The chemical behaviour of the six compositions is discussed with respect to basic dissolution-hydrolysis-polycondensation processes that occur in Na$_2$O-K$_2$O-Al$_2$O$_3$-SiO$_2$ systems.

INTRODUCTION

The formation of [M$_x$(AlO$_2$)x(SiO$_2$)$_y$.MOH.H$_2$O] gel, which essentially relies on the extent of dissolution of alumino-silicates materials, is a dominant step in geopolymerisation. The gel then diffuses outward from the particle surface into larger interstitial spaces between the particles, with precipitation of gel and concurrent dissolution of new solid. When the gel phase hardens, the separate alumino-silicate particles are therefore bound together with the gel which acts as binder[1]. Authors[2,3] described the reaction process of the gel formation indicating that Al-Si solid particles in alkaline solution lead to the formation of monomer ($^-$OSi(OH)$_3$ + Al(OH)$_4^-$. The successive reactions between these monomers and alkali ions and water result in the formation of dimer; and with concentrated silicate anion addition, the tetramer, pentamer, hexamer, octamer, nonamer, ... and their compounds will appear[4]. Geopolymers are then formed with tightly packed polycrystalline structure so as to give better mechanical properties.

The reaction mechanism involves the dissolution of Al and Si in the alkali medium, transportation of dissolved species, followed by polycondensation, forming a 3D network of alumino-silicate structure. Condensation occurs between alumino-silicate species or silicate species themselves, depending on the concentration of Si in the system. When Si/Al > 1, the silicate species formed as a result of hydrolysis of SiO$_2$, tend to condense among themselves to form oligometric silicates which condense with Al(OH)$_4$- forming rigid 3D geopolymeric structures[5,6,7]. Typically, better strength behavior are obtained for mixtures with SiO$_2$/Al$_2$O$_3$ ratios in the range of 1.65-2.10 with a Na$_2$O/SiO$_2$

ratio near 1. Higher amounts of hydroxyl ions facilitate the dissociation of different silicate and aluminate species, thus promoting further polymerization[8]. NaOH promtotes Na^+ ions with strong pair formation (better dissolution) of silicate oligomers. KOH promotes K^+ that favors the formation of larger silicate oligomers with which $Al(OH)_4$- prefers to bind. Therefore in KOH solutions more geopolymer precursors exist resulting thus in better setting and stronger compressive strength since K^+ would promote a high degree of condensation[9]. It was found that by combining NaOH and KOH, highly dissolved and cross linked samples of geopolymer materials can be obtained. The multiple alkali sources can act in a synergistic way to promote samples of optimal characteristics[3,4].

High-performance materials for construction, adhesives, coatings, hydroceramics and an ever-growing range of niche applications are produced by the reaction sequence described above. The mechanical strength result in rapid solidification within hours and rapid early strength development. The strength for these materials is believed to originate from the strong chemical bonding in the alumino-silicate gel formed, as well as the physical and chemical reactions occurring between the geopolymer gel, un- or partly reacted phases, and particulate aggregates. The compressive strength of geopolymer materials depends on a number of factors including gel phase strength, the ratio of gel phase/undissolved Al-Si particles sizes, the distribution and the hardness of the undissolved Al-Si particles the amorphous nature of geopolymers or the degree of crystallinity as well as the surface reaction between the gel phase and the undissolved Al-Si particles [10, 11]. After geopolymerisation, the undissolved particles remain bonded in the matrix, so that the hardness of the minerals correlates positively with the final compressive strength[11]. The significance of the Si/Al ratio during alkaline dissolution of the individual minerals indicates that compressive strength is acquired by complex reactions between the mineral surface, alumino-silicate and the concentrated sodium silicate solution.

The curing regime has a very important impact on the mechanical strength of geopolymer materials since it directly influences the rate of reactions that take place during geopolymerisation. geoolymer materials of the Na_2O-K_2O-Al_2O_3-SiO_2 system contain a relatively large amount of water in open pores available for evaporation, which would not result in capillary strain. This may account for the low temperature of dimensional stability[11]. When the freely evaporate water is removed from pores, the surface area of the gel structure increases as water is liberated from the surface of the gel and small pores resulting in shrinkage are observed. The gel contraction may be correlated with the reduction in surface area[12]. This fact indicates that during curing rapid drying should be avoided.

In this work, the assessment of the of the degree of dissolution and polycondensation phenomena is used to evaluate the packing behavior by using the compressive and bi-axial four point flexural strength values. Both measurements are in turn indicative of the structural and densification behavior which are dependent on the bonding and interlocking developed during geopolymerisation.

MATERIALS AND EXPERIMENTAL PROCEDURE

Materials

Metakaolin was used as the principal source of alumino-silicate. The choice was based on the fact that it is the cheapest and alumino-silicate that presents a good degree of purity. Metakaolin improves mechanical strength and reduces the transport of water and salts in the final product[13]. Metakaolin is important in the production of geopolymer materials for applications as adhesives, coatings and hydroceramics[12]. Two metakaolins were produced by the calcination at 700°C of two different kaolins for 4h. MK1 was obtained from a standard kaolin with SiO_2/Al_2O_3 = 1.2 with very low amount of impurities. MK2 was obtained from a kaolinitic clay with SiO_2/Al_2O_3 = 1.5, with 2.5 wt% of K_2O, 1% of Fe_2O_3 and 0.5% of TiO_2. 35 wt% concentrated solutions of NaOH and KOH (Aldrich, reagent grade) were used for all the experiments. The sodium silicate sodium had SiO_2/Na_2O molar ratio of 3.1 (Table I).

Table I. Chemical composition (wt%) of the starting raw materials used to prepare geopolymers.

Reactant	SiO₂	Al₂O₃	Na₂O	K₂O	Fe₂O₃	TiO₂	Supplier
Na silicate	27.00		8.71				Aco Sil Verona, Italy
Standard kaolin	44.37	36.12	0.10	0.82	0.97	1.52	Ceramic Co, Italy
Kaolinite clay	51.67	34.45	0.21	2.50	1.01	0.50	Ceramic Co, Italy

Geopolymer preparation

Solution of sodium hydroxide-potassium hydroxide-sodium silicate was prepared by volumetric mixing the reactant in the proportion 2:1:3. The solution was then agitated for homogenization for 5 min. The solution was used to dissolved six samples of geopolymers with SiO_2/Al_2O_3 ratios of 1.0, 1.2, 1.5, 1.65, 2.1 and 3.0 (samples GPM1 to GPM6, respectively) (Table II). Samples with SiO_2/Al_2O_3 ratio less to 1 did not give good results in terms of mechanical properties and were not considered for the rest of the study. Geopolymer materials were prepared by adding metakaolin to the respective solutions in the volume proportion of liquid:solid = 2:3. The mixtures were ball milled to the paste apparent density of ≈ 1.5 g/cm³.

The composition GPM5 with SiO_2/Al_2O_3 ratio = 2.1 has been chosen among the others for further testing due to its mechanical strength. This basic composition received different proportions of water representing respectively 2.5, 5, 7.5, 10, 12.5 and 15% of total volume of the geopolymer paste (Table I). The viscous pastes obtained were poured in different moulds:
- Samples with 40 mm of diameter x 10 mm thickness (for four-point bending tests);
- Samples with 40 mm of diameter x 80 mm thickness (for compressive strength tests).

Table II. Geopolymers compositions in the Na₂O-K₂O-Al₂O₃-SiO₂ system.

Compositions	SiO₂/Al₂O₃	Na₂O/K₂O	Liquid:Solid ratio
GPM1	1.00	2:1	2:3
GPM2	1.20	2:1	2:3
GPM3	1.50	2:1	2:3
GPM4	1.65	2:1	2:3
GPM5	2.10	2:1	2:3
GPM6	3.00	2:1	2:3

Biaxial bending strength tests: The piston-on-three-ball test

For the execution of the test a load is applied to the specimen centre by a right circular cylinder of hardened steel. The test specimen is a disk, with an average diameter of 40 mm and of suitable thickness. It is supported on three ball bearings with a diameter of 2.67 mm and positioned at 120°C relative to each other on a circle (9.5 mm diameter).

The specimens were loaded centrally at a rate of 3 mm/min, using a spherical indenter with a diameter of 1.66 mm. The testing machine used was of the type MTS 810, USA. The biaxial flexure strength of each disc was calculated according to the equation[14]:

$$\sigma_{max} = \frac{3P(1+\upsilon)}{4\pi t^2}[1+2\ln\frac{a}{b}+\frac{(1-\upsilon)}{(1+\upsilon)}\{1-\frac{b^2}{2a^2}\}\frac{a^2}{R^2}] \tag{1}$$

where P is the load, t the thickness, a the radius of the circle of the support points, b the radius of the region of uniform loading at the centre, R the radius of the discs, and υ Poisson's ratio.

Substituting values for a, b and the typical Poisson's ratio of brittle materials ≈ 0.23, the equation of the maximum strength becomes:

$$\sigma_{max} = .30\frac{P}{t^2}(5.88 + \frac{54}{R^2})$$
(2)

Compressive strength

The compressive strength was determined by using the same testing machine as for bi-axial four point flexural strength with modification of loading configuration. The end surface of specimens were polished flat and parallel to avoid the requirement for capping. The cylinders and prisms were centered in the compression-testing machine and loaded to complete failure. The compressive strength was calculated by dividing the maximum load (N) at failure by the average cross-sectional area (m^2).

All the values presented in the current work were an average of ten samples, with error reported as standard deviation from the mean.

RESULTS AND DISCUSSION

The role of silica/alumina content

Figures 1 (a and b) shows the influence of SiO_2/Al_2O_3 ratio on the bi-axial four point flexural and compressive strength of geopolymer materials based $Na_2O-K_2O-Al_2O_3-SiO_2$ system. Passing from GPM1 to GPM5, i.e. for ratio values of $SiO_2/Al_2O_3 = 1$ to 2.1, the compressive strength increases linearly to its maximum value, before decreasing again at the highest SiO_2/Al_2O_3 ratio value of 3.0. The bi-axial four point flexural strength reaches its maximum at GPM4 passing from 8 MPa for GPM1 to about 20 MPa for SiO_2/Al_2O_3 ratios of 1.65 (after 21 days curing time).

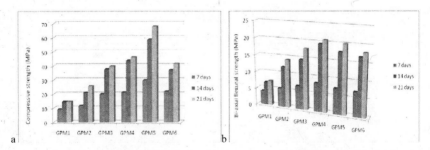

Figure 1. Influence of SiO_2/Al_2O_3 ratio on the compressive (a) and bi-axial four point flexural (b) strength of geopolymer materials based $K_2O-Na_2O-SiO_2-Al_2O_3$.

It can also be observed from Figure 1 that as SiO_2/Al_2O_3 increases, the difference in the mechanical properties with time is more important, that is to say that when the SiO_2/Al_2O_3 increases the time necessary to reach complete geopolymerisation increases. GPM1, with SiO_2/Al_2O_3 ratio of 1.0, achieved the maximum compressive strength after 14 days, while GPM5, with SiO_2/Al_2O_3 ratio of 2.1, present 30 MPa at as compressive strength after 7 days, 56 MPa after 14 days and 68 MPa after 21 days.

These observations indicate that there is a correlation between the SiO_2/Al_2O_3 ratio and the rate of geoplymerization and hence to the mechanical properties of the geoplymers. This indicates that reactions of polycondensation and hardening take more time when silica content is increased. The sialate (silicon-oxo-aluminate) network consists of SiO_4 and AlO_4 tetrahedra linked alternatively by sharing all oxygens. Polysialates are chain and ring polymers with bridging oxygens and their empiric formula is often related as $M_n(-(SiO_2)z-AlO_2)_n.wH_2O$ where z is 1, 2 or 3 and M the monovalent cation such as Na^+ or K^+ in the case under study, n the degree of polycondensation[15].

For the samples with SiO_2/Al_2O_3 ratio value near to 1 the lowest mechanical properties have been observed together with micro-fissures in a thick fragile layer formed at the surface. Such a surface layer, which has been identified to be high alkali content, can result from the migration of alkali ions wichi counterbalanced the charge of a number of free AlO_4 tetrahedra. These AlO_4 units are weakly linked to the network of sialate with as principal consequence the poor mechanical behavior. The increase of SiO_2/Al_2O_3 ratio in fact ameliorated the distribution of monomers of SiO_2 and AlO_4 in the network contributing to increase the bonds in the structure, as well as the packing of polycrystalline structure and by the way the mechanical strength (Figure 1 and 4).

The standard deviation of the bi-axial four point and compressive strength is about 5 MPa (Table III) which is indicative on the interval of variation of mechanical strength of geopolymer materials which do not have stoichiometric composition and comprise mixtures of amorphous to semicrystalline structure and crystalline aluminosilicate particles[15].

Table III. Mechanical properties of the geopolymers in the Na_2O-K_2O-Al_2O_3-SiO_2 system.

	Compressive strength (MPa)			Bi-axial four point flexural strength (MPa)		
	7 days	14 days	21 days	7 days	14 days	21 days
GPM1	9,1 ± 4,2	14,5 ± 4,4	14,5 ± 4,2	4,1 ± 4,8	6,9 ± 4,0	7,4 ± 4,1
GPM2	11,7 ± 4,9	20,9 ± 4,0	25,6 ± 4,8	5,5 ± 4,7	12,0 ± 4,6	14,2 ± 4,3
GPM3	20,0 ± 4,7	37,6 ± 4,8	39,7 ± 4,5	6,9 ± 4,2	14,7 ± 4,7	17,9 + 4,1
GPM4	21,3 ± 4,1	43,7 ± 4,9	46,2 ± 4,3	8,4 ± 4,3	19,6 ± 4,5	20,7 ± 4,2
GPM5	29,8 ± 4,5	58,5 ± 5,0	68,0 ± 4,4	7,5 ، 4,9	17,8 ± 4,4	20,0 ± 4,5
GPM6	21,9 ± 4,6	37,1 ± 4,8	41,5 ± 4,6	7,2 ± 4,9	16,9 ± 4,8	18,1 ± 4,7

Considering the empirical formula of geopolymer $M_n(-(SiO_2)z-AlO_2)_n.wH_2O$, the role of water during the production cycle of geopolymer materials is important. Under alkaline conditions, alumino-silicates are transformed into extremely reactive materials and it is generally believed that the dissolution process is initiated by the presence of hydroxyl ions with the formation of $[M_z(AlO_2)_x(SiO_2)_ynMOH\cdot mH_2O]$ gel which is essentially linked to the extent of the dissolution itself. It was in our objective to study the influence of water content on the mechanical properties of geopolymer materials using the composition GPM5 with SiO_2/Al_2O_3 ratio of 2.1 corresponding to the maximum mechanical properties. In Figure 2, the water progressively added to a basic composition of geopolymer did not substantially changed density and water absorption, being the first around 1.5-1.6 g/cm^3 and the second approximately 22-24 %.

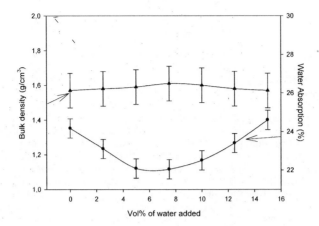

Figure 2. Influence of initial water added (%vol) on water absorption and bulk density of geopolymer based on $Na_2O-K_2O-Al_2O_3-SiO_2$ [3].

It has been observed that the amount of water strongly affects the viscosity of the paste giving the possibility to optimize the mechanical properties or ameliorate the characteristics of products such as pores dimensions and distribution. With insufficient amounts of water in the geopolymer compositions, they present very high viscosity with products showing large open pores. These large and open pores (Figure 3a) are failure precursors during mechanical tests and they are responsible for the poor mechanical properties of final products. By adding 5.0 to 7.5 vol% of water, the geopolymers obtained show better homogeneous microstructure with lower porosity (number and dimensions) than those obtained at lower water content, as an example 2.5 vol% in Figure 3b.

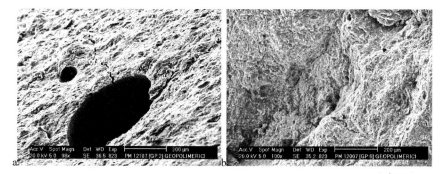

Figure 3. Micrographs of geopolymers GPM5 (SiO_2/Al_2O_3 = 2.1) with 2.5 vol% (a) and 7.5 vol% (b) water content; showing larger open pores source cracks (a) when no sufficient water is used for the appropriate viscosity of gel responsible for the better densification (b).

These compositions also showed the better mechanical strength (bi-axial four point flexural and compressive strength), as it can been observed in the Figure 4.

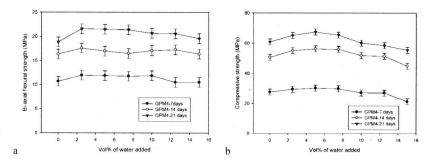

Figure 4. Influence of initial water added (vol%) on the bi-axial flexural (a) and compressive (b) strength of geopolymer GPM5 (SiO_2/Al_2O_3 = 2.1).

The better mechanical properties of geopolymers obtained with a water content between 5 and 7.5 vol% together with the lower water absorption (22%) and relatively high density (1.61 g/cm3) are indicative for the higher degree of reactivity and polycondensation. Time of hardening strongly affects the values of these properties, being almost linear for flexural strength while for compressive strength after 14 days almost complete geopolymerization is reached.

In the Figures 5 and 6 it can be observed that products of these compositions show better densification.

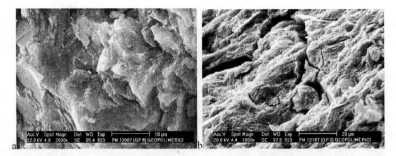

Figure 5. Micrographs of samples of geopolymer GPM5 with 7.5% water added showing fractured specimens after mechanical test with no cracks (a) and with intergranular cracks (b).

Figure 6. Etched surface of samples of Fig. 5(a) showing unreacted quartz grains (see arrow).

When the added water is in excess of 10 vol%, the viscosity of the geopolymer gel is progressively modified and the products obtained gradually loose their good mechanical properties. The strength is believed to originate from the strong chemical bonds in the alumina silicate gel formed, as well as the physical and chemical reactions occurring between the geopolymer gel, non or partly reacted phases, and particulate aggregates.

The process of geopolymerisation involves leaching, diffusion, condensation and hardening steps. These steps generally need time to go through and the difference in mechanical properties (Figure 4) of $Na_2O-K_2O-Al_2O_3-SiO_2$ geopolymer system with time is sufficient to understand the long term of these types of reactions. As indicated in Figure 1, the reaction velocity as well as hardening process decrease with time. Relatively faster during the first week, the progressively decrease during the second and third weeks of curing. Through microstructural investigations, it has been observed that the ratio of SiO_2/Al_2O_3 influence the homogeneity of the $Na_2O-K_2O-Al_2O_3-SiO_2$ geopolymers which in turn affects the bi-axial four point and compressive strengths that can be optimized by controlling the quality (viscosity, density, ...) of geolpolymer gel formed.

CONCLUSIONS

Geopolymer materials of the Na_2O-K_2O-Al_2O_3-SiO_2 system were studied. The influence of SiO_2/Al_2O_3 ratio and the water content of geopolymer gel were used to investigate on the bi-axial four point flexural and compressive strength behavior of obtained products which were cold cured after slip casting. It was observed that:

The bi-axial four point flexural and compressive strength of geopolymer material-based K_2O-Na_2O-SiO_2-Al_2O systems are strongly depending on the SiO_2/Al_2O_3 ratio: increasing the ratio from 1.0, the mechanical properties increases up to 2.1, point at which the increase in SiO_2 content decreases the mechanical properties. This behavior was ascribed to the influence of these ratios on the distribution of SiO_4 and AlO_4^- monomers during the polycondensation and the impact on the hardening of bonds formed.

The geopolymirisation process was found to be positive in the cold environment of curing even if in this case the reaction process take a long time.

The water content of geopolymer gels can be used to monitor the viscosity and density which are very important for the optimization of the final characteristics of geopolymer materials.

The respective values of 20 MPa and 68 MPa of bi-axial four point flexural and compressive strength of the obtained geopolymer materials of the Na_2O-K_2O-Al_2O_3-SiO_2 system are indicative for the possible applications of these materials with advantageous characteristics.

ACKNOWLEDGMENTS

Authors are particularly grateful to Dr. Mirko Braga, Laboratorio R.S.A., INGESSIL S.r.l., via dei Peschi, 13, 37141 Montorio (Verona), Italy for supplying of sodium silicate and to Dr. Elena Venturelli, Esmalglass-ITACA S.p.A., Viale Emilia Romagna 37, 41049 Sassuolo (Modena), Italy for supplying kaolin.

REFERENCES
[1] R. Cioffi, L. Maffucci, L. Sandoro, Optimization of geopolymer synthesis by calcination and polycondensation of kaolinite residue, Ressources, Conservation and re cycling, **40** 27-38 (2003).
[2] J.S.J. Van Deventer, J.L. Provis, P. Duxson, G.C. Lukey, Reaction mechanisms in the geopolymeric conversion of inorganic waste to useful products, J. Harz. Matls. **A139** 506-513 (2007).
[3] H. Xu, J.S.J. Van Deventer, The geopolymerisation of alumino-silicate minerals, Int. J. Miner. Process., 59, 247-266 (2000).
[4] W.M. Hendricks. A.T. Bell, C. J. Radke, Effect of organic and alkali metal cations on the distribution of silicate anions in aqueous solutions, J. Phys. Chem., **95** 9513-9518 (1991).
[5] L. Weng, K. Sagoe-Crentsil, T. Brown, Speciation and hydrolysis kinetics of aluminates in inorganic polymer systems, presented to geopolymer, International Conference on geopolymers 28-29 October, Melbourne, Austrialia (2002).
[6] M.R. Anseau, J.P. Leung, N.Sahai, T.W. Swaddle, Interactions of silicate ions with Zinc(II) and Aluminium(III) in alkali aqueous solution, Inorg. Chem., **44(22)** 8023-8032 (2005).
[7] M.R. North, T.W. Swaddle, Kinetics of silicate exchange in alkaline alumino-silicate solutions, Inorg. Chem., **39(12)** 2661-2665 (2000).
[8] J.S.J. Davidovits, Long term durability of hazardous toxic and nuclear waste disposals. In: Davidovits, J.S.J., Orlinski, J. (Eds.), Proceedings of the 1st International Conference on Geopolymers, Vol 1, Compiege, France, 1-3 June, PP. 125-134 (1988).
[9] J. W. Phair, J.S.J.Van Deventer, Effect of the silicate activator pH on the microstructural characteristics of waste-based geopolymers. Intl. J. Min. Processing, **66 (1-4)** 121-143 (2002).

[10]Van Deventer J.S.J., J.L. Provis, P.Duxson, G.C. Luckey, 2007. Reaction mechanisms in the geopolymeric conversion of inorganic waste to useful products. Journal of hazardous Materials A139, 506-513.

[11]K. Komnitsas, D. Zaharaki, Geopolymerisation: A review and prospects for minerals industry, Mineral Engineering, , **20** 1261-1277(2007).

[12] P. Duxson, G.C. Lukey, J.S.J. Van Deventer, Physical evaluation of Na-geopolymer derived from metakaolin up to 1000°C, J. Matls Sci, **42** 3044-3054 (2007).

[13] J. A. Kostuch, G. V. Walters, T. R. Jones, High performance concrete containing metakaolin-A review. In: Dhir R.K., Jones M.R (Eds.), Proceedings of the Concrete 2000 International Conference on Economic and Durable Concrete through Excellence. University of Dundee, Scotland, UK, 7-9 September, 2, pp. 1799-1811 (2000).

[14] D. K. Shetty, A.R. Rosenfield, P. M. Guire, P. Bansal, J. K. Winston, H. Duckeworth, Biaxial flexure tests for ceramics, Amer. Ceram. Soc. Bull. **59** 1193-1197 (1980).

[15] J.S.J. Davidovits, Geopolymers: Inorganic polymeric new materials. J.Therm. Anal. **37** 1633-1656 (1991).

A STUDY ON ALKALINE DISSOLUTION AND GEOPOLYMERISATION OF HELLENIC FLY ASH

Ch. Panagiotopoulou[1], T. Perraki[2], S. Tsivilis[1], N. Skordaki[1], G. Kakali[1]
[1] National Technical University of Athens, School of Chemical Engineering,
[2] National Technical University of Athens, School of Mining Engineering and Metallurgy,
Zografou Campus, 15773 Athens, Greece

ABSTRACT

This work concerns the use of fly ash (coming from the power station at Megalopolis, Greece), as raw material for the synthesis of geopolymers and it is part of a research project concerning the exploitation of Greek minerals and by-products in geopolymer technology.

The experimental part comprises two parts: i) the dissolution of fly ash in alkaline media and the investigation of the effect of the alkali ion (K or Na), the concentration of the solution (2, 5 and 10M) and time (5, 10 and 24 hours) on the dissolution rate of Al^{+3} and Si^{+4} and ii) the synthesis of fly ash based geopolymers and the investigation of the effect of curing conditions (T= 50, 70 and 90 °C, t= 24, 48 and 72 hours) and the Si/Al ratio (Si/Al= 1.75-4.5) on the development of the compressive strength. The dissolution rate of the studied fly ash, in alkaline media, shows that this material can be considered as potential raw material for geopolymerization. The optimal curing conditions are 48 hours at 70 °C. The increase in curing time leads to the decrease of compressive strength, while the increase in curing temperature causes a slight increase in compressive strength but it also favors the surface cracking of the specimens. The compressive strength of geopolymers is found to depend systematically on the Si/Al ratio, with the maximum being 45.5±2 MPa for a Si/Al ratio of 2.5

INTRODUCTION

The need for construction materials that possess improved fire-resisting properties led professor Joseph Davidovits to the synthesis of new materials which he named geopolymers [1]. These materials have excellent mechanical properties and high resistance to thermal and chemical attack, while their synthesis is based on the activation of aluminosilicate materials by an alkali metal hydroxide and an alkali metal salt and their transformation into a three-dimensional inorganic amorphous structure [2]. The synthesis and chemical composition of geopolymers are similar to those of zeolites, but their microstructure is amorphous to semi-crystalline.

Theoretically, any aluminosilicate material can undergo geopolymerisation under certain circumstances. Previous works have reported the formation of geopolymers from natural minerals [2-5], calcined clays [6,7], industrial by-products [8-11] or a combination of them [12-16].

The formation of geopolymers involves a chemical reaction between an aluminosilicate material and sodium silicate solution in a highly alkaline environment. The exact mechanism of this reaction is not yet fully understood, but it is believed to be a surface reaction consisting of four main stages: (1) the dissolution of solid reactants in an alkaline solution releasing Si and Al species, (2) the diffusion of the dissolved species through the solution, (3) the polycondensation of the Al and Si complexes with the added silicate solution and the formation of a gel and (4) the hardening of the gel that results to the final polymeric product. Stages (2) to (4) cannot be monitored since the procedures cannot be stopped and the products cannot be isolated. According to Hua Xu and Van Deventer [3] the mechanism of geopolymerisation is represented schematically according to Eqs.(1) and (2):

$$n(Si_2O_5.Al_2O_2) + 2nSiO_2 + 4nH_2O + NaOH \text{ or } (KOH) \rightarrow Na^+,K^+ + n(OH)_3\text{-}Si\text{-}O\text{-}Al^-\text{-}O\text{-}Si\text{-}(OH)_3$$

(Si-Al materials)

$$| \\ (OH)_2$$

(Geopolymer precursor)

(1)

$$n(OH)_3\text{-}Si\text{-}O\text{-}Al^-\text{-}O\text{-}Si\text{-}(OH)_3 + NaOH \text{ or } (KOH) \rightarrow (Na^+,K^+)\text{-}(\text{-}Si\text{-}O\text{-}Al^-\text{-}O\text{-}Si\text{-}O\text{-}) + 4nH_2O$$

$$| \\ (OH)_2 \qquad\qquad O \quad O \quad O \\ | \quad | \quad |$$

(Geopolymer backbone)

(2)

As indicated by the proposed mechanism of geopolymerisation, the presence of dissolved Si and Al species is necessary for the progress of the reaction, so the Si/Al molar ratio of starting materials is crucial since it causes significant structural differences which determine the final properties of geopolymers [17-19]. Low Si/Al molar ratio leads to the creation of a three-dimensional framework while high Si/Al ratio (>15) gives a polymeric character to the geopolymer [17]. Yet, the amount of Si and Al species in the gel, therefore the Si/Al ratio, depend not only on their presence in the starting materials but also on their dissolution from the aluminosilicate framework of raw materials. So, the extent of dissolution of the starting materials is of high significance for geopolymerisation.

Another factor that influences the properties of the produced geopolymers is the curing conditions, since the synthesis of geopolymers depends on the temperature of polycondensation and the time that the formatted gel remains in this temperature. In low temperatures amorphous or glassy structures are formed resulting in poor chemical properties. Temperatures from 35 °C to 85 °C lead to the formation of amorphous to semi-crystalline structures that possess good physical, thermal and mechanical properties, while curing in temperatures higher than 100 °C results in semi-crystalline structures that have excellent properties [20,21].

Fly ash is an industrial by-product, coming from power-supply plants, that causes deposit problems. Due to its high content in silica and alumina, fly ash can be used as raw material for geopolymer synthesis. This work concerns the use of fly ash (coming from the power station at Megalopolis, Greece), as raw material for the synthesis of geopolymers and it is part of a research project concerning the exploitation of Greek minerals and by-products in geopolymer technology.

EXPERIMENTAL

The experimental part comprises two stages:

i) The dissolution of fly ash in alkaline media and the investigation of the effect of the alkali ion (K or Na), the concentration of the solution (2, 5 and 10 M) and time (5, 10 and 24 hours) on the dissolution rate of Al^{+3} and Si^{4+}

ii) The synthesis of fly ash based geopolymers and the investigation of the effect of curing conditions (T= 50, 70 and 90 °C, t= 24, 48 and 72 hours) and the Si/Al ratio (Si/Al= 1.75-4.5) on the development of the compressive strength.

Fly ash comes from the power station at Megalopolis, Greece and its chemical composition is presented in Table I. This material consists mainly of quartz (SiO_2) and feldspars ($NaAlSi_3O_8$) while anhydrite, gehlenite, maghemite and calcite are found in smaller quantities. Fly ash was previously ground and its mean particle size (d_{50}) was approximately 10 μm. This is the typical fineness of fly ash used in construction technology (as main constituent in blended cements).

The leaching of fly ash was conducted by mixing 0.5 (± 0.0001) g of solid with 20 ml of alkaline solution for certain hours under continuous stirring. The variables studied are the kind of alkali metal (K, Na), the concentration of the alkaline solution (2, 5 and 10M) and the time of dissolution (5, 10 and 24 h). After filtering, the liquid part is diluted to 250 ml, the pH is adjusted to pH<1 by adding concentrated HCl acid and AAS is used in order to determine the Al and Si concentration. The solid part is examined by means of XRD in order to evaluate the effect of Si and Al leaching on the structure of the starting material.

Table I. Chemical composition of Hellenic fly ash (% w/w)

SiO_2	Al_2O_3	Fe_2O_3	CaO	MgO	K_2O	SO_3	L.O.I.
47.86	23.54	7.15	10.56	2.28	1.58	2.50	4.30

For the determination of the optimal curing conditions an aqueous activation solution containing sodium silicate and sodium hydroxide was used. Geopolymer samples were prepared according to the following molar ratios: Si/Al =1.9, H_2O/Na_2O = 12.19, Na_2O/SiO_2 = 0.27 and m_{solids}/m_{liquid}= 2.4, forming a homogenous slurry. This is a typical composition reported by other authors [21,22]. After mechanical mixing, the slurry was transferred to moulds which were mildly vibrated. Then the moulds were left for two hours at ambient temperature before they were cured in laboratory oven at 50 °C, 70 °C and 90 °C for 24, 48 and 72 hours. After cooling the specimens were transferred to sealed vessels and their compressive strength was measured after 7 days.

The last part of the experimental involves the investigation of the effect of Si/Al molar ratio on the development of compressive strength. The Si/Al ratio of Hellenic fly ash is 1.75. Sodium silicate solutions were prepared by dissolving amorphous silica in sodium hydroxide solutions so that the final Si/Al ratios would be from 1.75 to 3.0. Due to the high amount of the additional silica needed for Si/Al ratios from 3.5 to 4.5, the sodium silicate solutions were prepared by mixing commercial silica solutions containing 30% and 50% w/w silica with sodium hydroxide solutions. In all cases Al_2O_3/Na_2O=1 and the activation solutions were stored for a minimum of 24 hours prior to use. All geopolymer samples were synthesized using the same procedure and cured at the optimal curing conditions that were determined previously. Finally, the samples were examined by means of XRD and FTIR.

X-ray powder diffraction patterns were obtained using a Siemens D-5000 diffractometer, $CuK_{\alpha1}$ radiation (λ= 1.5405Å), operating at 40kV, 30mA. The IR measurements were carried out using a Fourier Transform IR (FT-IR) spectrophotometer (Perkin Elmer 880). The FTIR spectra in the wavenumber range from 400 to 4000 cm^{-1} were obtained using the KBr technique. The pellets were prepared by pressing a mixture of the sample and died KBr (sample: KBr approximately 1:200) at 8 tons cm^{-2}.

RESULTS AND DISCUSSION

Extent of dissolution- characterization of solid residue

Tables II and III present the concentration of Al and Si in the NaOH and KOH solutions, respectively, in relation to the time of leaching and the alkalinity of the leaching solution. These values are the measured concentrations, after the leaching of 0.5 g of fly ash and the dilution of filtered liquid to 250 ml.

Table II. Extend of Al and Si dissolution in NaOH solutions

Alkalinity (M)	2			5			10		
Time (h)	5	10	24	5	10	24	5	10	24
Al (ppm)	4.57	27.74	42.61	19.83	24.18	45.62	14.02	30.00	46.89
Si (ppm)	12.04	53.29	70.59	16.41	54.21	78.34	23.30	57.66	87.65

Table III. Extent of Al and Si dissolution in KOH solutions in relation to time and alkalinity

Alkalinity (M)	2			5			10		
Time (h)	5	10	24	5	10	24	5	10	24
Al (ppm)	2.68	20.13	20.91	5.41	19.78	29.82	5.35	25.68	41.40
Si (ppm)	7.22	33.97	23.78	10.06	10.10	54.33	10.71	58.81	75.64

Figures 1 and 2 present the % dissolved Al and Si after leaching, for 24 hours, in 10 M NaOH and 10 M KOH, respectively. As it is seen, although the fly ash consists of more than one -and not all amorphous- aluminosilicate phases, it appears to have a considerable solubility and therefore it is a potential raw material for geopolymerisation. As it has been previously reported, the thermal treatment of materials improves their reactivity, especially if their crystalline structure is modified in order to store energy [23,24].

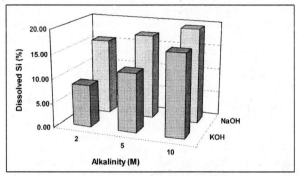

Figure 1. Dissolved Si (% w/w) in relation to alkalinity and alkali ion

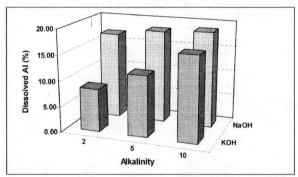

Figure 2. Dissolved Al (% w/w) in relation to alkalinity and alkali ion

According to Figures 1 and 2, Si and Al seem to have a synchronized leaching behaviour in both alkaline solutions. This indicates that Si and Al are probably dissolved in some kind of linked form, at least in concentrated alkaline solutions. Other researchers have also reported a simultaneous leaching of Al and Si in the case of pure mineral phases [24]. In the case of NaOH the increase of alkalinity has a slight effect on the dissolution of Al and Si, while in the case of KOH, the dissolution of both elements is proportional to the alkalinity of the solution. This may indicate that, in case of NaOH, equilibrium between dissolved species and solid is obtained at low alkalinity. The dissolution rate of Al and Si is higher in NaOH than in KOH. The same effect was also observed by other researchers and was associated with the smaller size of the Na^+ which can better stabilize the silicate monomers and dimmers in the solution, increasing in this way the dissolution rate of fly ash [3,5].

Figure 3 presents the XRD patterns of fly ash and its solid residue after leaching in 10 M NaOH for 24 h. As it is seen the solid residue contains mainly quartz while the other crystalline phases of fly ash were almost completely dissolved.

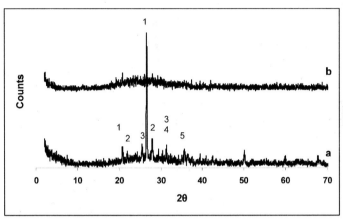

Figure 3. XRD patterns of fly ash (a) and solid residue after leaching in NaOH 10 M for 24 h (b).
1: quartz, 2: feldspars, 3: anhydrite, 4: gehlenite, 5: maghemite

Optimisation of curing conditions

Figure 4 presents the 7-day compressive strength of fly ash based geopolymers after curing at 50, 70 and 90 °C for 24, 48 and 72 hours. As it is seen the optimal curing time is 48 hours. The negative effect of longer curing time was also reported by other authors [24]. The increase of curing temperature from 50°C to 70°C induces a significant increase of the geopolymer compressive strength. Curing at higher temperature results in a slight increase of compressive strength, but causes the formation of surface cracking confirming that some water needs to be retained in the geopolymeric network in order to avoid cracking and maintain structural integrity.

Effect of Si/Al ratio

Figure 5 shows the ultimate compressive strength of the geopolymers in relation to the Si/Al ratio in the raw mixture. The specimens were cured at 70 °C for 48 days. All values presented are the average of three measurements with error reported as standard deviation. As it is seen, the increase of Si/Al ratio from 1.7 to 2.5 almost doubles the compressive strength. Further increase of the Si/Al ratio has a negative effect on the development of the compressive strength. The same trend has been reported in the case of metakaolin-based geopolymers [6,26,27]. It is thought that the initial increase of the Si content results in the formation of larger geopolymeric networks with higher structural integrity. However, the further increase of soluble Si content may inhibit the dissolution of fly ash by shifting the dissolution reaction to the left, leaving unreacted material and affecting the microstructure of the geopolymer. It must be noted that the geopolymers with Si/Al 4.0 and 4.5 required higher amounts of water in order to obtain a homogenous slurry and this is probably the reason for the very low compressive strength.

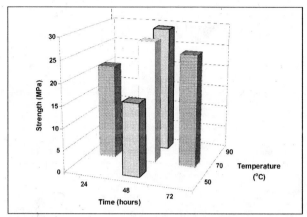

Figure 4. Compressive strength of geopolymers in relation to curing time and temperature

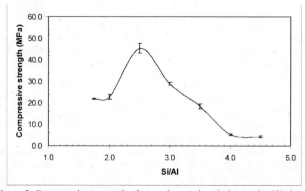

Figure 5. Compressive strength of geopolymers in relation to the Si/Al ratio

Figure 6 presents the XRD patterns of the geopolymers in relation to the Si/Al ratio. In the samples with Si/Al ratio higher than 2, the only crystalline phases are quartz and unreacted feldspars, while hydroxysodalite and faujasite are detected in the sample with Si/Al ratios 1.7 and 2.0, respectively. The formation of zeolitic phases in geopolymers with low Si/Al ratios has been also reported by other authors[28]. In samples with Si/Al ratio higher than 2, there was no formation of zeolitic phases. It seems that the increase in the Si content in the activation solution favors the formation of more disorded structures.

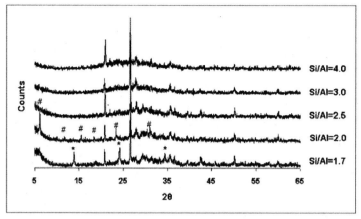

Figure 6. XRD patterns of geopolymers in relation to Si/Al ratio (*:hydroxysodalite, #: faujasite)

Figure 7 shows the FTIR spectra of fly ash and the prepared geopolymers in relation to the Si/Al ratio. The spectrum of fly ash is broad and relatively featureless, due to the glassy nature and heterogeneity of this material. The main aluminosilicate phases in fly ash (both crystalline and amorphous) have overlapped peaks in the region 800-1300 cm^{-1}. This broad hump is the main feature in both the fly ash and geopolymer spectra and is associated with the Si-O-T (T: tetrahedral Si or Al) asymmetric stretching vibrations. As it is seen in Figure 7, the maximum of this hump, in the case of fly ash, is around 1110 cm^{-1}, while in the case of geopolymers this band becomes narrower and shifts to lower wavenumbers. As the Si/Al ratio in geopolymers increases, this peak shifts progressively to higher wavenumbers (1000, 1011, 1025 and 1050 cm^{-1} for 1.7, 2.5, 3.5 and 4.5 ratios, respectively). The position of this band gives an indication of the length and angle of the bonds in a silicate network. For amorphous silica, this peak occurs at approximately 1100 cm^{-1}. The shift of this band to lower wavenumbers, in the case of geopolymers, indicates an increase in the substitution of tetrahedral Al in the geopolymeric network. As the Si/Al ratio increases, the peak shifts back to higher wavenumbers due to the higher Si/Al ratio in the binder. Any peaks in the region 630-760 cm^{-1} generally correspond to units such as the aluminosilicate ring and cage structures and indicate the presence of crystalline zeolitic phases. The vibrations at approximately 740 and 650 cm^{-1}, in the geopolymer with Si/Al=1.7, indicates the presence of hydroxysodalite. In the geopolymers with higher Si/Al ratios there are not any distinct peaks in the zeolite region. These observations confirm the XRD measurements and are in agreement with the relative literature [28,29].

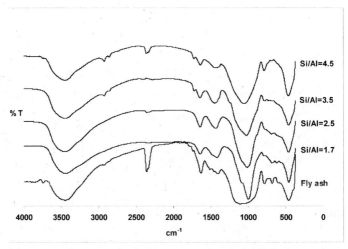

Figure 7. FTIR spectra of fly ash and geopolymers

CONCLUSIONS

The following conclusions can be drawn from the present study:

- Fly ash coming from Megalopolis, Greece is a promising raw material for geopolymer synthesis
- The dissolution of fly ash is higher in NaOH than in KOH, while Si and Al seem to have a synchronized leaching behaviour in both alkaline solutions
- The optimal curing conditions are 48 hours at 70 °C. The increase of curing time leads to the decrease in compressive strength, while the increase of the curing temperature causes a slight increase in the compressive strength but it also favors the surface cracking of the specimens.
- The compressive strength of geopolymers is found to depend systematically on the Si/Al ratio, with the maximum being 45.5±2 MPa for a Si/Al ratio of 2.5

REFERENCES

1   J. Davidovits, Geopolymers: Inorganic polymeric new materials, *J. Mater Edu*, **16**, 91(1994)
2   J. Davidovits, Geopolymers: Inorganic polymeric new materials, *J. Therm Anal Calorim*, **37**, 1633-1656 (1991)
3   Hua Xu, J.S.J. van Deventer, The geopolymerisation of aluminosilicate materials, *Int J Miner Process*, **59**, 247-258 (2000)
4   Hua Xu, J.S.J. van Deventer, The effect of alkali metals on the formation of polymeric gels from alkali feldspars, *Colloids Surf A Physicochem Eng Asp*, **216**, 27-44 (2003)
5   Hua Xu, J.S.J. van Deventer, G.C Lukey, Effect of alkali metals on the peferential geopolymerisation of Stilbite/ Kaolinite mixtures, *Ind Eng Chem Res*, **40**, 3749-3756 (2001)
6   M. Rowles, B. O' Connor, Chemical optimisation of the compressive strength of aluminosilicate geopolymers synthesized by sodium silicate activation of metakaolinite, *J Mater* Chem **13(5)**, 1161-1165 (2003)
7   M. Schmucker, K.J.D. Mackenzie, Microstructure of sodium polysialate siloxo geopolymer, *Ceram Int*, **31**, 433-437 (2005)

8   J.W. Phair, J.S.J. van Deventer, Characterisaton of fly ash based geopolymeric binders activated with sodium aluminate, *Ind Eng Chem Res*, **41**, 4242-4251 (2002)

9   A.M. Fernandez- Jimenez, A. Palomo, M. Criado, Microstructure development of alkali activated fly ash cement: a descriptive model, *Cem Concr Res*, **35**, 1204-1209 (2005)

10 T. Backarev, Geopolymeric material prepared by using Class F fly ash and elevated temperature curing, *Cem Concr Res*, **35**, 1224-1232 (2005)

11 W.K.W. Lee, J.S.J. van Deventer, G.C Lukey, Effect of anions on the formation of aluminosilicate gel in geopolymers, *Ind Eng Chem Res*, **41**, 4550-4558 (2002)

12 Hua Xu, J.S.J. van Deventer, Geopolymerisation of multiple minerals, *Miner Eng*, **15**, 1131-1139 (2002).

13 J.C. Swanepoel, C.A. Strydom, Utilisation of fly ash in geopolymeric materials, *Appl Geochem* **17**, 1143-1148 (2002)

14 P.S. Singh, M. Trigg, I. Burgar, T. Bastow, Geopolymer formation processes at room temperature studied by Si and Al MAS- NMR, *Mate Sci Eng A*, **396(1-2)** 392 (2005)

15 R.A. Flecher, C.L. Nicholson, S. Shimada, K.J.D. Mackenzie, The composition of aluminosilicate geopolymers, *J Eur Ceram Soc*, **25**, 1471-1477 (2005)

16 D. Feng, H. Tan, J.S.J. van Deventer, Ultrasound enhanced geopolymerisation, *J Mater Sci*, **39**, 571-580 (2004)

17 J. Davidovits, Chemistry and geopolymeric systems terminology, *Proccedings, Geopolymer* **99**, 9-40 (1999)

18 J. Davidovits, Geopolymers on the first generation: SILIFACE process, *Proccedings, Geopolymer* **88**, 49-68 (1988)

19 Hua Xu, J.S.J. van Deventer, Effect of source materials on Geopolymerisation, *Ind Eng Chem Res*, **42**, 1698-1706 (2003)

20 J. Davidovits, Structural Characterisation of Geopolymeric Materials with X-Ray Diffractometry and MAS-NMR Spectrometry, *Proceedings, Geopolymer* **88**, 149-166 (1998)

21 V.F.F. Barbosa, K.J.D. Mackenzie, C.D. Thaumaturgo, Synthesis and Characterisation of sodium polysialate inorganic polymer based on alumina and silica, *Proccedings, Geopolymer* **99**, 65-78 (1999)

22 J. Davidovits, Mineral polymers and methods of making them, *United States Patent* 4,349,386, (1982)

23 Ch. Panagiotopoulou, E. Kontori, Th. Perraki, G. Kakali, Dissolution of aluminosilicate minerals and by-products in alkaline media, *J Mater Sci*, **42**, 2967–2973 (2007)

24 Hua Xu, Geopolymerisation of aluminosilicate minerals, *pHD Thesis, The University of Melbourne* (2002)

25 J.G.S. van Jaarsveld, J.S.J. van Deventer, G.C. Lucey, The effect of composition and temperature on the properties of fly ash and kaolinite based geopolymers, *Chemical Engineering Journal* **89**, 63-73 (20002)

26 P. Duxson, S.W. Mallicoat , G.C. Lukey , W.M. Kriven , J.S.J. van Deventer, The effect of alkali and Si/Al ratio on the development of mechanical properties of metakaolin-based geopolymers, *Colloids and Surfaces A: Physicochem. Eng. Aspects* **292**, 8-20 (2007)

27 P. Duxson, J.L. Provis, G.C. Lukey, S.W. Mallicoat, W.M. Kriven, Jannie S.J. van Deventer, Understanding the relationship between geopolymer composition, microstructure and mechanical properties, *Colloids and Surfaces A: Physicochem. Eng. Aspects* **269**, 47-58 (2005)

28 C.A.. Rees, J.L. Provis, G.C. Lukey, J. S. J. van Deventer, Attenuated Total Reflectance Fourier Transform Infrared Analysis of Fly Ash Geopolymer Gel Aging, *Langmuir* **23**, 8170-8179 (2007)

29 C.A.. Rees, J.L. Provis, G.C. Lukey, J. S. J. van Deventer, In Situ ATR-FTIR Study of the Early Stages of Fly Ash Geopolymer Gel Formation, *Langmuir* **23,**, 9076-9082 (2007)

ROLE OF OXIDE RATIOS ON ENGINEERING PERFORMANCE OF FLY-ASH GEOPOLYMER BINDER SYSTEMS

Kwesi Sagoe–Crentsil
CSIRO Materials Science and Engineering
PO Box 56, Highett, Victoria 3190, Australia

ABSTRACT

This study examines specific roles of various constituent oxides on the hydrolysis and condensation reactions that underpin engineering performance of Geopolymer binder systems. Fly-ash Geopolymer systems characterised by high Si/Al ratios, i.e., $SiO_2/Al_2O_3 \geq 3$, provides an ideal system for this form of analysis given its widespread consideration for emerging mainstream engineering applications. Specific emphasis is placed on the roles of silica and alkali species present in the feedstock material and their impact on mechanical properties such as strength development.

It is noted that alongside other chemical variables discussed for fly-ash based Geopolymer systems, high alkali content ($Na_2O/Al_2O_3 = 1.0$) systems are shown to be characterised by high strength, low porosity and dense microstructures. Such properties may be correspondingly achieved with high silica and low alumina contents ($SiO_2/Al_2O_3 = 3.5-3.8$), beyond which strength deterioration can occur. The observed synthesis parameters suggest potential beneficial and novel applications for a variety of fly-ash sources as raw feedstock material for building product and civil construction applications.

INTRODUCTION

The properties of Geopolymer binder systems are largely controlled by the reaction chemistry of SiO_2, Al_2O_3 and other minor oxides present in its highly alkaline environment. It is these characteristic properties, notably the high compressive and tensile strengths, thermal performance and acid resistance properties compared to equivalent Ordinary Portland cement (OPC) systems that have generated interest in Geopolymer systems in recent years.[1-3] Clearly, the factors controlling geopolymer binder performance hinge on materials selection and process route adopted for geopolymer synthesis. In particular, the type and chemical composition of oxide components of the feedstock material – typically fly ash or metakaolin, and concentration of alkali silicate activator, water content, and cure conditions[4,5] play a major role in both microstructure development and tailoring of engineering properties of the Geopolymer binder product.

As noted by several researchers[6,7] the basic mechanisms of the geoplymerisation reaction involves an initial dissolution step in which Al and Si ions are released in the alkali medium. Transport and hydrolysis of dissolved species are followed by a polycondensation step, forming 3-D network of silico-aluminate structures. According to Davidovits[1], these structures can be of three types: Poly (sialate) (-Si-O-Al-O-), Poly (sialate-siloxo) (Si-O-Al-O-Si-O), and Poly (sialate-disiloxo) (Si-O-Al-O-Si-O-Si-O).

The chemical processes governing polymerisation reactions of Al_2O_3 and SiO_2 in these systems are largely controlled by stability of the respective speciated phases. X-ray diffraction (XRD) analysis shows geopolymers to be largely amorphous[8,9] although there is published evidence of occurrence of nanocrystalline particles, within the geopolymer matrix structure.[10,11] Correspondingly, in the alkaline aqueous solutions of geopolymers, aluminium is present mostly as monomeric aluminate ions $(Al(OH)^{4-})$.[12,13] Thus all the aluminium present in solution is in IV-fold coordination irrespective of the coordination of the aluminium in the precursor. Silicon, by contrast, forms a variety of oligomeric ions particularly at high concentrations and high SiO_2/M_2O (M = Na, K) ratios.[14] Swaddle and co-workers[13] list 25 such silicate oligomers in linear, cyclic and three dimensional forms.

Unlike the well understood roles of oxide components comprising the hydrated gel phases present in $CaO-Al_2O_3-SiO_2$ systems i.e., Portland and pozzolanic cements, the equivalent contributions of oxide components governing polymerisation reactions and, hence, geopolymer properties are now only beginning to emerge.[15,16] Accordingly, the reaction pathways required to achieve desired engineering performance of geopolymer systems is becoming increasingly important. Especially so, since the properties of geopolymer systems are particularly dependent on reaction chemistry given its characteristic multiphase structure. While aspects of physical and chemical property relationships of generic geopolymer systems have been investigated,[17] the need exists to extend such studies to cover raw materials selection, process conditions through to large scale production issues.

This study draws on experimental and theoretical studies to examine the relationships between chemical formulation, microstructure and mechanical properties of selected fly-ash based geopolymer systems to explore the effects of feedstock oxide compositional limits and mineralogy. Analysis of the compressive strength development characteristics of fly ash formulations provides a basic index to assess the interrelationships of key oxide components and their implications on geopolymer performance.

EXPERIMENTAL

Geopolymer mixtures were prepared from fly ash, alkali silicate solution, alkali hydroxide and distilled water. The Class F fly ash, used was mostly amorphous, however it also contained approximately 10% quartz, 20% mullite and 5% ferrite spinel, with the majority of particles passing the 45μm sieve; supplied by Pozzolanic Enterprises, Australia. The composition by weight percent of oxides is: 47.9% SiO_2, 29.79% Al_2O_3, 13.93% Fe_2O_3, 3.29% CaO with minor amounts of other oxides.

The alkali silicate solution was supplied by PQ Australia and had the following composition: sodium silicate (8.9 wt% Na_2O, 28.7 wt% SiO_2, and 62.5 wt% H_2O). Several geopolymer formulations were prepared with the varying proportions of ingredients selected to allow the effect of alkali and silica contents to be assessed; ccolloidal silica (Ludox HS-40) provided supplementary silica in formulations.

Sample preparation involved initial mixing of alkali hydroxide, alkali silicate solution and fly ash. After mixing samples were immediately poured into preheated moulds which were then sealed. Samples were cured at $85^{\circ}C$ for 2 hours, after which they were demoulded and cooled in a refrigerator to arrest reaction kinetics. Compressive strengths were measured on 25.4 mm cubes. A minimum of two(usually three) were tested for each formulation, and at times increased to six cubes to obtain acceptable standard deviation on measurements Compressive strength testing was performed using Baldwin testing apparatus. SEM was performed on fractured surfaces on a Philips XL30FEG field emission SEM with all images representing averaged parts of samples.

The chemical composition of formulated geopolymer systems with varying Si loading described in this study may be nominally represented as: $x Na_2O \cdot 3SiO_2 \cdot Al_2O_3 \cdot 10H_2O$ with the Na_2O/Al_2O_3 values of x = 0.6, 0.8, 1.0 and 1.2. A second series of mixtures with a composition of $Na_2O \cdot y SiO_2 \cdot Al_2O_3 \cdot 10H_2O$ and SiO_2/Al_2O_3 ratio equivalent to y = 2.7, 3.0, 3.5 and 3.9, is also reported. Analysis of the effects of molar concentrations of Na_2O and SiO_2 were performed with reference to microstructural and compressive strength testing. The latter series with a compositional ratio range of SiO_2/Al_2O_3 = 2.7 to 3.9, differed in the proportion of silica added in solution compared to that added as solid fly ash.

RESULTS
Effect of alkali content

The microstructure of geopolymer systems of varying Na_2O/Al_2O_3 (0.6 and 1.2) are shown in Fig 1. The basic structure consists of dense gel phase of rounded growths with bridging between them (Fig. 1). As observed in Fig 1, the respective gel phase matrices not only varied in densification but also the amount of bridging material between spherical particles. Therefore, the samples may be viewed as composites comprising of gel phase and partially reacted and unreacted spherical fly ash particles.

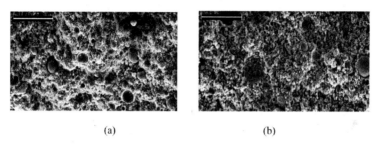

(a)   (b)

Fig 1. Microstructures of samples containing different alkali levels.[16] (a) Na_2O/Al_2O_3 = 0.6 (b) Na_2O/Al_2O_3 = 1. 2. Scale bar represents 50 μm

The strength of the unreacted particles and the interfaces between them and the geopolymer matrix can be expected to have a significant bearing on the overall strength of the material. In general, the microstructures of geopolymer samples appear similar to images previously published for these materials.[5,17] Despite the range of compressive strengths observed for various formulations, only minor differences exist in the respective microstructures. A slight decrease in the number of unreacted particles and an increase in the size of spherical-shaped particles with increasing alkali content Na_2O/Al_2O_3 = 0.6 through to 1.2 was evident as seen from Fig 1.

The compressive strengths at 24hrs after steam cure at $85^{\circ}C$ for 2hrs, show a progressive increase for samples with Na_2O/Al_2O_3 = 0.6, 0.8 and 1.0 then a decrease at ratios of 1.2, as shown in Fig 2. As the role of the alkali in geopolymers is partially to balance the charge of the aluminate groups in the tectosilicate, it is not unexpected to find that the compressive strength goes through a maximum when the alkali and alumina concentrations are equal. The other role of alkali in the system is to increase the solubility of the aluminosilicate, as indicated akin to alkali attack of silicate shown in Equation 1.

In alkaline media (e.g. containing sodium ions) a conventional silica depolymerisation reaction likely occurs, with any excess NaOH tending to disrupt internal Si-O-Si links of the silica tetrahedron after the acidic hydroxyl groups at the surface of the silica have been neutralized, as indicated by:

$$Si\text{-}O\text{-}Si + 2NaOH = Si\text{-}O\text{-}Na^+ + Na^+\text{-}O\text{-}Si + H_2O \tag{1}$$

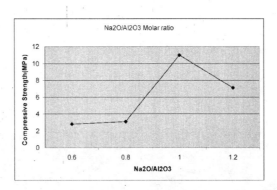

Fig 2. Effect of varying alkali concentrations i.e. Na_2O/Al_2O_3 = 0.6, 0.8, 1.0, 1.2 on compressive strength (tested 24hrs after Steam cure at 85^OC for 2hrs)

Effect of silica content

The microstructures of geopolymer systems with SiO_2/Al_2O_3 = 2.7 and 3.9 are shown in Fig 3. Mixtures of varying Si contents were produced to determine the effect of the silica concentration on the structure and strength of geopolymers represented by; $Na_2O \cdot ySiO_2 \cdot Al_2O_3 \cdot 10H_2O$ [y = 2.7, 3.0, 3.5 and 3.9]. The fineness of the texture and the density of the geopolymer increases greatly from SiO_2/Al_2O_3 = 2.7 to 3.9. The images reveal higher silica samples contain more unreacted particles, and also that crystals have grown on a number of the unreacted fly ash particles, particularly in the higher silica mixture.

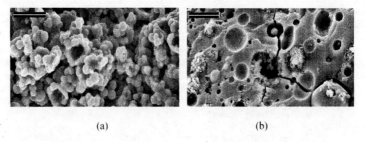

(a)   (b)

Fig 3. High magnification images of geopolymers containing different amounts of silica (a) SiO_2/Al_2O_3 = 2.7 (b) and SiO_2/Al_2O_3 =3.9. Scale bar represents 1 µm.

From the corresponding compressive strength at 24hrs (after Steam cure at 85^OC for 2hrs) plots of Fig 4, it can be observed that the difference in strengths observed between SiO_2/Al_2O_3 = 3.0 and 3.5 was greater than that between ratios of 2.7 and 3.0 or 3.5 and 3.9, as would be expected given the range of silica concentrations. Also, the compressive strengths of the Si series of samples followed expected trends in the microstructure, i.e. the densest, finest grained samples being also the strongest as shown in Fig 4.

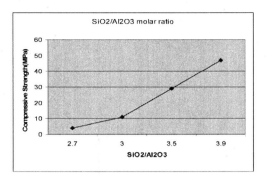

Fig 4. Effect of varying silica content on compressive strength of geopolymer systems

Compared to previous observations for varying alkali levels, the corresponding change in strength with silica content is rather dramatic. In particular, the small change (11%) from SiO_2/Al_2O_3 = 3.5 to 3.9 gave a 62% strength increase as shown in Fig 4. Specimens with SiO_2/Al_2O_3 = 2.7 3.0 3.5 and 3.9 showed trends of increasing strength with increased silica concentration in agreement with expectations based on microstructures of the respective geopolymers.

Effect of alumina content

Varying the alumina concentrations ranging from Al_2O_3/Na_2O=0.7 to 1.1 show that low alumina formulation give strong and dense microstructures. The microstructure of low Al geopolymer compositions typically shows a large proportion of unreacted particles and crystallinity. These two factors could be expected to weaken the system, as the strength is likely to be derived from the amorphous gel rather than crystalline or unreacted components. This may counteract the effect of the extra strength given to the binding phase by its density and fine grain. At an alumina concentration around that of Al_2O_3/Na_2O=0.9, a strength maximum is reported between these two counteracting factors.[16]

The Al component of geopolymers mostly governs condensation processes during synthesis. Thus, the Si/Al ratio in geopolymer systems has an important influence on the structure and mechanical properties. A higher Al content assists in condensation occurring more feasibly, and tends to lead to a denser network structure due to the removal of more hydroxyl groups. Other researchers such as Palomo et al.[18] showed that Al solubility is generally much higher than Si under alkaline conditions, thus aluminate anions are likely to react with silicate species from alkaline silicate solutions. Fig 5 below shows the setting time characteristics of geopolymer systems with varying SiO_2/Al_2O_3 content.

As shown by previous analysis,[19,20] $[Al(OH)_4]^-$ species largely regulate the condensation reaction rate and, consequently, setting rates. The concentration of aluminate species in solution such as $[Al(OH)_4]^-$ can be adjusted by varying pH value of the solution. Hence lower pH solutions may result in lower concentration of $[Al(OH)_4]^-$ species and longer setting time.

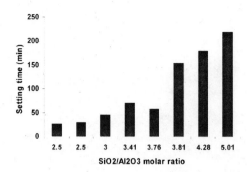

Fig 5. Setting time characteristics of geopolymer systems with varying SiO_2/Al_2O_3 content

Effect of water content

The effect of varying the water content of fly ash inorganic polymers has not been widely reported in the past. For instance, formulations with a water/fly ash ratio of 0.34 achieve around twice the strength of equivalent ratio of 0.75 for various alkali concentrations has been reported.[16]

The compressive strengths of geopolymer formulations based on water to fly ash mass ratios varying from 0.15 to 0.57 suggest that the lower water content, the stronger the mix.[16] However, there are practical limitations to low water contents. In practice, it is these limitations which would limit the strength that could be achieved by water reduction. Low water content greatly raises the viscosity of the liquid component reducing dispersal and ease of mixing. Effectively, trends in compressive strength indicate lower water content formulations give relatively higher strengths, a finding that logically follows the trend in porosity.

Essentially higher water content geopolymers generally set slowest. This is partly because more time is available for mobility of ions in the system as the solids dissolve. This provides a possible reason for observed low level of unreacted material. A porosity increase is to be expected, as there will be more water-filled pores in the product with increase in water content akin to high water/binder ratio in hydraulic cement systems. The practical importance of the viscosity of geopolymer formulations has been commented upon previously by other authors.[15,17]

Alternatively, increasing overall liquids/solids (L/S) to improve mix workability but keeping the relative proportions of the liquid components fixed should not significantly alter compressive strengths. However, Silverstrim et al.[21] found that the strength of inorganic polymer mortars substantially increased with increasing proportions of activator. Palomo et al.[18] on the other hand found that there was little difference between formulations with 20% and 23% activator (L/S = 0.25 and 0.30). While these proportions are fairly similar, the data of Silverstrim et al.[21] implies that such small increments would make a measurable difference.

Effect of pH

The alkalinity of the activator governs the rate of condensation between aluminate and silicate species, besides such factors as temperature and the nature of the feedstock as shown previously by calorimetric results[20]. However, the highly alkaline nature of the mixtures, with initial pH in excess of 13.9 makes silicate and aluminosilicate oligomeric ions less stable than at pH<12.[13] Phair et al,[22] found that geopolymer condensation was favoured in conditions of high pH in which most of the dissolved silicon was monomeric. By comparison, zeolitic reactions which have similar chemistries to geopolymers are also formed in conditions favouring monomeric or small oligomeric silicate ions. However the speciation of aluminosilicate solutions at high pH and high silica concentration is not still evolving.[23]

Effect of Si/Al ratio

Considering the facts that aluminium component of geopolymer mixtures tends to dissolve more easily than the silicon component, [16,24] it is plausible that more $Al(OH)_4^-$ species and relatively less Si species are available for condensation in the system with low SiO_2/Al_2O_3 ratios. Therefore condensation is more likely to occur between aluminate and silicate species producing poly (sialate) polymer structures. With increasing Si content, more silicate species are available for condensation and reaction between silicate species, resulting oligomeric silicates, becomes dominant. And further condensation between oligomeric silicates and aluminates result in a rigid 3D net works of poly (sialate-siloxo), and poly (sialate-disiloxo) 3-D rigid polymeric structures.

DISCUSSION

The mixing stage of proportioned solid and liquid feedstock components of geopolymer systems initiate an immediate dissolution process. Depending on the pH regime and oxide concentrations, the resultant species in the liquid phase may comprise monomeric $[Al(OH)_4]^-$, $[SiO_2(OH)_2]^{2-}$ and $[SiO(OH)_3]^-$ or similar. [11,24] These subsequently condense with each other. It should be noted that the condensation between Al and Si species occurs more readily due to the characteristic high activity of species such as $[Al(OH)_4]^-$. For $[SiO(OH)_3]^-$ and $[SiO_2(OH)_2]^{2-}$, although the latter species is more capable of condensing with $[Al(OH)_4]^-$ since there exists a larger attraction, they are likely to produce only small aluminosilicate oligomers as discussed by Weng et al.[20] The above discussions are summarised in the synthesis pathway as given below:

1. At the onset of mixing, solid aluminosilicate components dissolve releasing aluminate and silicate ions into solution, with concurrent hydrolysis reactions of dissolved ions
2. The aluminate and silicate species subsequently begin the condensation process, initially giving aluminosilicate monomers and perhaps oligomers. These ions further condense with one another to produce a gel phase while the mixture starts to set.
3. Condensation reactions continue within the gel phase with the silicate/aluminate ions continuing to dissolve from the solid and onset of initial hardening.
4. Re-dissolution of the gel and precipitation of less soluble and more stable aluminosilicate species may occur while the Geopolymer hardens completely as condensation reactions rapidly escalates.
5. Over a long period of time, the condensation reactions continue but at a decreasing rate. The rigidity of the gel and reduced free water greatly reduce the rate of dissolution of the original aluminosilicate solid.

The variety of complex microstructures that characterise Geopolymer systems depends on selected mix composition. It is apparent that there is a maximum SiO_2/Al_2O_3 ratio which is favourable in producing high strength geopolymers. Accordingly, the most favourable SiO_2/Al_2O_3 molar ratio for geopolymer strength is about 3.8. For this, Na_2O/Al_2O_3 ratio is about unity.

Given that aluminate anions for the reaction are solely derived from the dissolution of mineral oxides under alkaline conditions, monomeric $[Al(OH)_4]^-$ ions are probably the only aluminate species existing under high alkaline conditions.[24] On the other hand, silicate species come from both soluble alkaline silicates and the dissolution of mineral oxides. In the specific case of fly ash systems, the silicate species from the dissolution of particles are difficult to predict because the hydrolysis process of amorphous silica is kinetically dependent on various factors, such as the activity of the particles, temperature, time, and the concentration and pH value of alkaline silicate solutions.

Higher Al composition also suggests that condensation occurs more feasibly, and leads to a denser network structure due to the removal of more hydroxyl groups. Therefore, as expected, geopolymers composed of a higher Al component may be brittle, and have high hardness. Moreover, geopolymers with a low Al component will be better binders since more hydroxyl groups exist in the structure, resulting in improved bond characteristics primarily from existing hydrogen bonds. These observations suggest that the condensation process in these systems occurs in two stages: (a) quick condensation between aluminate and silicate species; followed by (b) a slow condensation stage solely involving silicate species.[20]

Silicon solubility, on the other hand, is generally much less than aluminium under alkaline conditions. Thus, aluminate anions are likely to react with silicate species from alkaline silicate solutions. Alkaline (mainly Na and K) silicate solutions used in the production of geopolymers, typically have high SiO_2 concentrations above 5M, and M_2O/SiO_2 ratios of 0.66~0.83 by the addition of NaOH. These alkaline silicate solutions have a high content of polymeric silicate species, as demonstrated by Barbosa et al.[8] Therefore, the condensation in geopolymers will likely occur between monomeric $[Al(OH)_4]^-$ and a variety of silicate species, including monomers and oligomers.

Compressive strengths in general can be related to both composition and microstructure in a logical way. These general observations and their implications on feedstock material selection may be best understood by examining the fundamental dissolution and condensation reactions occurring during synthesis. Compared with hydration reactions in ordinary Portland cements, condensation in geopolymers appears to be more complex due to the existence of a variety of oligomeric silicate species. The process may be further complicated by the overlap of dissolution processes, hydrolysis and condensation reactions, which regulates the speciation conditions for condensation. In this context, the continuously changing environment of concentration of various species and the pH value of the liquid phase also remains critical.

CONCLUSIONS

Observations described in this study broadly follow expected theoretical correlations between geopolymer mix composition, microstructure and strength. It is observed that high strength formulations are characterised by low porosity and dense, fine grained microstructures. Such structures are found in geopolymers with high alkali contents ($Na_2O/Al_2O_3 = 1.0$). High silica and low alumina contents ($SiO_2/Al_2O_3 = 3.5$-3.8) also produced this structure. The observed synthesis parameters suggest potential beneficial mix design methods can be adopted for different raw feedstock materials to produce a variety of building product and civil construction applications.

Setting time of geopolymer formations is mainly controlled by the amount of Al available for the reaction while increasing SiO_2/Al_2O_3 ratio leads to longer setting times. Correspondingly, high SiO_2/Al_2O_3 ratios are responsible for the high strength gains at later stages.

REFERENCES
[1] J. Davidovits, Geopolymer chemistry and properties. In: Davidovits, J, Orlinski, J. (Eds.), Proceedings of the 1st International Conference on Geopolymer '88, vol. 1, Compiegne, France, 1–3 June, 25–48(1988)
[2] P.V. Krivenko, G.Yu Kovalchuk, Directed synthesis of alkaline aluminosilicate minerals in a geocement matrix. J. Mater. Sci., 42, 2944–52(2007).
[3] F. Pacheco-Torgal, J. Castro-Gomes, and S. Jalali, Investigations about the effect of aggregates on strength and microstructure of geopolymeric mine waste mud binders, Cem. Con. Res., 37, 933–41(2007)

[4] H. Rahier, W. Simons, B. van Mele, and M. Biesemans, Low temperature synthesised aluminosilicate glasses. Part III: Influence of the composition of the silicate solution on production, structure and properties, *J. Mater. Sci.* **32**, 2237-47(1997).

[5] A. Fernandez-Jimenez and A. Palomo, Composition and microstructure of alkali activated fly ash binder: Effect of activator *Cem. Concr. Res.*, **35**, 1984-92(2005)

[6] M. Rowles, and B. O'Connor, Chemical optimisation of the compressive strength of aluminosilicate geopolymers synthesised by sodium silicate activation of metakaolinite *J Mat. Chem.*, **13**. 1161-65(2003).

[7] A. Fernández-Jiménez, A. Palomo, M.Criado, *Cem. Concr. Res.*, **35**, Issue: 6, 1204-09(2005).

[8] V.F.F. Barbosa, K.J.D. MacKenzie, and C. Thaumaturgo, Synthesis and characterisation of materials based on inorganic polymers of alumina and silica: sodium polysialate polymers, *Intl J. of Inorg. Mat*, **2**, 309-317(2000).

[9] J.G.S. Van Jaarsveld, J.S.J. van Deventer, and A. Schwartzman, The potential use of geopolymeric materials to immobilise toxic metals: Part I. Theory and applications, *Min. Eng.*, **10(7)**, 659-669(1999).

[10] J.L. Provis, G.C. Lukey, and J.S.J. Van Deventer, Do Geopolymers actually contain nanocrystalline zeolites? A re-examination of existing results, *Chem. Mater.*, **17**, 3075-85(2005).

[11] R.A. Fletcher, J.D. MacKenzie, C.L. Nicholson, and S. Shimada, The composition range of aluminosilicate geopolymers *J. Europ. Cer. Soc.*, 25, 1471-77(2005).

[12] C.J. Brinker and G.W. Scherer, *Sol-Gel Science: The Physics and Chemistry of Sol-Gel Processing*, Academic Press, Boston (1990).

[13] T.W.Swaddle, J. Salerno, and P.A. Tregloan, Aqueous aluminates, silicates and aluminosilicates, *Chemical Society Reviews*, 319-325(1994).

[14] Iler, R.K. *The Chemistry of Silica*, Wiley-Interscience, New York.

[15] P.Duxson, J.L. Provis, G.C. Lukey, S.W. Mallicoat, W.M. Kriven, and J.S.J. van Deventer, Understanding the relationship between geopolymer composition, microstructure and mechanical properties, *Col. and Surf. A: Physicochem. Eng. Aspects*, **269**, 47-58, (2005).

[16] M. Steveson, and K. Sagoe–Crentsil, Relationship between composition, structure and strength of inorganic polymers Part I – Metakaolin-derived inorganic polymers *J Mater. Sci.*, **40**, 2023-36, (2005). Relationship between composition, structure and strength of inorganic polymers Part II – Flyash-derived inorganic polymers *J. Mater. Sci.*, **40**, 4247-59(2005).

[17] J.W. Phair, and J.S.J. Van Deventer, Effect of silicate activator pH on the leaching and material characteristics of waste-based inorganic polymers. *Miner. Eng.* **14** 289–304(2001).

[18] Palomo, A., Grutzeck, M.W. and Blanco, M.T. Alkali-activated fly ashes, A cement for the future, *Cem. Concr. Res.*, **29**, 1323-29(1999).

[19] P. De Silva, K. Sagoe-Crenstil, and V. Sirivivatnanon, Kinetics of geopolymerization: Role of Al_2O_3 and SiO_2. *Cem. Concr. Res.* **37** (4), 512–518(2007).

[20] L. Weng, and K. Sagoe-Crenstil, Dissolution processes, hydrolysis and condensation reactions during geopolymer synthesis: Part 1 – Low Si/Al ratio systems. *J. Mater. Sci.*, **42**, #9. 2997–06(2007).

[21] T. Silverstrim, H. Rostami, B. Clark, and J. Martin, in Proceedings of the 19th International Conference on Cement Microscopy, Cincinnati, USA, 355–373(1997).

[22] J.W. Phair, S.J. van Deventer, and J.D. Smith, Mechanism of polysialation in the incorporation of zirconia into fly ash-based geopolymers, *Ind. Eng. Chem. Res.*, **39**, 2925-34(2000).

[23] J.G.S. Van Jaarsveld, J.S.J. van Deventer, and A. Schwartzman, The potential use of geopolymeric materials to immobilise toxic metals: Part II. Material and leaching characteristics, *Miner. Eng*, **12(1)**, 75-91(1999).

[24] K. Sagoe-Crenstil, and L. Weng, Dissolution processes, hydrolysis and condensation reactions during geopolymer synthesis: Part II – High Si/Al ratio systems *J. Mater. Sci.* **42**, #9, 3007-14 (2007).

ALKALINE ACTIVATION OF VOLCANIC ASHES: A PRELIMINARY STUDY

B. Varela[1], A. Teixeira-Pinto [2], P. Tavares[2], T. Fernandez[1], A. Palomo[3]

[1] Rochester Institute of Technology, New York, USA
[2] Universidade de Trás-os-Montes e Alto Douro, Vila Real, Portugal
[3] Instituto de Ciencias de la Construcción Eduardo Torroja, Madrid, Spain

ABSTRACT

In recent years increasing work has been done towards the development of cementitious materials based on aluminosilicates, with the goal of an environmentally conscious alternative to ordinary Portland cement. Aluminosilicates such as metakaolin, fly-ashes or blast furnace slags are currently being used as raw materials for alkaline activation, providing products of high mechanical strength and enhanced durability. Until now no research has been pursued using volcanic ashes as an aluminosilicate source for alkali activation. Volcanic ashes are abundant in certain parts of the world, as a consequence of past and present volcanic activity. These materials seem to offer an interesting potential for alkaline activation when one considers not only their unique chemical composition and fineness but also thermal history. In this work ash samples from North Dakota and the Cape Verde Islands are utilized for alkaline activation. The resultant material is characterized, studied and tested under different conditions.

Keywords: - alkaline activation, volcanic ash, geopolymerization

1. INTRODUCTION

Environmental issues put some urgency in changing several aspects of current industrial practice. Up to recent years, the exploitation of natural resources has continued with practices based on the idea of an inexhaustible supply and there has been little concern for the impact that these practices could impart to the well being of present and future generations. The cement industry faces serious criticism, partially because of their necessary involvement with quarrying which permanently disturbs the surrounding area and also for the hazardous emission of CO_2 to the atmosphere [1]. It is estimated that every year, the Portland cement industry is responsible for more than 1.5 billion tons of CO_2. Therefore, research is being carried out with the goal of the development of alternative binding systems that could overcome these disadvantages. Alkaline activation is a rediscovered technique which promotes chemical reactions among the main oxides in an aluminosilicate (SiO_2, Al_2O_3, Na_2O and K_2O) under strong alkaline conditions, giving place to final products that present high mechanical strength and chemical stability [2].

Aluminosilicates are decidedly abundant, since they represent more than 75% of the solid inorganic compounds of the terrestrial crust. The level of calcium content of aluminosilicates is generally very low, therefore, they do not release CO_2 during the heating and curing processes. The calcination temperature is generally much lower than the one necessary for producing clinker, thus a substantial saving in energy is achieved. Moreover, the thermal treatments provoke a release of constituent water so that the final products are generally amorphous. This particular state of disorder is vulnerable to chemical combination with other species.

The use of aluminosilicates like coal-ash or slag generally presents a major advantage: - they are produced as waste in other industrial processes and are considered residues, not involving any spending of energy or any special treatment before the alkaline activation. By using industrial by-products, the industry and the environment gains are two fold, a reduction of the need for disposal of industrial waste, and the needed for quarrying of new raw materials [3]. In what concerns to the use of volcanic ashes (most of them aged for thousands of years) one could think that this material is ready

185

to use. Yet, after that long period of time it is expected that a great part of the volcanic ashes are partially re-hydrated. If this is the case, a thermal treatment could be necessary.

In this work the activation of two different volcanic ashes, from the USA and the Islands of Cape Verde is studied. First the chemical composition was determined by atomic absorption. Both materials were characterized by means of XRD (X'Pert Pro, detector X'Celerator CuKα) and Infra-Red Spectroscopy (FTIR Unicam Research Series). Samples were tested for mechanical trength. Results of alkaline activation are discussed.

2. EXPERIMENTAL

2.1 Materials

This work studied two sets of volcanic ashes: - one from Linton, North Dakota, United Sates of America (USA), ranging in age from 7 to 20 million years believed to have originated from volcanoes in south-central and western Montana, as well as northern Wyoming. That tuff is thought to be the result of single or related air fall events that likely deposited a thin layer of volcanic dust across a wide area. In turn this volcanic dust was transported by wind and rain into streams and redeposited in local basins as thick tuffs. Because of this process, an inch or less of volcanic dust may result in 20 feet or more of tuff. The second material under study came from the Cape Verde island of Santo Antão (CV), the western most island of the archipelago. This yellowish volcanic ash, locally named "pozzolan of Cape Verde", is being sold to cement factories abroad. The chemical compositions of USA and CV ashes are presented in Table 1.

Table 1 – Chemical composition of USA and CV ashes in wt%

Oxides	SiO_2	Al_2O_3	Fe_2O_3	CaO	MgO	Na_2O	K_2O	TiO_2	MnO	P_2O_2	LoI
USA	68.89	15.44	2.86	2.05	1.34	2.55	3.13	0.31	0.06	0.12	1.51
CV	54.32	21.65	2.55	1.39	1.08	8.68	6.08	0.21	0.22	0.20	3.42

The Cape Verde ashes are very alkaline, considering the mutual action of sodium and potassium oxides. Therefore, attention has to be given during the activation to guard against the possibility of excess of Na^+ and K^+ ions on the paste to avoid the occurrence of efflorescence. Most of the compositions included granitic sand from a local quarry, in a two to one proportion with the ashes. The alkaline activator was prepared by a two to one mixture of sodium silicate (Soda Solvay D-40, analytical grade) and sodium hydroxide solution of 20 molal concentration.

2.2 Methods and Techniques

Initially, both materials were ground in a Retsch Ball Mill and sieved through a 125μm mesh in order to make them homogeneous. The North Dakota Ashes and Pozzolans of Cape Verde were activated under several conditions: - (i) direct (ii) direct with chemical correction (iii) calcination and direct activation (iv) calcination and chemical correction. The correction of alumina and sodium content was made with sludge from a water treatment plant (rich in aluminium hydroxide) and sodium carbonate. The calcination of both ashes, with the aim of preparing the species for activation, was obtained in an oven at 800°C during 2 hours.

The overall conditions of mixing were the same for all experiments: - the paste obtained in every experiment, after being mixed and homogenized for 5 minutes in a lab mixer (Controls 65-L0006/A) was poured into prismatic plastic moulds (4 cm x 4 cm x 16 cm) and compacted on a vibrating table. Three samples for each experiment were considered. Different ratios of binder/aggregate, liquid/solid components and curing conditions were evaluated. After curing, the samples were tested in flexure (three point bending) and compression in a Seidner 15-250 machine until failure. Only the samples that gave better mechanical results are reported.

3. RESULT
3.1 Characterization

XRD spectra of the materials (raw and calcined) is shown on Figures 1 and 2. The spectra shows a common visible sign of amorphism (between 10 and 38° 2θ), despite several crystalline phases which were mainly albite (A), quartz (mullite and cristobalite) (Q), dicktite (D) in USA ashes, and phillipsite (P) and chabbazite (C), two well known zeolites in CV ashes. The USA ashes present a smaller amount of the crystalline phases and a well defined amorphism hallo, even in the raw material. The Cape Verde ashes because of the re-hydration process over a long time, present the previous zeolite forms. After calcination these forms were observed to almost fully disappear because of the evaporation of constituent water.

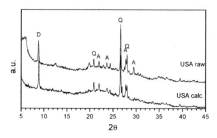

Fig. 1 – XRD spectrum of USA ashes

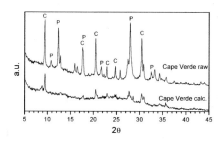

Fig. 2 - XRD spectrum of CV ashes

Figures 3 and 4 present the FTIR spectra for both ashes before calcination. These spectra confirm the presence of the OH groups at the band of 3500 cm^{-1} due to the water of re-hydration. The peak near the 1000 cm^{-1} corresponds to asymmetrical stretching of Si-O-Si and Al-O-Si bonds, while the peaks in the lower wavenumbers correspond to symmetrical stretching of the same bonds [4]. In both cases, the spectrum is similar to the one found in aluminosilicates, as can be expected from their chemical composition and nature of the material. However, a more detailed study particularly in the far infrared region is required for further identification of these bands.

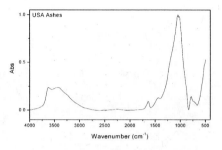

Fig. 3 - FTIR spectrum of USA ashes

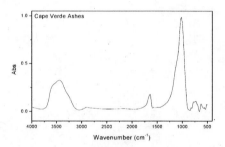

Fig. 4 - FTIR spectrum of CV ashes

3.2 Mechanical Results

The samples prepared with raw ashes (USA and CV) did not harden under any curing condition, consequently, a calcination process was necessary to adequate the species for activation. Two sets of compositions were studied: - (i) direct formulations without correction of chemical composition and (ii) formulations where silica, alumina and sodium corrections were introduced. Sodium was corrected using sodium carbonate (SC), but silica, and alumina contents were corrected via the use of glass powder (GP) and calcined sludge recovered from a water treatment plant (SWT) respectively. Table 2 presents the results and compositions of the first set of experiments. The mechanical results are the average of three samples. These samples were cured at 65°C for 48 hours with a 98% relative humidity.

When tested in flexure the samples showed a partially wet fracture. For this reason, all samples were dried at room temperature for 24 hours before testing. The second set of tests included the addition of calcined sludge, sodium carbonate and glass powder to modify the SiO_2 / Al_2O_3, Na_2O/SiO_2 and Na_2O/Al_2O_3 theoretical ratios. Tables 3 and 4 present the composition and the mechanical results obtained after these corrections.

Table 2 – Direct activation of calcined ashes. Composition and mechanical results

	Ashes (g)	Sand (g)	Activator (g)	Flexure (MPa)	Compression (MPa)
USA	100	200	85	2.42	3.81
CV	100	200	75	1.10	2.98

Table 3 – Final theoretical ratios in wt% after addition of correctors

	SiO_2/Al_2O_3	Na_2O/SiO_2	Na_2O/Al_2O_3
USA	5.98	0.32	1.93
CV	5.86	0.32	1.84

Table 4 – Activated calcined ashes with chemical corrections. Composition and mechanical results

	Ashes (g)	SWT (g)	GP (g)	SC (g)	Sand (g)	Activator (g)	Flexure (MPa)	Compression (MPa)
USA	140	30	-	30	400	170	5.06	8.96
CV	120	18	36	18	380	140	3.21	8.20

4. CONCLUSIONS

Both types of ashes, from the plains of North Dakota and from the Island of Santo Antão in Cape Verde are of the volcanic origin and have the potential for alkaline activation. In both cases, a heat treatment to remove the water of re-hydration was necessary to improve the mechanical strength. Strength was further improved by the addition of chemical corrector such as sodium carbonate, the sludge from a water treatment plant, which is rich in Alumina and recycled glass powder. Mechanical tests showed that compressive strength for both samples is very similar, however, the flexural strength in the USA ashes is remarkable. This strength represents 56% of the compressive strength. It seems, from our observations, that curing is a very important process in this type of materials. After 48 hours under 65°C the samples were not dried completely, therefore, it was necessary to let them dry at room temperature for a day before testing. Preliminary observations show that the mechanical strength can be improved by curing for 72 hours at 95 °C, but this requires further study.

This preliminary study shows that it is possible to make a cementitious material by the alkaline activation of volcanic ashes. However, further studies and larger samples are necessary to clarify some aspects connected with curing conditions that have emerged during this research, as well as to further characterize the mechanical properties and failure modes. One last word of caution for future researches with these materials. Some volcanic ashes might contain Erionite, a mineral who might provoke cancer in humans. For this reason, an analysis to detect the presence of this mineral is necessary before attempting any work with volcanic ashes.

5. ACKNOWLEDGEMENTS

The authors would like to thank Mr. Edward C. Murphy, State Geologist from the North Dakota Geological Survey for providing the samples from Linton, North Dakota and Mr. Herculano Lizardo, Landscape Architect from the Republic of Cape Verde for providing the samples from Santo Antão for this research.

6. REFERENCES
[1] Hardjito, D., Wallah, S.E., Sumajouw, D.M.J, and Rangan, B.V. Geopolymer concrete: Turn waste into environmentally friendly concrete, INCONTEST 2003.
[2] Palomo, A., Macías, A., Blanco, M.T. and Puertas, F. Physical, chemical and mechanical characterization of geopolymers. 9th International Congress on the Chemistry of Cement, pp. 505-511.
[3] Xu, H. and Van Deventer, J.S.J. The geopolymerization of alumino-silicate minerals. Int. J. Miner. Process. 59(2000) 247-256.
[4] The Infrared Spectra of Minerals. Farmer, V.C., ed. London: Mineralogical Society 1974.

Multifunctional Ceramics

THE EFFECT OF DOPING WITH TITANIA AND CALCIUM TITANATE ON THE MICROSTRUCTURE AND ELECTRICAL PROPERTIES OF THE GIANT DIELECTRIC CONSTANT CERAMIC CaCu$_3$Ti$_4$O$_{12}$

Barry A. Bender, Ed Gorzkowski, and Ming-Jen Pan
Naval Research Lab
Code 6351
Washington, DC 20375

ABSTRACT

Small amounts (1-5 mole%) of TiO$_2$ and CaTiO$_3$ were added to the giant dielectric constant ceramic CaCu$_3$Ti$_4$O$_{12}$ (CCTO) in the attempt to lower dielectric loss without sacrificing high permittivity. The undoped and doped ceramics had similar microstructures consisting of primarily large grains in the range of 35 to 40 microns. Doping CCTO with TiO$_2$ lead to an increase in the dissipation factor of CaCu$_3$Ti$_4$O$_{12}$ from 0.049 to a high of 0.078, while its permittivity increased from 43949 to 77585. Doping with CaTiO$_3$ followed a similar trend as the tan δ increased to a high of 0.303 and the dielectric constant at 1 kHz increased to a high of 75687. Doping at these levels also led to a 50% drop in electrical breakdown voltage.

INTRODUCTION

In its development of the all-electric ship the US Navy has made significant outlays in the technology of power electronics. Passive components, especially filter capacitors, remain a limiting factor in power converter design. This is due to their low volumetric efficiency which causes them to be responsible for occupying 50 to 60% of the volume associated with today's state-of-the-art power converter. The ideal ceramic filter capacitor would consist of a high dielectric ceramic with good stability over a range of temperatures and frequencies. Commercial dielectric oxides such as BaTiO$_3$ typically sacrifice high permittivity for temperature stability. Recently, a new dielectric oxide, CaCu$_3$Ti$_4$O$_{12}$, has been uncovered with the potential to have high permittivity (single crystal dielectric constant is 80,000) that is stable over a wide range of temperatures and frequencies.[1,2] Also the material can be engineered into an internal barrier layer capacitive-like (IBLC) dielectric via one-step processing in air[3] at modest sintering temperatures of 1050 to 1100°C and it is environmentally-friendly since CCTO is a lead-free dielectric.

However, the dielectric loss properties of CaCu$_3$Ti$_4$O$_{12}$ have to be improved if this material is going to be used commercially. Dissipation factors as low as 0.05 to 0.06 (20°C, 10 kHz) have been reported for undoped CaCu$_3$Ti$_4$O$_{12}$ but are very sensitive to temperature.[4,5] At temperatures as low as 40°C dielectric loss values begin to climb leading to losses that exceed 0.10 before 60°C is reached.[4,6-8] To improve CCTO's dielectric loss properties the nature of the giant permittivity of CaCu$_3$Ti$_4$O$_{12}$ has to be fully comprehended. The consensus of most researchers is that the high permittivity of CaCu$_3$Ti$_4$O$_{12}$ is extrinsic in nature and is the result of the formation of insulating layers around semiconducting grains. This creates an electrically inhomogeneous material that is similar to internal barrier layer capacitors (IBLCs).[3] However, the exact nature of the insulating boundaries and semiconducting grains is still under scientific debate. Electrical measurements show that the insulating boundaries are electrostatic potential barriers that can be best described using a double Schottky barrier (DSB) model.[9,10] The electrical properties of electroceramics that contain these type of DSB barriers can be very sensitive to the presence of dopants and oxygen.[11-13] Electrical measurements show that the grains are n-semiconductors,[14] but why the grains are semiconducting is unclear. Li et al.[15] believe that cation nonstoichiometry occurs during processing resulting in the replacement of Ti ions on Cu ion sites. They believe that small increases in Ti ion concentration as low as 0.0001 can account for the measured semiconductivity in CCTO grains. Researchers have added TiO_2 and $CaTiO_3$ in amounts of 10% or higher and have shown that it leads to dramatic reductions in dielectric loss.[16-18] However, it also results in substantial 75% decreases in permittivity too. At the moment, no research has been reported on the addition of small additions of TiO_2 or $CaTiO_3$ which according to Li's model[15] may have significant impact on the dielectric and electrical properties of CCTO. This paper reports on the attempt to decrease the loss of CCTO ceramics by doping them with small amounts of TiO_2 and $CaTiO_3$. The effects of doping on the resultant microstructure, dielectric properties, and electrical breakdown of $CaCu_3Ti_4O_{12}$ are reported.

EXPERIMENTAL PROCEDURE

$CaCu_3Ti_4O_{12}$ was prepared using ceramic solid state reaction processing techniques. Stoichiometric amounts of $CaCO_3$ (99.98%), CuO (99.5%) and TiO_2 (99.5%) were mixed by blending the precursor powders into a purified water solution containing a dispersant (Tamol 901) and a surfactant (Triton CF-10). The resultant slurries were then attrition-milled for 1 h and dried at 90°C. The standard processed powder, STD, was calcined at 900°C for 4 h and then 945°C for 4 h. After the final calcination the STD powders were attrition-milled for 1 h to produce finer powders. The titania-doped powders, T, were fabricated by mixing various amounts of titania with the calcined STD powder (0.95 mole% (95T), 1.9 mole% (190T) and 2.8 mole% (280T)). The $CaTiO_3$-doped powders, C, were made by mixing various amounts of $CaTiO_3$ with the calcined STD powder (0.95 mole% (95C), 1.35 mole% (135C), 2.7 mole% (270C), and 5.4 mole% (540C)). A 2% PVA binder solution was mixed with the powders and they were sieved to eliminate large agglomerates. The dried powder was uniaxially pressed into discs typically 13 mm in diameter and 1 mm in thickness. The discs were then placed on platinum foil and sintered in air for three hours at 1100°C.

Material characterization was done on the discs and powders after each processing step. XRD was used to monitor phase evolution for the various mixed powders and resultant discs. Microstructural characterization was done on the fracture surfaces using scanning electron microscopy (SEM). To measure the dielectric properties, sintered pellets were ground and polished to achieve flat and parallel surfaces onto which palladium-gold electrodes were sputtered. The capacitance and dielectric loss of each sample were measured as a function of temperature (-50 to 100 °C) and frequency (100 Hz to 100 KHz) using an integrated, computer-controlled system in combination with a Hewlett-Packard 4284A LCR meter. Electrical breakdown was measured on samples typically 1 mm in thickness with gold electrodes at an applied rate of voltage of 500 volts per second.

RESULTS and DISCUSION

Effect of Ti-Doping on the Microstructure and Dielectric Properties of $CaCu_3Ti_4O_{12}$
Ti-doping had a mixed effect on the microstructure of CCTO. The microstructure of undoped CCTO was bimodal consisting of pockets of small grains distributed randomly throughout a matrix of coarser grains (see Fig. 1a). The average grain size of the large grain was 37 microns (see Table I) while the smaller grains were typically 2 to 4 microns in size and occupied less than 5 vol%. Fractographs indicated that the fracture was basically transgranular and showed evidence of a possible very thin submicron grain boundary phase (Fig. 1b). The only overt evidence of a second phase observed to be present was a Cu-rich phase whose presence was always detected in the mix of the smaller grains (see Fig. 1c). Ti-doping did not lead to the presence of any other overt second phases either as detected by XRD or SEM. However, higher resolution techniques like TEM need to be used to clarify the nature of the grain boundaries and the presence and nature of second phases. Ti-doping did lead to mixed results in microstructure and transgranular fracture (see Fig. 1). The average grain size of 190T and 280T had similar values (Table I) but at the 99% confidence level the 95T CCTO had

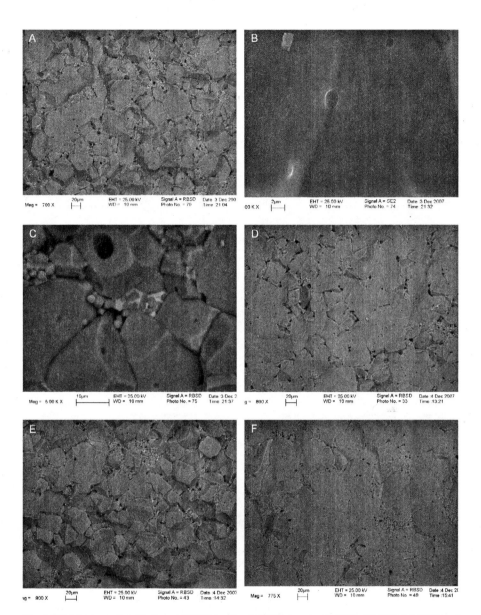

Fig. 1 SEM fractographs of undoped CCTO (a-c) and CCTO doped with TiO$_2$- (d) 95T, (e) 190T, and (f) 280T. Fig. (a) and (c-f) are back-scattered SEM micrographs.

a grain size that was 15% smaller. It is unclear how doping with titania at this level could reduce the grain size but not at the higher dopants levels. The only other major difference between all the samples was that the 280T sample fractured in a very flat, highly transgranular mode which is often indicative of a stronger grain boundary (see Fig. 1f).

Table I. Effect of Doping on Grain Size, Dielectric, and Electrical Properties of CaCu$_3$Ti$_4$O$_{12}$

Sample	Grain Size (μm)	Permittivity (1 kHz)	Dissipation Factor	Breakdown Voltage (V/mm)
STD	37 (9)	43939	0.049	1100
95T	31 (10)	49100	0.060	750
190T	39 (11)	48249	0.049	740
280T	40 (10)	77585	0.078	520
95C	32 (8)	65000	0.193	415
135C	35 (9)	70683	0.303	400
270C	40 (10)	75687	0.231	450
540C	38 (9)	67420	0.287	950

(xx) is the standard deviation of the measured grain size

However, doping CCTO with titania did have an effect on the dielectric properties. Though the trends were mixed, adding titania to CCTO increased its dielectric constant (20°C- 1 kHz) from 43939 to 77585 for the 280T sample (see Table I and Fig. 2). It also led to an increase in the dissipation factor as it increased from 0.049 for the undoped sample to 0.078 for the 280T sample. The 190T sample was an anomaly in regards to dielectric properties as the permittivity and dielectric loss both decreased in value as compared to the 95T. However, this result provided clues to what the effect of doping with small amounts of titania has on the CaCu$_3$Ti$_4$O$_{12}$ system. The dielectric constant in an IBLC or material with a Schottky barrier layer can be represented as[10]

$$\varepsilon' = \varepsilon_B\,(d/t_B) \qquad (1)$$

where ε_B is the permittivity of the boundary or barrier layer, d can be approximated by the grain size, and t_B is the thickness of the boundary or barrier layer. Since the dielectric loss is very similar for the STD and 190T sample, at both 20 and 80°C, they should have similar values of ε_B. Grain sizes are similar for the two samples so that means that the effective barrier or boundary thickness for the Ti-doped system must have decreased. This would also explain the increase in permittivity for the other two samples (95T and 280T). The dielectric data also provides other clues to the effect of Ti-doping. In an IBLC the approximation for the loss tangent is[19]

$$\tan \delta = 1/(\omega R_{gb}C) + \omega R_g C \qquad (2)$$

where ω is the angular frequency, R_{gb} is the resistance of the boundary layer, R_g is the resistance of the semiconducting grains, and C is the capacitance. This means that the low frequency loss is dominated by R_{gb} while higher frequency loss is dominated by R_g. Looking at Fig. 2 it can be determined for 95T and 280T that the change in loss from the STD ceramic at 100 Hz is 3 to 4 fold while at 100 kHz the change in loss is only 1.1 to 1.5. This implies that Ti-doping effects R_{gb} causing the grain boundaries to be more conductive resulting in higher dielectric losses. How the presence of extra Ti ions affects the nature of the boundary or barrier layer is unknown. It is well known that Ti segregates to the grain boundaries in BaTiO$_3$ and SrTiO$_3$ ceramics.[20] Also Ti has been measured by STEM analysis to segregate to the grain boundaries in CCTO ceramics.[21] Electrical measurements by Zang et al.[10] have shown that Schottky barriers exist in CCTO. They and Marques et al.[22] postulate that the barrier layer consists of adsorbed oxygen atoms. Since Ti is a known oxygen getter it is possible that the presence of extra Ti segregating to the boundary changes the chemistry of the barrier layer and leads to less adsorbed oxygen. This would lead to a thinner boundary or barrier layer and a lower barrier height. From I-V measurements on $SrTiO_3$ and impedance spectroscopy (IS) on Ti-doped CCTO, barrier heights and R_{gb} has been directly linked with changes in permittivity and dielectric loss so that lower barrier heights lead to a decrease in R_{gb} which leads to higher permittivity and higher loss which is what we observe in the Ti-doped CCTO samples.[23,16]

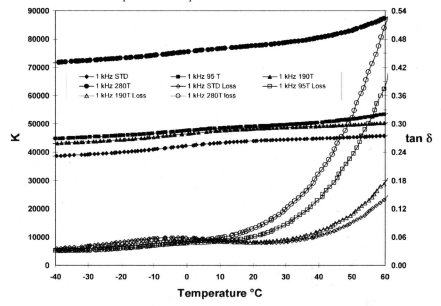

Fig. 2 Temperature dependence at 1 kHz of the dielectric constant and dissipation factor of undoped CCTO (STD) and TiO_2-doped CCTO (95T, 190T, and 280T).

Effect of $CaTiO_3$-Doping on the Microstructure and Dielectric Properties of $CaCu_3Ti_4O_{12}$

$CaTiO_3$-doping led to results that were very similar in nature to the effects that titania-doping had on the microstructure of CCTO. Again the sample with the least amount of $CaTiO_3$, 95C, showed a 15% reduction in grain size (see Table I) while the other $CaTiO_3$-doped samples had grain sizes similar to that of the STD sample (see Fig. 3). Also the $CaTiO_3$-doped samples tended to fracture in a more transgranular fracture as compared to the undoped sample with the 270C sample showing the flattest fracture (Fig 3- c vs. d). XRD and SEM picked up no evidence of any second phases but their presence can not be ruled out until higher resolution TEM analysis is done on these ceramics too.

Doping with $CaTiO_3$ led to mixed results in its effect on the dielectric properties of $CaCu_3Ti_4O_{12}$. All the doped samples had significantly higher dielectric constants (see Table I and Fig. 4) as the permittivity increased by 75% from 43939 to 75687 when doped with 2.7 mole% $CaTiO_3$. All the samples showed significant increases in dielectric loss as the tan δ increased 6 fold going from 0.049 to 0.303 when doped with 1.35 mole% $CaTiO_3$. Again there were anomalies present in the data. The 270C data showed a significant drop in dissipation factor (see Table I) which was counter to the trend of the rest of the data, while the 540C showed a drop in dielectric constant which was counter to the trend of increasing permittivity with increasing doping amounts of CaTiO$_3$. These results indicate that there may be possible different mechanisms or effects that small amounts of CaTiO$_3$ have on the defect chemistry of the semiconducting grains and insulating boundaries in CCTO dielectrics.

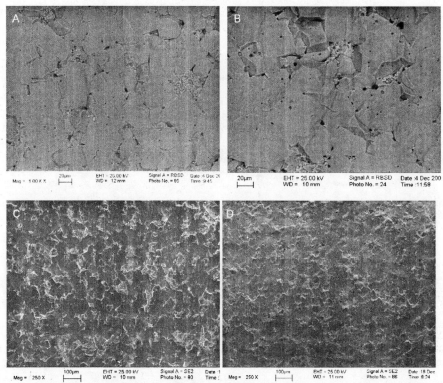

Fig. 3 SEM fractographs of undoped CCTO doped with CaTiO$_3$- (a) 95C, (b) 135C, (c) 270C, (d) 540C. Fig. (a) and (b) are back-scattered SEM micrographs.

In CCTO ceramics doped with large amounts of CaTiO$_3$ (x> 0.1, Ca$_x$Cu$_3$Ti$_{4+x}$O$_y$), the addition of CaTiO$_3$ led to 75% drops in permittivity and an order of magnitude drop in tan δ.[17,18] The researchers attributed the drop in dielectric constant due to the formation of CaTiO$_3$ in CCTO. Since CaTiO$_3$ is a low loss (tan $\delta = 0.001$) dielectric of 180 that is used to lower dielectric loss and the temperature dependence of loss in titanates it is not unexpected a composite containing CaTiO$_3$ would have lower permittivity and dielectric loss.[24,25] However, as shown from the data in Table I this is not the case for when small amounts of CaTiO$_3$ were added to CCTO, which resulted in the opposite general trend of higher permittivity and higher losses as compared to the undoped material. Similar to Ti-doped CCTO, detailed examination of the dielectric spectra (Fig. 4) reveals that the changes in loss as compared to the undoped sample versus the doped sample are greater at 100 Hz then at 100k Hz (i.e. for 135C- 33 fold (100 Hz) vs. 2.5 fold at (100 kHz)). This shows that the addition of small amounts of $CaTiO_3$ is lowering R_{gb}. This is in contrast to Yan *et al.*[17] who showed using IS that large $CaTiO_3$ additions increased the resistance of the boundary layers. It is postulated that with low amounts of added $CaTiO_3$, that $CaTiO_3$ is segregating to the boundary layer. This is quite feasible as Ti, Ca, and $CaTiO_3$ have been reported to segregate to the grain boundaries.[21,26,27] There it can react with the boundary layer changing its chemistry and defect nature to make the effective boundary layer thinner and less insulative which leads to a higher dielectric constant and higher losses as seen for 95C and 135C. In the case of 540C it is possible that the segregation has reached its saturation limit and $CaTiO_3$ is now forming as a second phase. Since the dielectric constant of $CaTiO_3$ is much lower than CCTO (43939 vs. 180)[24] and the permittivity is much lower too (0.0490 vs. 0.001)[24] it is expected by the rules of mixture that the resultant doped-CCTO would start to see a decrease in its permittivity and tan δ as compared to 135C.

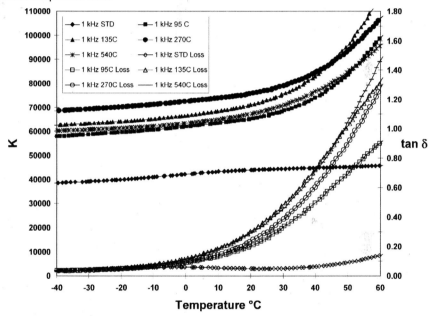

Fig. 4 Temperature dependence at 1 kHz of the dielectric constant and dissipation factor of undoped CCTO (STD) and $CaTiO_3$-doped CCTO (95C, 135C, 270C, and 540C).

Effect of Doping on the Electrical Breakdown Voltage of $CaCu_3Ti_4O_{12}$

High breakdown voltage (E_b) is a desirable characteristic for a capacitive material. Commercial ferroelectric $BaTiO_3$-based capacitor materials have high breakdown voltages in the range of 30 kV/cm.[28] However, varistors and IBLCs have a much lower E_b due to the presence of Schottky barriers. Typical values for varistors range from 400 for TiO_2-based varistors to 1300 V/cm for ZnO and SnO_2-based varistors.[29-31] For a $SrTiO_3$-based IBLC E_b has been measured to range from 200 to 400 V/cm.[32] For CCTO materials there has been no thorough study on its breakdown voltage but there is scattered data available in the literature. Breakdown voltages for undoped CCTO have been reported to be 400, 600, 1300, and 1300 v/cm.[33-35,14] However, E_b is sensitive to processing as the breakdown voltage drops from 1300 to 300 V/cm due to a 10-fold increase in grain size and E_b increases from 400 to 570 V/cm when heat-treated in flowing oxygen.[14,33] Table I data indicates that the breakdown voltage for CCTO is sensitive to doping. In the samples doped with TiO_2 E_b drops from 1100 V/cm for the undoped dielectric to as low as 520 V/cm for the 280T sample. As discussed earlier it was postulated that Ti doping leads to effectively a thinner boundary layer which leads to a lower boundary energy and a less insulative boundary layer. This should result in a lower E_b because Chung et al.[36] have done IS research on Sc-doped CCTO and have shown a direct correlation between changes in R_{gb} and E_b.

Doping with $CaTiO_3$ also lowered the breakdown voltage for CCTO. However, the data trend was mixed in a similar fashion as the dielectric data was (see Table I). The E_b dropped from 1100 for the STD sample to a low of 400 for the 135C sample and started to increase with further doping to a value of 950 for 540C. As postulated for the dielectric data it is believed that the doping CCTO at the lower amounts of $CaTiO_3$ led to a decrease in R_{gb} and as discussed above this would lead to lower electrical breakdown voltages as observed for 95C and 135C. However, when doped with 5.4 mole% $CaTiO_3$ it was believed that $CaTiO_3$ started to develop as a second phase at the grain boundaries. This could lead to the observed increase in E_b for 540T because a 1 part $CaCu_3Ti_4O_{12}$:2 part $CaTiO_3$ composite saw a dramatic increase in E_b to 6500 V/cm due to the large amounts of $CaTiO_3$ present in the composite.[26]

CONCLUSIONS

Unlike the results of reduced loss and permittivity reported for doping CCTO with large amounts of TiO_2 and $CaTiO_3$, addition of small amounts of these dopants led to higher dissipation factors and higher dielectric constants. Instead of 10-fold reduction of tan δ the dissipation factors were increased from about 0.05 to a high of 0.08 for the Ti-doped material and 0.3 for the $CaTiO_3$-doped material. This increase in tan δ was accompanied by a 75% increase in permittivity as the dielectric constant increased from about 44000 to 77000 for both doped-materials. Analysis of the dielectric spectra indicated that doping led to a decrease in the resistivity of the insulative boundary layers. It also indicated that the effective thickness of the layers was narrower. These results can also account for the 50% drop in electrical voltage breakdown from doping CCTO dielectrics with TiO_2 or $CaTiO_3$. Impedance spectroscopy and TEM analysis are needed to shed further insight on the complex role of these dopants at low levels. With this information the nature of the boundary layers should be clearer which will allow for material engineering CCTO in order to obtain low-loss CCTO dielectrics without sacrificing the giant permittivity of $CaCu_3Ti_4O_{12}$.

REFERENCES

[1] C. C. Homes, T. Vogt, S. M. Shapiro, S. Wakitomo, A. P. Ramirez, "Optical Response of High-Dielectric-Constant Perovskite-Related Oxide," Science, 293, 673-76 (2001).

[2] M. A. Subramanian, D. Li, N. Duan, B. A. Reisner, and A. W. Sleight, "High Dielectric Constant in $ACu_3Ti_4O_{12}$ and $ACu_3Fe_3O_{12}$ Phases," J. of Solid State Chem., 151, 323-25 (2000).

[3] T. B. Adams, D. C. Sinclair, and A. R. West, "Giant Barrier Layer Capacitance Effects in $CaCu_3Ti_4O_{12}$ Ceramics," Adv. Mater., 14, 1321-23 (2002).

[4] B.A. Bender and M.-J. Pan, "The Effect of Processing on the Giant Dielectric Properties of $CaCu_3Ti_4O_{12}$," Mat. Sci. Eng. B., 117, 339-47 (2005).

[5] E.A. Patterson, S. Kwon, C.-C. Huagn, and D.P. Cann, "Effects of ZrO_2 Additions on the Dielectric Properties of $CaCu_3Ti_4O_{12}$," Appl. Phys. Lett., 87, 182911-1-3 (2005).

[6]T.-T. Fang, L.-T. Mei, H.-F. Ho, "Effects of Cu Stoichiometry on the Microstructures, Barrier-Layer Structures, Electrical Conduction, Dielectric Responses, and Stability of CaCu$_3$Ti$_4$O$_{12}$," *Acta Mat.*, **54**, 2867-75 (2006).

[7]R.K. Grubbs, E.L. Venturini, P.G. Clem, J.J. Richardson, B.A. Tuttle, and G.A. Samara, "Dielectric and Magnetic Properties of Fe- and Nb-doped CaCu$_3$Ti$_4$O$_{12}$," *Phys. Rev. B*, **72**, 104111-1-11 (2005).

[8]T.-T. Fang and H.K. Shiau, "Mechanism for Developing the Boundary Barrier Layers of CaCu$_3$Ti$_4$O$_{12}$," *J. Am. Ceram. Soc.*, **87**, 2072-79 (2004).

[9]T.B.Adams, D.C. Sinclair, A.R.West, "Characterization of Grain Boundary Impedances in Fine- and Coarse-Grained CaCu$_3$Ti$_4$O$_{12}$ Ceramics", *Phys. Rev. B*, **73**, 094124-1-9 (2006).

[10]G. Zang, J. Zhang, P. Zheng, J. Wang, and C. Wang, "Grain Boundary Effect on the Dielectric Properties of CaCu$_3$Ti$_4$O$_{12}$ Ceramics," *J. Phys. D. Appl. Phys.*, **38**, 1824-27 (2005).

[11]D.R. Clarke, "Varistor Ceramics," *J. Am. Ceram. Soc.*, **82**, 485-502 (1999).

[12]R.C. Buchanan, *Ceramic Materials for Electronics*, Marcel Dekker, New York, pp. 377-431 (2004).

[13]G.V. Lewis, C.R. Catlow, and R.E. Casselton, "PTCR Effect in BaTiO$_3$," *J. Am. Ceram. Soc.*, **68**, 555-58 (1985).

[14]S.Y. Chung, I.-D. Kim, and S.J.L. Kang, "Strong Nonlinear Current-Voltage Behaviour in Perovskite-Derivative Calcium Copper Titanate," *Nature Materials*, **3**, 774-78 (2004).

[15]J. Li, M.A. Subramanian, H.D. Rosenfeld, C.Y. Jones, B.H. Toby, and A.W. Sleight, "Clues to the Giant Dielectric Constant of CaCu$_3$Ti$_4$O$_{12}$ in the Defect Structure of SrCu$_3$Ti$_4$O$_{12}$," *Chem. Mater.*, **16**, 5223-25 (2004).

[16]Y.-H. Lin, J. Cai, M. Li, C.-W. Nan, and J. He, "High Dielectric and Nonlinear Electrical Behaviors in TiO$_2$-Rich CaCu$_3$Ti$_4$O$_{12}$ Ceramics," *Appl. Phys. Lett.*, **88**, 172902-1-3 (2006).

[17]Y. Yan, L. Jin, L. Feng, and G. Cao, "Decrease of Dielectric Loss in Giant Dielectric Constant CaCu$_3$Ti$_4$O$_{12}$ Ceramics by Adding CaTiO$_3$," *Mat. Sci. Eng. B*, **130**, 146-50 (2006).

[18]W. Kobayashi and I. Terasaki, "CaCu$_3$Ti$_4$O$_{12}$/CaTiO$_3$ Composite Dielectrics: Ba/Pb-free Dielectric Ceramics with High Dielectric Constants," *Appl. Phys. Lett.*, **87**, 032902-1-3 (2005).

[19]Y. Yan, L. Jin, L. Feng, and G. Cao, "Decrease of Dielectric Loss in Giant Dielectric Constant CaCu$_3$Ti$_4$O$_{12}$ Ceramics by Adding CaTiO$_3$," *Mat. Sci. Eng. B*, **130**, 146-50 (2006).

[20]Y.-M. Chiang and T. Takagi, "Grain-Boundary Chemistry of Barium Titanate & Strontium Titanate: I, High-Temperature Equilibrium Space Charge," *J. Am. Ceram. Soc.*, **73** [1], 3278-85 (1990).

[21]C. Wang, H. J. Zhang, P.M. He, and G.H. CaO, "Ti-rich and Cu-Poor Grain-Boundary Layers of CaCu$_3$Ti$_4$O$_{12}$ Detected by X-ray Photoelectron Spectroscopy," *Appl Phys. Lett.*, **91**, 052910 (2007).

[22]V.P.B Marques, A. Ries, A.Z. Simoes, M.A. Ramirez, J.A. Varela, and E. Longo, "Evolution of CaCu$_3$Ti$_4$O$_{12}$ Varistor Properties During Heat Treatment in Vacuum," *Ceram. Intl.*, **33**, 1187-90 (2007).

[23]P. Prabhumirashi, V. P. Dravid, A.R. Lupini, M.F. Chisholm, and S.J. Pennycook, "Atomic-Scale Manipulation of Potential Barriers at SrTiO$_3$ Grain Boundaries, *Appl. Phys. Lett.*, **87**, 121917-1-3 (2005).

[24]W. Qin, W. Wu, J. Cheng, and Z. Meng, "Calcium-Doping for Enhanced Temperature Stability of (Ba$_{0.6}$Sr$_{0.4}$)$_{1-x}$Ca$_x$TiO$_3$ Thin Film Dielectric Properties," *Mat. Lett.*, **61**, 5161-63 (2007).

25R.K. Dwivdi, D. Kumar, and O. Parkash, "Valence Compensated Perovskite Oxide System Ca$_{1-x}$La$_x$Ti$_{1-x}$Cr$_x$O$_3$," *J. Mat. Sci.*, **36**, 3641-48 (2001).

[26]M.A. Ramirez, P.R. Bueno, J.A. Varela, and E. Longo, "Non-Ohmic and Dielectric Properties of a Ca$_2$Cu$_2$Ti$_4$O$_{12}$ Polycrystalline System," *Appl. Phys. Lett.*, **89**, 212202-1-3 (2006).

[27] S.F. Shao, J.L. Zhang, P. Zheng, C.L. Wang, J.C. Li, and M.L. Zhao, "High Permittivity and Low Dielectric Loss in Ceramics with the Nominal Compositions of CaCu$_{3-x}$La$_{2x/3}$Ti$_4$O$_{12}$," *Appl. Phys. Lett.*, **91**, 042905-1-3 (2007).

[28]B.-C. Shin, S.-C. Kim, C.-W. Nahm, and S.-J. Jang, "Nondestructive Testing of Ceramic Capacitors by Partial Discharge Method," *Mat. Lett.*, **50**, 82-86 (2001).

[29]A.B. Gaikwad, S.C. Navale, and V. Ravi, "TiO$_2$ Ceramic Varistor Modified with Tantalum and Barium," *Mat. Sci. Eng. B*, **123**, 50-52 (2005).

[30]C.-W. Nahm, "Microstructure and Electrical Properties of Tb-Doped Zinc Oxide-Based Ceramics," *J. Non-Cryst. Sol.*, **353**, 2954-57 (2007).

[31]S.R.D. Hage, V. Choube, and V. Ravi, "Nonlinear I-V Characteristics of Doped SnO2." *Mat. Sci. Eng. B*, **110**, 168-71 (2004).

[32]S.-M. Wang and S.-J. L. Kang, "Acceptor Segregation and Nonlinear Current-Voltage Characteristics in H2-Sintered SrTiO3, "*Appl. Phys. Lett.*, **89**, 041910-1-3 (2006).

[33]V.P.B Marques, P.R. Bueno, A.Z. Simoes, M.Cilense, J.A. Varela, E. Longo, and E.R. Leite, "Nature of Potential Barrier in CaCu$_3$Ti$_4$O$_{12}$ Polycrystalline Perovskite," *Sol. State Comm.*, **138**, 1-4 (2006).

[34]S.-Y. Chung, S.-Y. Choi, T. Yamamoto, Y. Ikuhara, and S.-J. L. Kang, "*Site-Selectivity of 3d Metal Cation Dopants and Dielectric Response in Calcium Copper Titanate*," *Appl. Phys. Lett.*, **88**. 091917-1-3 (2006).

[35]M. Guo, T. Wu, T. Liu, S.-X. Wang, and X.-Z. Zhao, "Characterization of CaCu$_3$Ti$_4$O$_{12}$ Varistor-Capacitor Ceramics by Impedance Spectroscopy," *J. Appl. Phys.*, **99**, 124113-1-3 (2006).

[36]S.-Y. Chung, S.-I. Lee, J.-H. Choi, S.-Y. Choi, "Initial Cation Stoichiometry and Current-Voltage Behavior in Sc-Doped Calcium Copper Titanate," *Appl. Phys. Lett.*, **89**, 191907-1-3 (2006).

DIFFUSE PHASE TRANSITION IN THE La and Ga DOPED BARIUM TITANATE

D.D. Gulwade and P. Gopalan
Department of Metallurgical Engineering and Materials Science
Indian Institute of Technology, Bombay, Powai, Mumbai 400 076, India

ABSTRACT

Various approaches have been adopted to modify properties of pure compound for potential options as lead free, high k memory material. In the present study, small amounts of co-doping La and Ga on the A and B site of $BaTiO_3$ respectively, resulting in a solid solution of the type $Ba_{1-3x}La_{2x}Ti_{1-3x}Ga_{4x}O_3$ have been investigated. The compounds have been prepared by conventional solid-state reaction. The X-ray Diffraction (XRD) of calcined powder shows the presence of tetragonal (P4/mmm) phase only. The XRD data has been analyzed using the FULLPROF Rietveld refinement package. The sintered pellets have been characterized by dielectric spectroscopy between room temperature and 200°C. The resulting compounds exhibit a remarkable decrease in Curie temperature as well as a significant enhancement in the dielectric constant.

The high temperature x-ray diffraction (HTXRD) of few compositions having higher diffuseness have been performed. The tetragonal phase persists over a wide range of temperature, and tetragonality decreases steadily with increase in temperature, which corroborates with the observed diffuse phase transition.

INTRODUCTION

The simpler crystallographic structure of ferroelectric perovskites is continually being explored since the discovery of $BaTiO_3$. Pure $BaTiO_3$ exhibits a ferroelectric to paraelectric transition at 120°C accompanied by a sudden change in dielectric constant and a crystallographic transformation from tetragonal to the Cubic phase. The properties are modified by various dopants; the widely investigated dopants are Sn,[1,2] Sr,[3] La,[4-6] Zr,[7,8] Ce,[9-13] Ca, Pb, and Y.[14,15] A majority of the doped material exhibit a diffuse phase transition or relaxor behavior. The high-k materials exhibiting diffuse phase transition are of great interest for memory applications as well as the various EIS (Electronic industry standards) capacitors. We have recently reviewed a literature on the pure and doped $BaTiO_3$.[16]

The diffuse phase transition is characterized by a deviation from Curie-Weiss (CW) law and a transition occurs over a temperature range rather than at an unique temperature. Unlike $BaTiO_3$, the transition is not followed by a sudden change in dielectric constant. The various explanations provided in literature for the diffuse phase transition are compositional inhomogenity, formation of polar nano domains and inter- and intra-grain stresses. Also, the diminished enthalpy is believed to result in non-vanishing thermodynamic probability of existence of the ferroelectric phase over a wide temperature range around the transition temperature. All these explanations are not mutually exclusive and possible existence of the ferroelectric phase over a range of temperature is believed to cause diffuseness of the transition. The present work is an attempt to investigate the experimental evidence of the crystallographic changes in materials exhibiting a diffuse transition.

In the present investigation we synthesized a family of compositions synthesized by using ball mill, which helped achieve a densification in excess of 95%. We further studied these compositions using HTXRD to observe the crystallographic changes. The materials in this investigation exhibited a highly diffuse transition and HTXRD confirmed that the transition temperature reflected in Dielectric constant and temperature plot (T_{max}) does not corresponds to the crystallographic phase transition and tetragonality exist well above T_{max}.

EXPERIMENTAL

Nominal compositions of the type $Ba_{1-3x}La_{2x}Ti_{1-3x}Ga_{4x}O_3$ for x=0.002, 0.004, 0.006, 0.008 and 0.01 were synthesized by conventional solid-state reaction. Stoichiometric amounts of the starting materials were mixed in a ball-mill with zirconia as a grinding media. The powder was calcined at 1100°C for 12h, followed by repetitive stages of ball milling and calcinations. The powder was characterized by X-Ray diffraction. The lattice parameters were extracted using the FULLPROF least square refinement software.[17] Further, the calcined powder was characterized by HTXRD between room temperature and 200 °C.

The calcined powder was ball-milled, dried, pelletized using a 10mm diameter tungsten carbide die and sintered at 1350°C for 4 h. The dielectric measurements were recorded in the temperature range between room temperature and 300°C at different frequencies using a HP impedance analyzer (4192A).

RESULTS AND DISCUSSION

The XRD patterns for all the compositions in Figures 1 [(a)-(d)] exhibit a single phase. The least square fit of the pattern assigned using P4/mmm symmetry in FULLPROF yield the lattice parameters. The tetragonality (c/a) is plotted in Figure 2; the tetragonality decreases with increase in dopant concentration. The change in tetrgonality is consistent with the corresponding change in tolerance factor. The dielectric constant as a function of temperature, at a constant frequency (10 KHz) for all compositions are plotted in Figure 3.

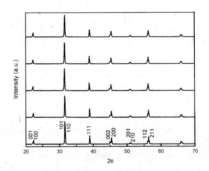

Figure 1. XRD pattern for $Ba_{1-3x}La_{2x}Ti_{1-3x}Ga_{4x}O_3$ compositions.

The transition temperature decreases from 135°C to 40°C in going from x = 0 ($BaTiO_3$) to x = 0.008 (see Table I); in good agreement with corresponding decrease in tetragonality (Figure 2). The transition temperature for x=0.01 composition is below room temperature. The data for dielectric constant at room temperature and T_c, at a constant frequency 10 KHz for all compositions is provided in Table I.

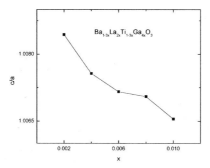

Figure 2. Variation in tetragonality as a function of dopant concentration (x).

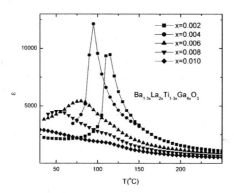

Figure 3. Dielectric constant at 10KHz as a function of temperature for all the compositions.

The diffuseness of the transition increases with increase in dopant concentration, this may be observed qualitatively in Figure 3. Further, the normalized dielectric constant as a function of normalized temperature is plotted in Figure 4; the diffuseness increases with increase in dopant concentration. The compositions exhibited a deviation from the C-W law. Therefore, the modified C-W law provided below (eqn 1) is used for further analysis [18]

$$\frac{1}{\varepsilon} - \frac{1}{\varepsilon_{max}} = \frac{(T - T_{max})^{\gamma}}{C} \tag{1}$$

In eqn 1, C and γ are constants, and ε_{max} is the maximum dielectric constant at transition temperature T_{max}. The constant γ varies between 1 (normal ferroelectric) and 2 (relaxor). The constant γ represents the slope of graph between log ($1/\varepsilon-1/\varepsilon_{max}$) and log ($T-T_{max}$) and is an indicator of diffuseness of transition. The constant γ increases with increase in dopant concentration (see Table I); this leads us to the same conclusion that the diffusivity increases in going from x = 0.002 to 0.008.

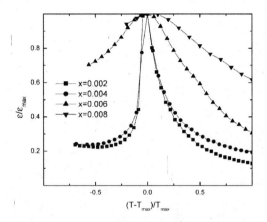

Figure 4. Plot of normalized dielectric constant at 10 KHz as a function of normalized temperature.

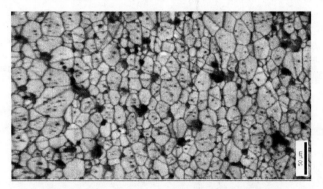

Figure 5. Image quality map for x=0.004 composition.

The representative OIM plots is exhibited in Figure 5, none of the compositions exhibit preferred orientation. The data points below the confidence index value of 0.1 are excluded from the analysis. The grain size decreases from ~100 μm for pure BaTiO₃ to ~2 μm for the x=0.008 composition. In general, the average grain size obtained from OIM data decreases with increase in dopant concentration. However, the grain size increases for x=0.004 (20 μm), this results in a higher dielectric constant relative to that for the x=0.002 (5 μm) compositions (see Table I). The latter composition exhibits higher grain average misorientation, indicating incomplete sintering and hence a finer grain size relative to x=0.004 composition. However, grain average misorientations for all other samples are within the accuracy of the OIM. Satisfactory data could not be recorded for x=0.01 composition, as smaller grain size is possibly comparable to electron-beam spot diameter, resulting in higher overlap of diffraction patterns and weak patterns.

The tetrgonality decreases with the increase in dopant concentration. However, the composition x=0.01 exhibit the tetragonal splitting at room temperature, although its transition temperature is below room temperature. This established an existence of tetragonal phase above T_{max} in the compositions exhibiting diffuse phase transition. Further, in order to interpret crystallographic changes, HTXRD experiments were performed between room temperature and 250 °C. The tetragonal (301)-(310) peak splitting at a higher angle (~75°) was observed qualitatively, prior to a rigorous Rietveld analysis. The tetragonal splitting disappered for pure BaTiO$_3$ abruptly at T_{max}, unlike doped compositions.

The changes in volume and lattice parameter as a function of temperature are plotted in Figure 6. Pure BaTiO$_3$ exhibits a sharp change in volume at T_{max}, unlike doped compositions. However, the change in lattice parameters for x=0.006 and 0.008 are continuous and the structure is tetragonal well above T_{max}(see Figure 6); this corroborates with the observed diffuse phase transition and a higher γ (see Table I). The higher value of γ represents the deviation from C-W law and is indicative of formation of polar domains. The wider difference between the crystallographic transition temperature (established from HTXRD) and T_{max} is proportional to the value of γ, namely the diffuseness of the transition. The present investigation establishes the existence of ferroelectric phase at temperatures well above T_{max} and is an experimental evidence of the existence of tetragonality above T_{max} for materials exhibiting diffuse phase transition. The present investigation corroborates with our earlier work on doped BaTiO$_3$,[19] which includes investigation with two complimentary techniques namely High temperature Raman spectroscopy and XRD.

Table I. Transition temperature and dielectric constant for all the Ba$_{1-3x}$La$_{2x}$Ti$_{1-3x}$Ga$_{4x}$O$_3$ compositions.

x	T_{max}	ε at T_{max}	ε at RT	λ
0.002	115	9449	2182	1.08
0.004	95	12150	2945	1
0.006	80	5422	3810	1.5
0.008	50	4532	4272	1.7
0.01	-	-	2914	-

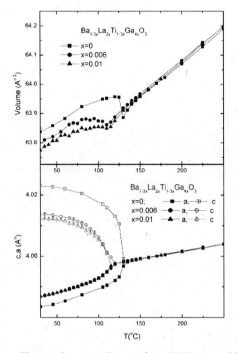

Figure 6. Image quality map for x=0.004 composition.

CONCLUSIONS

In the present investigation, the changes in dielectric properties as a result of co-doping extremely small amounts of La and Ga in barium titanate are addressed. The doping results in an increase in the ease of transition, the transition persists over a wide temperature range and hence the diffuseness.

The La and Ga co-doped compositions meet highly demanding material characteristics, including a high dielectric constant at room temperature, a small change in dielectric constant in vicinity of room temperature, low frequency dispersion as well as a high resistivity. This appears to be the first $BaTiO_3$ based compound at such a low doping level exhibiting high diffuseness and a high dielectric constant.

The present investigation is experimental evidence that the transition observed in dielectric constant as a function of temperature does not correspond to a crystallographic phase change. The existence of ferroelectric phase at higher temperature explains the observed diffuseness.

REFERENCES

1. R. Vivekanandan and T. R. N. Kutty, *Ceram. Int.*, **14**, 207 (1988).

2. V. V. Shvartsman, W. Kleemann, J. Dec, Z. K. Xu and S. G. Lu, *J Appl. Phys.*, **99**, 124111 (2006).

3. B. S. Chiou and J. W. Liou, *Mater. Chem. Phys.*, **51**, 59 (1997).

4. F.D. Morrison, D.C. Sinclair, and A. R. West, *J Appl. Phys. Soc.*, **86**, 6355 (1999)

5. F.D. Morrison, D.C. Sinclair, J.M.S. Skakle, and A.R. West, *J Am. Ceram. Soc.*, **81**, 1957 (1998)

6. F.D. Morrison, D.C. Sinclair, and A.R. West, *J Am. Ceram. Soc.*, **84**, 474 (2001)

7. B. Jaffe, W.R. Cook, and H. Jaffe "Piezoelectric Ceramics", Academic Press Inc. London, (1971).

8. D. Hennings, A. Schnell and G. Simon, *J. Am. Ceram. Soc.*, **65**, 539 (1982).

9. D. Makovec and D. Kolar, *J Am. Ceram. Soc.*, **80**, 45 (1997).

10. D. Hennings, B. Schreinemacher, and H. Schreinemacher, *J. Euro. Ceram. Soc.*, **13**, 81 (1994).

11. Chen, Y. Zhi, J. Zhi, P.M. Vilarinho and J. L. Baptista, *J. Euro. Ceram. Soc.*, **17**, 1217 (1997).

12. Chen, J. Zhi and Y. Zhi, *J Phys: Condens. Mat.*, **14**, 8901 (2002).

13. Z. Yu, C. Ang, Z. Jing, P.M. Viarinho and J. L. Baptista, *J Phys: Condens. Mat.*, **9**, 3081 (1997).

14. Z. Jing, C. Ang, Z. Yu, P. M. Vilainho and J. L. Batptista, *J Appl. Phys.*, **84** 983 (1998).

15. J. Zhi, A. Chen, Y. Zhi, P. M. Vilainho and J. L. Batptista, *J Am. Ceram. Soc.*, **82** 1345 (1999).

16. D.D. Gulwade, S.M. Bobade, A.R. Kulkarni and P. Gopalan, *J. Appl. Phys.*, **97**, 074106 (2005); S.M. Bobade, D.D. Gulwade A.R. Kulkarni and P. Gopalan, *J. Appl. Phy.*, **97**, 074105 (2005).

17. J.R. Carvajal, "FULLPROF" Version 2 K, Laboratoire Leon Brillouin CEA–CNRS, (2000).

18. K. Uchino and S. Nomura, Integr. Ferroelectr. **44**, 55 (1982).

19. D.D.Gulwade and P.Gopalan, *Solid State comm.* Submitted (2007)

PRESSURELESS SINTERING OF TITANIUM DIBORIDE POWDERS

Michael P. Hunt[1] and Kathryn V. Logan[1]
[1]Virginia Polytechnic Institute and State University
Blacksburg, VA, USA

ABSTRACT

Key parameters that affect the densification of pure TiB_2 were being determined. The pure TiB_2 powders used in the determination of the key parameters were obtained from commercial sources. Taguchi methods were used to study the significant factors in achieving high green density from pressed TiB_2 pellets. The results of the study showed that pressing pressure and powder dryness account for ~66% of the factors that effect TiB_2 green density. A confirmation study confirmed that the optimal configuration of the factors studied was pressing with a dual action configuration at 20,000 psi with powder that was not dried. Pressed samples were fired between 1800 - 2100°C using a reducing atmosphere. Preliminary results indicated that densification of these materials may begin at 1800°C. Samples that were fired at 1800°C showed evidence of early densification stages while samples fired at 2100°C showed varying degrees of densification.

INTRODUCTION

Titanium diboride (TiB_2) is the most stable form of all Ti – B compounds but is not found in nature. Traditionally, TiB_2 is produced using carbothermal reduction of TiO_2 and B_2O_3.[1] The powders are then generally formed and sintered simultaneously using hot pressing. Alternatively, the powders may be formed by pressing and pre-sintering to provide some structural strength and then hot isostatically pressing (HIP'ing) to full density. The second approach allows for the possibility of machining the pre-sintered part before its final densification.[1,2]

Properties of Technical Ceramics

As a technical ceramic, TiB_2 has a high compressive strength, high oxidation resistance, high hardness, and a high melting point.[3,4,5] However, TiB_2 is unique among ceramics because of its thermal conductivity, electrical conductivity, and fracture toughness properties in comparison with other ceramics as shown in Table 1.[1,6,7]

I. Property Comparison Chart for Selected Technical Ceramics[1,6,7]

Property	TiB_2	Al_2O_3	SiC
Density (g/cm^3)	4.52	3.99	3.19
Compressive Strength (MPa)	3388 – 3736	2470 – 2730	3047 – 3360
Hardness Vickers (GPa)	15 – 36	15 – 20	23 – 26
Fracture Toughness (MPa*m$^{1/2}$)	6.7 – 8.0	5.7 – 6.3	4.28 – 4.72
Electrical Resistivity (μohm*cm)	8.99 – 17	$1 \times 10^{21} – 1 \times 10^{22}$	$1 \times 10^9 – 3.16 \times 10^{10}$
Thermal Conductivity (W/m*K)	24 – 26	28.8 – 31.2	90 – 110
Melting Point (°C)	2916 – 3045	2004 – 2096	2151 – 2249

* Calculated using the composite rule of mixtures.

Material Uses

The properties of TiB_2 make it an excellent candidate for many potential applications such as thermal protection for hypersonic vehicles, tank armor, and radiation shielding for space craft. However, each of these applications requires that these properties be consistent from part to part.[1,5,8] Currently TiB_2 is primarily being used in the aluminum industry in electrically heated crucible materials due to its electrical conductivity and chemical resistance to molten aluminum.[7] It has also been tested as armor plating and used, to a lesser extent, in the microelectronics and semiconductors industries.[4,5,9,10]

Methods of Powder Production

The raw materials serve as the foundation of these experiments and may be obtained by commercial means and SHS reaction. The SHS reaction powders are produced using the Logan technique.[1] The commercial powders are generally produced by the carbothermic process. The carbothermic process relies on the oxidation reduction reaction of TiO_2 and B_2O_3 with carbon to produce TiB_2 as shown in equation [1]. As a result, producing TiB_2 by the carbothermic process generally leaves small amounts of carbon and other impurities in the final product. These powders have particle sizes between 3 and 10 μm on average.[1,10]

$$2TiO_2 + 2B_2O_3 + 5C \rightarrow 2TiB_2 + 5CO_2 \qquad [1]$$

The SHS process makes use of the heat released during an exothermic oxidation-reduction reaction to initiate a series of reactions. The self-propagating reaction continues until there are no un-reacted particles left. The SHS reaction to produce pure TiB_2 powder is accomplished by reducing TiO_2 and B_2O_3 with Mg to produce TiB_2 and MgO according to the formula shown in equation [2]. The concurrently formed MgO powder produced by the SHS reaction is then removed by leaching in a nitric acid solution to remove the MgO.[1] The resulting pure TiB_2 powder has a particle size between 0.1-10 μm (0.5 μm avg.).[1,8]

$$TiO_2 + B_2O_3 + 5Mg \rightarrow TiB_2 + 5MgO \qquad [2]$$

Processing Challenges

The forming and densification of TiB_2 is accomplished commercially by hot pressing. The use of pressure in the densification process reduces the temperatures needed to sinter the material.[1,3,12] As with most ceramics, there is a balance between sintering and grain growth. The ability to control the sintering and the grain growth partially determines the uniformity of the microstructure.[3,4,12] Since the sintering process for TiB_2 has not been studied in depth, there is a great deal of inconsistency in the final microstructure which consequently affects the final properties.

The pressure that is provided with hot pressing, whether great or small, increases the production cost of TiB_2 which has limited its widespread use.[3,7] The material shows a great deal of promise but in most applications it is impractical to replace existing materials with TiB_2 because of cost. However, each of the applications requires that the properties be consistent from part to part.[1,8] Since the final properties are ultimately controlled by the processing of the material, it is imperative to be able to control the processing, specifically the sintering or densification, of the TiB_2.[3,4,12]

Justification of Research

The purpose of this research is to provide information leading to a better understanding of the sintering process for TiB_2 and, particularly, to learn what factors control the final microstructure and properties of the material. With this knowledge it may be possible to produce high density parts with the lowest possible cost and optimize the material properties. Statistics may be used to design experiments that test which factors are significant and how well the process is understood. The use of Taguchi methods allows the possibility of drawing a large number of conclusions from a smaller data set than traditional single factor testing. Also, these conclusions can be reached in a significantly shorter amount of time than would be possible with single factor testing. The use of Taguchi methods can provide the possibility of studying important parameters which can not be studied with single factor testing. All and all, the results of this work should provide the route to potential cost savings for processing and manufacture of TiB$_2$ based materials.[13,14]

Analysis of Variance (ANOVA) is an important technique used in statistics to determine important parameters in a model by analyzing the variance in each parameter. ANOVA helps researchers to determine how well they understand a system or process. While using ANOVA it is possible to quantify the affect of any parameter being tested and to calculate the certainty of a parameters affect on the outcome of the system or process.[14]

Taguchi methods rely on ANOVA and orthogonal matrices to determine what factors affect a particular outcome without having to test all possible combinations of factors and levels. Also, the data that is generated using an experiment designed in conjunction with Taguchi methods provides valuable information on the interactions between different parameters. In sintering studies, it becomes possible to determine which parameters are important to each property individually by characterizing the material afterwards. Therefore, the Taguchi method of approaching experimentation allows for the factors and levels to be tailored to specific applications that require specific properties.[14]

EXPERIMENTAL METHODS
Pressing Experiments

Carbothermically produced powders were obtained from H.C. Starck and were used throughout the duration of these experiments. An L4 Taguchi array, as shown in Table II, was used to determine if pressing pressure, action mode, or drying the powder was significant in affecting the green density of pressed TiB$_2$ pellets. A three piece punch and die assembly was used in conjunction with a hydraulic hand operated press to uniaxially press powder samples. In single action pressing, only one of the

II. Taguchi L4 Array for Pressing Experiments

Experiment #	Pressing Pressure (psi)	Action Type	Powder Dried
1	10000	Single	Yes
2	10000	Double	No
3	20000	Single	No
4	20000	Double	Yes

punches was in motion and in dual action pressing both punches were in motion as shown in figure 1. Half of the samples were dried in air at 110°C for two hours before pressing. The diameter, thickness, and mass of each sample were measured to determine the green density of each pellet pressed. In order to remove statistical error and improve accuracy, complete randomization was used and samples were run in an order determined by a random number generator. Results were analyzed and a confirmation experiment was run.

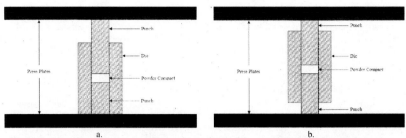

a. b.

Figure 1 – Diagrams for uniaxial pressing setup for a. single action and b. dual action pressing.

Preliminary Sample Firing

Samples were heated to 1800°C and 2100°C in flowing He at an average heating rate of 10°C/min and were held at temperature for one hour. Samples were placed onto a graphite disc within the high temperature furnace. The samples were allowed to furnace cool (~3-4°C/min). Upon cooling the samples were measured for density and analyzed using an SEM with EDS capability.

Density Measurements

The sintered samples were placed in a drying oven, under vacuum, overnight to remove all moisture from the open porosity. Once the samples were dry, they were weighed using a balance accurate to four decimal places. The mass of the dried sample was recorded in a spreadsheet. The samples were then placed in a vacuum chamber pumped down to -30 psi and then submerged in water. The submerged samples were weighed suspended in water, in accordance with Archimedes method, and the results were entered into the spreadsheet. The water temperature was measured to account for the changes of water density with temperature.

Once the suspended weight was determined, excess surface moisture was removed with a moist paper towel. The saturated sample was weighed and the results recorded. The density, apparent pore volume, and percent porosity of the samples were calculated. The samples were then dried under vacuum, to remove all water from their pores.

SEM Microscopy

The samples were mounted on a conductive sample holder using carbon tape and small amounts of conductive carbon paste. The samples were placed into the SEM chamber and micrographs of the surface morphology and structure were taken and analyzed. Some samples were subjected to EDS analysis of particular areas of interest.

RESULTS AND DISCUSSION

The pressed sample densities averaged 2.35 g/cm^3 (52.04% theoretical density), as shown in Table III. The samples pressed using configuration 1 averaged 2.19 g/cm^3 (48.38% theoretical density), configuration 2 averaged 2.35 g/cm^3 (51.98% theoretical density), configuration 3 averaged 2.50 g/cm^3 (55.34% theoretical density), and configuration 4 averaged 2.37 g/cm^3 (52.37% theoretical density). From the data in Table III, ANOVA was completed which showed a variance of 0.142 for pressing pressure, 0.001 for action mode, and 0.107 for powder dried as shown in Table IV.

III. TiB$_2$ Designed Experiment Raw Data

Configuration	Run 1 (g/cm^3)	Run 2 (g/cm^3)	Run 3 (g/cm^3)	Run 4 (g/cm^3)	Run 5 (g/cm^3)	Average (g/cm^3)	% Theoretical
1	2.22	2.08	2.07	2.28	2.28	2.19	48.38%
2	2.25	2.26	2.44	2.39	2.40	2.35	51.98%
3	2.54	2.51	2.45	2.52	2.49	2.50	55.34%
4	2.25	2.37	2.44	2.37	2.42	2.37	52.47%

The data suggests that the action mode may not have been significant to the green density. According to the ANOVA performed, both pressing pressure and powder drying have a 99% confidence interval for significance as shown in Table V. In addition, the optimal setting was found to be a pressing pressure of 20,000 psi using a double action pressing mode and powder that had not been dried. From this analysis, a density of ~2.52 g/cm^3 (55.71% theoretical density) should be achieved using these settings as shown in Table V.

IV. ANOVA Table for TiB$_2$ Pressing Experiments

Source	Sum of the Squares	Degrees of Freedom	Variance	Calculated F-Value	Table F-Value Exceeded	Confidence Interval	Influence Percentage
Pressing Pressure	0.142	1	0.142	22.21	5.29	99%	37.99%
Action Mode	0.001	1	0.001	0.21	-	-	-
Powder Dried	0.107	1	0.107	16.80	5.29	99%	28.19%
Error	0.102	16	0.006	-	-	-	-
Total	0.352	19	-	-	-	-	66.19%

Based on these results, it is plausible that the moisture content of powder before pressing plays a significant role in producing high green densities from pressing, as has been demonstrated in the literature. Further experimentation would be needed to determine the extent to which moisture content affects green densities for TiB$_2$ pressing. The total influence percentage was found to be 66.19% which means that 43.81% of the parameters affecting green density of pressed TiB$_2$ pellets have not been found.

V. Predicted Process Averages for Pressing TiB$_2$ Pellets

Pressing Pressure (psi)	Action Mode	Powder Dried	Predicted Density (g/cm^3)	Predicted Percent Theoretical
20,000	Dual	No	2.52	55.71%
20,000	Single	No	2.50	55.34%
20,000	Dual	Yes	2.37	52.47%
20,000	Single	Yes	2.36	52.10%
10,000	Dual	No	2.35	51.98%
10,000	Single	No	2.33	51.62%
10,000	Dual	Yes	2.20	48.75%
10,000	Single	Yes	2.19	48.38%

The data from the confirmation experiment showed a green density average of 2.484 ± 0.0399 g/cm³ as shown in Table VI. The prediction from Table V falls within the process average confidence interval range. Since the prediction falls within the range, the prediction was good and the model was validated despite the fact that the system was not completely understood. Since the model was validated, the predicted optimal setting will produce the highest green densities from these combinations of factors and levels.

Table VI. Process Average and Confidence Interval for Pressing TiB₂

Green Density Average	2.484 g/cm³
Standard Deviation	0.0231 g/cm³
Conference Interval (CI)	0.0399 g/cm³
CI Minimum	2.444 g/cm³
CI Maximum	2.524 g/cm³
Predicted Green Density	2.518 g/cm³

The green microstructure did not appear entirely uniform; however, the particles seemed to be tightly packed as shown in figure 2. As expected, there was no evidence of any physical fusion or bonding between the particles as a result of pressing. The powder compact did not show any evidence of why the particles were holding together or evidence of impurities.

Figure 2 – SEM micrographs of the general green microstructure of carbothermic TiB₂.

Preliminary Sintering Results

The samples that were fired at 1800°C had an average density of 2.50 g/cm³ (55.30% theoretical density), as shown in Table VII. The samples seemed to be fairly consistent in density with the exception of CT0100208002 which showed a substantially higher density. There were no noticeable differences between this sample and the rest. It is possible that the position of the sample on the sample holder may have impacted the density, however, there is not an obvious trend with this

Table VII. – Densities of Samples Fired at 1800°C

Sample ID	Density (g/cm³)	% Theoretical
CT122107010	2.49	55.10%
CT122107017	2.49	54.98%
CT010208001	2.46	54.49%
CT010208002	2.61	57.67%
CT010208003	2.45	54.24%
AVERAGE	2.50	55.30%

data. Initially, the observed porosity appeared to be open porosity at the surface. However, after further investigation using the SEM, it was discovered that the porosity was networked throughout the sample as shown by figure 3. The SEM images showed evidence of the beginning stages of sintering such as necking. There is a possibility that if left at these temperatures for longer times that more

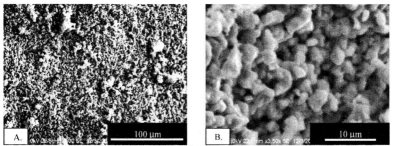

Figure 3 – Micrographs of a. general porosity and b. evidence of densification in carbothermic TiB₂ fired at 1800°C.

significant densification might be observed. The results of EDS at certain unique regions of the microstructure revealed the presence of carbon and oxygen which produced very interesting microstructural features as shown in figure 4. Normally, these regions would be considered as artifacts, however, they were observed in all of the TiB₂ samples that were fired. Since the samples

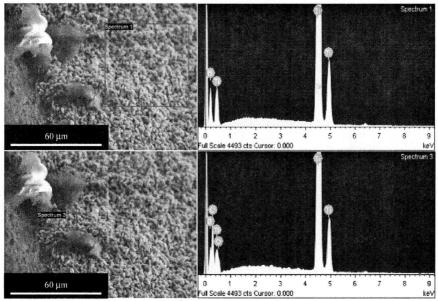

Figure 4 – Micrographs and EDS spectra near an impure region within the carbothermic TiB₂ microstructure.

were made from carbothermic TiB$_2$ and both elements were found in the composition report included with the powder, these formations may be a result of impurities from the carbothermic process.

After firing at 2100°C, the samples had an average density of 2.65 g/cm^3 which is 58.65% of the theoretical density as shown in Table VIII. There was an apparent difference in the final densities of the two samples that were fired, however, there was no noticeable difference between the microstructures or physical appearance of the two samples. Further study may be required to determine the source of differences between the samples. There was still a lot of porosity within the samples. The microstructure of these samples appeared to be very different from the samples that were fired at 1800°C, as shown in figure 5.

Table VIII. – Densities of Samples Fired at 2100°C

Sample ID	Density (g/cm^3)	% Theoretical
CT113007002	2.55	56.42%
CT113007003	2.75	60.89%

While observing these samples using the SEM, there were several areas that showed masses larger than the initial particle size of the powders as shown in figure 6. Upon further investigation these masses showed significant evidence that sintering was taking place within the sample. However, the sample surface also showed evidence of formations with similar appearance and EDS traces as

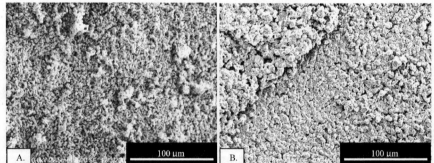

Figure 5 – SEM Micrographs of the carbothermic TiB$_2$ microstructure after being fired at a. 1800°C and b. 2100°C.

those fired at 1800°C as shown in Figure 7. After EDS and further observation, these formations had the same general form and EDS spectra. One major difference between the formations was that the impurity formations found in the samples fired at 1800°C were only found on the edges, however, the formations were found throughout the surface of the samples fired at 2100°C.

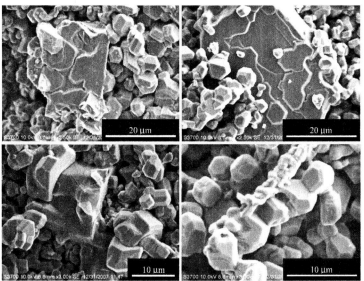

Figure 6 – SEM micrographs of sintered carbothermic TiB₂ in various areas of a sample fired at 2100°C.

It is possible that the higher temperature, while improving sintering conditions, also provided the necessary energy for the formation of defect structures throughout the bulk of the material. Since sharp edges on a material have a higher surface energy than the rest of the free surface, it is reasonable that the formations would appear along the edges of the material first. Also, these edge formations would begin to show up at lower temperatures than the formations on other surfaces.

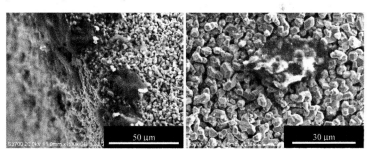

Figure 7 – Impurity formations in carbothermic TiB₂ samples fired at A. 1800°C and B. 2100°C

CONCLUSIONS

The TiB₂ in all of the discussed results was produced using the carbothermic process and according to the ANOVA performed, both pressing pressure and powder drying have a 99% confidence interval for significance. In addition, the optimal pressing settings were found to be a pressing pressure of ~20,000 psi using a double action pressing mode and powder that had not been dried. It is plausible that the moisture content of the powder before pressing plays a significant role in producing high green densities from pressing, as has been demonstrated in the literature. The total influence percentage was found to be 66.19% which means that 43.81% of the parameters affecting green density of TiB_2 pellets have not been found. The optimal setting prediction was good and the model was validated despite the fact that the system was not completely understood. Since the model was validated, the predicted optimal setting will produce the highest green densities from these combinations of factors.

The SEM images of the samples fired at 1800°C showed evidence of the beginning stages of sintering such as necking. The results of EDS at certain unique regions of the microstructure revealed the presence of carbon and oxygen which produced very interesting microstructural features. Since both elements were found in the composition report that was included with the powder, these formations may be a result of impurities from the raw carbothermic TiB_2 powder. The microstructure of the samples fired at 2100°C appeared to be very different from the samples that were fired at 1800°C. The samples fired at 2100°C showed significant evidence that sintering was taking place within the sample.

The surfaces of samples fired at 2100°C showed evidence of formations with similar appearance and EDS traces as those fired at 1800°C. The major difference between the formations was that the impurity formations found in the samples fired at 1800°C were only found on the edges while the formations were found throughout the surface of the samples fired at 2100°C. It is possible that the higher temperature, while improving sintering conditions, also provided the necessary energy for the formation of defect structures throughout the bulk of the material.

ACKNOWLEDGEMENTS

I would like to thank the Dr. Wallace Vaughn and Mr. Craig Leggette for their assistance and training on the use of high temperature furnace equipment used in these experiments. Mr. Jim Baughman for his aid in obtaining the SEM micrographs and EDS plots, the NASA Langley Research Center for use of their facilities, and the National Institute of Aerospace for their funding (Grant Award# VT-03-01) of this work.

REFERENCES
[1]Logan, K. V., *Elastic-Plastic Behavior of Hot Pressed Composite Titanium Diboride/Alumina Powders Produced using Self-Propagating High Temperature Synthesis.* 1992, Georgia Institute of Technology: Atlanta, GA.
[2]Pastor, H., *Metallic Borides: Preparation of Solid Bodies – Sintering Methods and Properties of Solid Bodies,* in *Boron and Refractory Borides,* V.I. Matkovich, Editor. 1977, Springer-Verlag: New York. p. 457-493 (Hardcopy).
[3]Callister, W.D., *Materials Science and Engineering: An Introduction.* 6th ed. 2003, New York: John Wiley & Sons, Inc.
[4]Richardson, D.W., *Modern Ceramic Engineering: Properties, Processing, and Use in Design.* 3rd ed. 2006, Boca Raton, FL: CRC Taylor & Francis. 707.
[5]Holt, S. and Logan, K.V. *Preliminary Results in the Experimental Determination of a Possible Eutectic in Composite $Al_2O_3 – TiB_2$.* in Materials Science and Technology Conference 2007. Detroit, MI.
[6]Ashby, M. F., *CES Selecter.* 2005, Granta Design Limited: Cambridge. Software program for finding material properties, comparing materials, and deciding which materials to use by user defined criteria.
[7]Kaufmann, E., *Characterization of Materials.* Vol. 1. 2003, Hoboken, NJ: Wiley-Interscience. 1392.
[8]Adams, J.W., et al., *Microstructure Development of Aluminum Oxide/Titanium Diboride Composites for Penetration Resistance.* Journal of the American Ceramic Society (Hardcopy).
[9]Keller, A.R. and M. Zhou, *Effect of Microstructure on Dynamic Failure Resistance of Titanium Diboride/Alumina Ceramics.* Journal of the American Ceramic Society, 2003. **Vol. 86** (3): p. 449-457

[10]Toon, J. (1995) *Company Begins Production of Improved High-Performance Materials for Cutting Tools, Dies, and Electrodes.* Georgia Tech Research News. Website: http://gtresearchnews.gatech.edu/newsrelease/TIB2.html

[11]Lok, J. *Acid Leaching of SHS Produced MgO/TiB$_2$.* 2006, Virginia Polytechnic Institute and State University: Blacksburg, VA.

[12]Rahaman, M.N., Ceramic Processing. 2007, Boca Raton, FL.: CRC/Taylor & Francis. 473.

[13]Ashby, M.F., *Materials Selection in Mechanical Design.* 3rd ed. 2005, Oxford: Elsevier Buttersworth-Heinemann.

[14]Antony, J. *Design of Experiments for Engineers and Scientists.* 2005, Oxford: Elsevier Buttersworth-Heinemann.

THE RELATION BETWEEN PEIERLS AND MOTT-HUBBARD TRANSITION IN VO$_2$ BY TUNNELING SPECTROSCOPY

Changman Kim[1], Tomoya Ohno[1], Takashi Tamura[1], Yasushi Oikawa[1],
Jae-Soo Shin[2] and Hajime Ozaki[1]

[1]Department of Electrical Engineering and Bioscience, Waseda University, Tokyo, Japan
[2]Department of Advanced Materials Engineering, Daejeon University, Daejeon, South Korea

ABSTRACT

Tunneling spectroscopy has been performed on W-doped VO$_2$ single crystal near the Metal-Insulator transition temperature. The tunneling energy gap was in good agreement with band calculations and optical measurements near the transition temperature. We have found by tunneling spectroscopy an additional density of states in the low-temperature phase. With increasing temperature, from room temperature to just below the transition temperature, an additional increase in the density of states was observed in the conduction band and it shifted downward to the bottom of conduction band. When the front of the additional density of states approaches the bottom of conduction band, edges of the tunneling energy gap becomes blurred, and the VO$_2$ turns into the high-temperature phase. A model for the mechanism of the Metal-Insulator transition in VO$_2$ is proposed.

INTRODUCTION

Vanadium oxides of magneli phase are expressed by V$_n$O$_{2n-1}$ (n=4⊓8,∞) and several vanadium oxides of them undergo a Metal-Insulator Transition (MIT) at their transition temperatures. Among them, vanadium dioxide (VO$_2$) has been received most attention because of not only the dramatic reversible changes of electrical resistivity and infrared transmission, but also the transition temperature (T_t) which is close to room temperature, T_t=340K[1]. For these characteristics, VO$_2$ has the possibility of applications to new electronic devices such as "Thermochromic Smart Windows", "Mott-Transition Field-Effect Transistor" etc[2]. The phase transition mechanism of VO$_2$ has often been the topic under debate whether it is Peierls type or Mott-Hubbard type.

The early qualitative aspects of the electronic structure in the low temperature phase of VO$_2$ were explained by Goodenough[3]. The d states of the V atoms are split into lower lying t$_{2g}$ state and higher lying e$_g$ state because of O octahedral crystal field. The tetragonal crystal field further splits the multiple t$_{2g}$ state into d$_\parallel$ and π^* states. In the low-temperature phase of VO$_2$, there are two structural components to the lattice distortion, namely a pairing and a twisting of V atoms out of the rutile axis c$_r$. The pairing and twisting of the V atoms result in two effects on the electronic structure. First, the π^* band is pushed higher in energy, due to the tilting of the pairs which increases the overlap of these states with O states. Second, the d$_\parallel$ band is split into a lower-energy bonding combination and a higher-energy anti-bonding combination. The band gap exists between the bottom of π^* band and the top of bonding d$_\parallel$ band. On the other hand, for such peierls-like band gap. Zylbersztejn and Mott[4], and

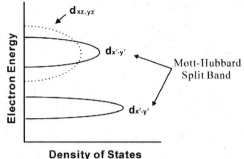

Density of States

Figure 1. Schematic illustration of VO$_2$ in the low temperature monoclinic structure. The upper and lower d$_{x2-y2}$ bands are Mott-Hubbard split bands. The band gap exists between the bottom of d$_{xz, yz}$ band and the top of the lower d$_{x2-y2}$ band.

Rice et al.[5] suggested that a crystallographic distortion is not sufficient to open up an energy gap, and that the electron-correlation effects play an important role in opening the energy gap. Zylbersztejn and Mott also suggested that the role of the crystallographic distortion is only to provide an empty π^* bands. Shin et al.[6] have estimated the energy band gap as about 0.7eV from UPS + reflectance measurements.

The recent study of band calculation via local density approximation plus Hubbard U (LDA+U) has estimated the band gap as about 0.7 eV[7]. Figure 1 shows a schematic illustration for the density of states calculated within LDA+U method (The d$_{xz, yz}$ and d$_{x2-y2}$ bands correspond with π^*, d$_\parallel$ bands in Goodenough's expression, respectively).

In our previous studies, we used tunneling spectroscopy in order to investigate the change in the electronic structure at the MIT in VO$_2$ doped with W[8-9].

In the present study, the tunneling spectroscopy results are explained for the onset of Metal-Insulator Transition in VO$_2$ in relation to the band diagram by the local density approximation plus Hubbard U calculation.

EXPERIMENTS

The crystal growth of VO$_2$ was performed using VO$_2$ and V$_2$O$_5$ powders and with WO$_3$ powder for W doping. The well mixed powders were sealed in quartz tube with 90 mm (long) ×10 mm (diameter) under 1×10^{-3} Pa. The sealed quartz tube was placed vertically in an electric furnace. The temperature of the furnace was kept at 1000°C for 5 hours and then decreased at a rate of 2.7 °C/hr to 800°C. At 800°C, the quartz tube was inverted in the furnace so as to separate the useless solution from the crystals which were grown in the melt at the bottom of the quartz tube. The crystals were annealed for 2 hours at 800 °C in the inverted tube, and then, the heater of the furnace was switched off. The typical size of the crystals obtained was $3 \times 1 \times 1$ mm^3. The crystalline c-axis of high-temperature

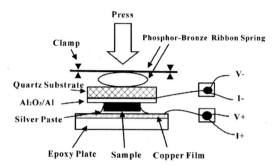

Figure 2. Schematic planar contact tunnel unit employed in this study.

rutile type lies along the length of the crystal.

The W concentration in the crystal was determined by a wavelength dispersive spectrometer electron probe microanalyser (WDS-EPMA), using JAX-8600 (JEOL). The resolution of the W content was ±0.1%.

In this study, the planar-contact structure, as shown in figure 2, was employed as the tunnel junction, instead of using an insulator evaporated on the sample surface, because in the latter case, the rigid contact and sometimes atomic diffusion between the insulator and the sample tends to suppress or modify the structural change associated with the MIT near the surface of the sample. The Al₂O₃/Al structure was fabricated as follows. Al was evaporated onto a clean quartz substrate. Then, it was heated in the evaporation chamber at about 100 °C for 1 hour in O₂ atmosphere of 1 atm to oxidize the Al surface. For the back electrode, sample was bonded to copper plate using silver paste. The surface of Al₂O₃ was pressed to the sample surface using phosphor-bronze ribbon spring to form a stable contact tunnel junction. By this planar-contact method, the tunnel junction resistance can be adjusted by controlling the pressure from the top of the apparatus through a rotating shaft with fine pitch screw. To avoid the influence of the series resistance by the lead wire on the tunneling spectroscopy, a quasi four-probe method was employed in measuring the bias voltage V. The tunneling spectroscopy was performed using the ac modulation technique. The modulation bias and frequency were 1 mV and 1 kHz, respectively.

RESULTS AND DISCUSSION

The W composition x in W$_x$V$_{1-x}$O$_2$ was estimated by EPMA. The relationship between the starting composition and the substituted one is shown in figure 3. The substituted composition is linearly proportional to the starting composition. The segregation coefficient is 0.69.

The temperature dependences of the electrical resistivities are shown in figure 4 (a). The W composition dependence of T_t is linear with coefficient -27.8K/at.% as shown in figure 4(b), in which

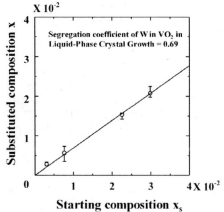

X 10⁻²

Segregation coefficient of W in VO₂ in Liquid-Phase Crystal Growth = 0.69

Starting composition x_s

Figure 3. The relationship between starting composition x_s and substituted one x in $W_xV_{1-x}O_2$ by EPMA. The plots show the average of five measured points for each sample.

the T_t is plotted for the heating process. In table 1, the transition characteristics are listed for various W composition x. Although $W_xV_{1-x}O_2$ samples for various W composition x are in different ρ-T characteristics, we can find the similarity of ρ-T profiles between the nondoped VO₂ and $W_xV_{1-x}O_2$ except for x=0.0153. It suggests that the electronic structure was not changed basically by the W doping for x ⊔0.01.

Figure 5(a) shows the tunneling dI/dV vs. V characteristics for $W_xV_{1-x}O_2$ with x=0.006 in the temperature region near the T_t. W was doped to reduce the electrical resistivity in the low temperature phase, for the sake of minimizing the potential drop across the bulk of VO₂, and thus, minimizing the spectroscopic error. Over the whole temperature range of measurements, the tunneling junction was unchanged. Figure 5(b) shows the curves shown in figure 5(a) shifted vertically for easy to see each curve. In figure 5, curves for 323.8 K and 327.3 K are in the high-temperature phase and those for 320.7 K and lower temperatures are in the low-temperature phase. The valence and conduction bands lie in the negative and positive bias region, respectively.

In the low-temperature phase, there appears an energy gap structure with diminished electronic density of states. The apparent residual density of states in the gap region might be due to some non-tunneling components. For temperatures above T_t, the gap structure disappears and the curves show a metallic state.

Figure 6(a) shows the temperature dependence of energy gap, estimated by the separation of biases between the maxima of $|d^2I/dV^2|$ in positive and negative biases in $|V| < 0.5V$. The tunneling energy gap at lower temperature, ~0.7eV, is in good agreement with those by optical studies[6].

In figure 5(b), a remarkable change in the density of states is seen in the conduction band

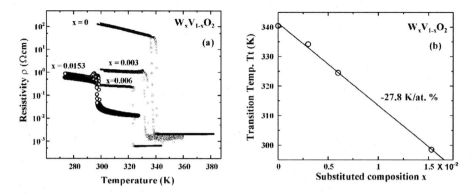

Figure 4. (a) Temperature dependences of electrical resistivities in W$_x$V$_{1-x}$O$_2$ for various W composition x. (b) MIT temperature T_t versus W composition x.

Table 1. Some characteristic parameters in the temperature dependence of electrical resistivity in tungsten doped VO$_2$. $T_{I \to M}$ and $T_{M \to I}$ are the midway temperatures of MIT during heating and cooling, respectively. $(\rho_I/\rho_M)_{Tt}$ is the ratio of electrical resistivities on both sides of the transition from insulator to metal. Hysteresis is the separation of temperatures between heating and cooling at the midway of transition.

Sample	W composition x	$T_{I \to M}$ (K)	$T_{M \to I}$ (K)	$(\rho_I/\rho_M)_{Tt}$	Hysteresis (K)
1	Pure	340.4	337.3	1.6×10^4	3.2
2	0.003	334.2	331.7	7.8×10^2	2.5
3	0.006	324.5	323.2	3.4×10^2	1.3
4	0.0153	298.5	297.1	1.7×10	1.4

region. At the measured lowest temperature, there appears an additional increase in the density of states above +0.8 V. With increasing the temperature, this increase in the density of states shifts toward lower bias voltage. When the front of the increase approaches the conduction band edge, the band gap structure becomes blurred, and then, the sample turns to the high-temperature phase. Figure 6(b) shows the temperature dependence of energy difference between the front of the additional increase in the density of states and the bottom of conduction band in the low-temperature phase.

Now, we would like to consider about this behavior in the change of density of states, which seems to lead to the MIT, in relation to our band diagrams in figure 7. At low temperature ($T \leq 315$K), the band gaps, shown in figure 5, might correspond to the energy separation between the d$_{xz, yz}$ and the

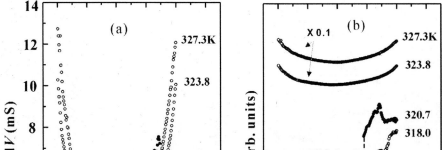

Figure 5. (a) Temperature dependence of tunneling spectroscopy for W$_x$V$_{1-x}$O$_2$ with x=0.006 in the temperature region around the T_i=323K. Measurements were carried out from high temperature to low temperature. (b) Temperature dependence of tunneling spectroscopy in (a), shifted vertically.

lower d$_{x^2-y^2}$ bands in figure 7(a), (b). Thermal carriers are excited across the band gap and the bottom of d$_{xz, yz}$ band is provided with electrons. As the upper band, where the electrons are provided, is not the Mott-Hubbard split band (upper d$_{x^2-y^2}$), the effect of the carriers in depressing the gap formation is less direct compared with the case the upper band is the Mott-Hubbard split band.

    Now, we might conjecture that the increase of density of states above the bottom of upper band is the Mott-Hubbard split band (upper d$_{x^2-y^2}$). The position of the increase shifts to lower bias with increasing temperature as shown in figure 7(a), (b). When the front of the increase (bottom of upper d$_{x^2-y^2}$) approaches the bottom of the d$_{xz, yz}$ band, the electrons in the latter band transfer to the former (figure 7(c)), and thus, the carriers begin to act on depressing the Mott-Hubbard gap, leading to the metallic state (figure 7(d)). When the sample is doped with W, there more carriers exist in the d$_{xz,yz}$ band in the low temperature phase. Thus, in increasing the temperature, the M-I transition occurs earlier than the undoped case.

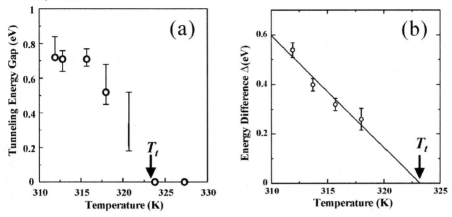

Figure 6. (a) Temperature dependence of tunneling energy gap, (b) Temperature dependence of energy difference between the bottom of the $d_{xz,yz}$ band and the bottom of the upper $d_{x^2-y^2}$ band.

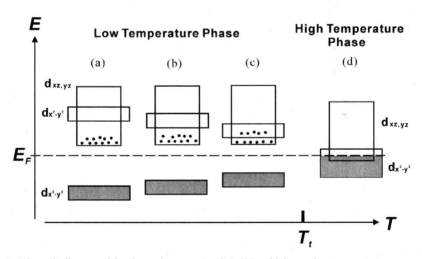

Figure 7. Schematic diagram of the change in energy bands in VO₂ with increasing temperature.

CONCLUSION

In the low-temperature phase, we observed the upper d_{x2-y2} band in the conduction band. This upper d_{x2-y2} band is one of the Mott-Hubbard split bands. We suggest that the cause of Mott-Hubbard Transition in VO$_2$ from insulator to metallic state is the electron transfer from the $d_{xz,yz}$ band to the Mott-Hubbard split upper d_{x2-y2} band. A crystallographic distortion by Peierls-transition-like effects brings the upper d_{x2-y2} band close to the bottom of conduction band $d_{xz,yz}$, which yield the electron transfer to the former one.

ACKNOWLEDGEMENT

This work is supported by a grant from the Marubun Research Promotion Foundation.

REFERENCES

[1] F. G. Morin, *Phys. Rev. Lett.*, **3**, 34 (1959).

[2] D. Yin, N. Xu, J. Zhang and X. Zheng, *J. Phys. D:Appl. Phys.*, **29**, 1051 (1996).

[3] J. B. Goodenough, *J. Solid State Chem.*, **3**, 490 (1971).

[4] A. Zylbersztejn and N. Mott, *Phys. Rev. B*, **11**, 4383 (1975).

[5] T. M. Rice, H. Launois and J. P. Pouget, *Phys. Rev. B*, **73**, 3042 (1994).

[6] S. Shin, S. Suga, M. Taniguchi, M. Fujisawa, H. Kanzaki, A. Fujimori, H. Daimon, Y. Ueda, K. Kosuge and S. Kachi, *Phys. Rev. B*, **41**, 4993 (1990).

[7] A. Liebsch, H. Ishida and G. Bihlmayer, *Phys. Rev. B*, **71**, 085109 (2005).

[8] C. Kim, Y. Oikawa, J. S. Shin and H. Ozaki, *J. Phys.:Condens. Matter*, **18**, 9863 (2006).

[9] C. Kim, J. S. Shin and H. Ozaki, *J. Phys.:Condens. Matter*, **19**, 096007 (2007).

INFLUENCE OF Yb_2O_3 AND Er_2O_3 ON $BaTiO_3$ CERAMICS MICROSTRUCTURE AND CORRESPONDING ELECTRICAL PROPERTIES

V.V. Mitic, Z.S. Nikolic, V. Paunovic, D. Mancic, Lj. Zivkovic
Faculty of Electronic Engineering, University of Nis
Nis, Serbia

V.B.Pavlovic
Faculty of Agriculture, University of Belgrade
Belgrade, Serbia

B.Jordovic
Faculty of Technical Sciences, University of Kragujevac
Cacak, Serbia

ABSTRACT

In this article the additives Yb and Er oxides are used as doping materials for $BaTiO_3$-based multilayer devices. The amphoteric behavior of these rare-earth ions leads to the increase of dielectric permittivity and decrease of dielectric losses. $BaTiO_3$-ceramics doped with 0.01 up to 0.5 wt % of Yb_2O_3 and Er_2O_3 were prepared by conventional solid state procedure and sintered up to $1320^{\circ}C$ for four hours. In $BaTiO_3$ doped with a low level of rare-earth ions the grain size ranged from 10-60μm. With the higher dopant concentration the abnormal grain growth is inhibited and the grain size ranged from between 2-10 μm. The measurements of capacitance and dielectric losses as a function of frequency and temperature have been done in order to correlate the microstructure and dielectric properties of doped $BaTiO_3$-ceramics. The temperature dependence of the dielectric constant as a function of dopant amount has been investigated.
Keywords: $BaTiO_3$-ceramics, dopant, dielectric constant, microstructure

INTRODUCTION

Due to their high dielectric constant, thermal stability and low losses barium titanate based materials are one of the most common ferroelectrics, with extensive use as a dielectric materials for multilayer ceramic capacitors (MLCCs), embedded capacitance in printed circuit boards, thermal imaging and actuators, dynamic random access memories (DRAM) in integrated circuits [1-4]. Since donor and acceptor type additions are basic components of dielectric materials based on $BaTiO_3$, extensive studies have been carried out on their effect on the defect structure and related properties of $BaTiO_3$ [5-7]. According to them, two types of dopants can be introduced into $BaTiO_3$: large ions of valency 3+ and higher, can be incorporated into Ba^{2+} positions, while the small ions of valency 5+ and higher, can be incorporated into the Ti^{4+} sublattice [8-10].Bassically, the extent of the solid solution of a dopant ion in a host structure depends on the site where the dopant ion is incorporated into the host structure, the compensation mechanism and the solid solubility limit [11]. For the rare-earth-ion incorporation into the $BaTiO_3$ lattice, the $BaTiO_3$ defect chemistry mainly depends on the lattice site where the ion is incorporated [12]. It has been shown that the three-valent ions incorporated at the Ba^{2+} -sites act as donors, which extra donor charge is compensated by ionized Ti vacancies ($V_{Ti}^{'''}$), the

231

three-valent ions incorporated at the Ti^{4+} -sites act as acceptors which extra charge is compensated by ionized oxygen vacancies (V$_O^{..}$), while the ions from the middle of the rare-earth series show amphoteric behavior and can occupy both cationic lattice sites in the BaTiO$_3$ structure [11]. As a result, the abnormal grain growth and the formation of deep and shallow traps at grain boundaries influenced by the presence of an acceptor-donor dopant can be observed. Taking into account that optimisation of the electrical properties of these materials requires microstructures of high density and homogeneous grains in this article the influence of Yb$_2$O$_3$ and Er$_2$O$_3$ on BaTiO$_3$ ceramics microstructure and corresponding electrical properties.

EXPERIMENTAL PROCEDURE

Samples were prepared from BaTiO$_3$ commercial powders (MURATA) , and with small amounts of Yb$_2$O$_3$ and Er$_2$O$_3$ from 0.01 up to 0.5 wt. %. The samples were prepared by conventional solid state procedure and sintered in the tunnel furnace type CT-10 MURATA at the temperature of 1320°C for 2 hours. Microstructure characterizations for various samples have been carried out by scanning electron microscope of the JEOL-JSM-T20 type, which enables the observation of samples surface by enlarging to 35000 times, with the resolution of 4.5 nm. The grain size distribution and porosity of the samples were obtained by LEICA Q500MC Image Processing and Analysis System. The linear intercept measurement method was used for estimating the grain size values, as well as the pores volume ratios. The capacitance and loss tangent were measured using HP 4276A LCZ meter in the frequency range from 1-20 KHz.

RESULTS AND DISSCUSSION

Microstructure development

The consolidation of ceramics powders on the base of barium-titanate has a great importance, especially from the point of view of further prognosis and properties design of these ceramics. Our investigations showed that for the sintering temperature of 1320°C ceramic densities varied from 72% of theoretical density (TD), for high doped samples, to 89%TD for the low doped samples, being higher for Yb doped ceramics. Microstructure investigations of the samples sintered with Er$_2$O showed that the grains were irregulary polygonaly shaped (Fig. 1).

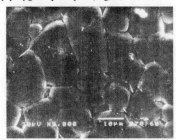

Figure 1. SEM micrograph of BaTiO$_3$ doped with 0.5% of Er$_2$O$_3$

For the lowest concentration, the size of the grains was large (up to 30 μm), but by increasing the dopant concentration the grain size decreased. As a result, for 0.5 wt% of Er$_2$O the grain was about 10 μm, while for the samples doped with 1wt% of Er$_2$O$_3$ grain size drastically decreased to the value of only few μm. Spiral concentric grain growth which has been noticed for the samples sintered with

0.01 wt% of Er₂O₃ disappeared when the concentration increased up to 1 wt% of Er₂O₃. For these samples the formation of the "glassy phase" indicated that the sintering was done in liquid phase. This is in accordance with the EDS analysis which has been shown that for the small concentration of Er dopant was uniformly distributed, while the increase of dopant concentration led to the coprecipitation between grains (Fig.2).

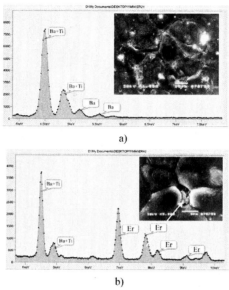

a)

b)

Figure 2. SEM/EDS spectra of BaTiO₃ doped with a) 0,1 wt% Er₂O₃ and b) 0,5 wt%Er₂O₃.

The similar microstructure development has been noticed for the samples sintered with Yb₂O₃ as well. The polygonal shaped grains were rather large (up to 30 μm) for the samples sintered with 0.01 wt% of Yb₂O₃, while their size decresed with the increase of dopant concentration (Fig. 3). The average grain size of 10 μm has been observed for the samples sintered with 1 wt% of Yb₂O₃. Spiral concentric grain growth has been also noticed.

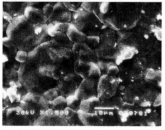

Figure 3. SEM micrograph of BaTiO₃ doped with 0,5 wt% of Yb₂O₃.

Dielectric characteristics

The dielectric properties evaluation has been made by capacitance and dielectric loss measurements in the frequency range from 100 Hz to 20 kHz. According to the obtained results (Fig. 4), the dielectric permittivity in both types of specimens maintains almost the same value for the entire frequency range (100 Hz-20 kHz).

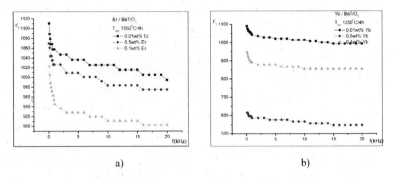

a)   b)

Figure 4. Dielectric constant vs. frequency for a) Er/BaTiO$_3$ and b) Yb/BaTiO$_3$.

After a slight higher value of ε_r at low frequency, dielectric constant becomes nearly constant at frequency greater than 5 kHz. The dielectric constant of the investigated samples ranged from 650 to 1200 at room temperature (Tab. I). For 0.01 wt% Er doped BaTiO$_3$ dielectric constant is 1200 and for 0.5 wt%Yb-BaTiO$_3$ dielectric constant is 650. In general, Er-BaTiO$_3$ samples exhibit greater dielectric constant compared with Yb-BaTiO$_3$ samples.

Table I

Sample Er or Yb in wt%	ε_r at 300K	ε_r at Tc	Tc [^{0}C]	Curie constant [K]	(T$_0$) [^{0}C]	γ
0.01 Er	1110	4505	127	1.93· 10^5	98	1.12
0.5 Er	1099	4100	128	1.24· 10^5	88	1.19
0.1 Er	1022	3750	127	1.46· 10^5	96	1.17
0.01 Yb	1090	5450	127	2.29· 10^5	85	1.19
0.5 Yb	614	1660	124	1.55· 10^5	26	1.15
0.1Yb	947	4500	127	1.91· 10^5	85	.1.15

The influence of additive type and microstructural characteristics on the dielectric behavior of Er and Yb-doped BaTiO$_3$ can be evaluated through permittivity-temperature response curves (Fig 5.). The greatest change in dielectric constant vs. temperature for low doped samples (0.01 wt%) is observed in Yb doped BaTiO$_3$ for which the dielectric constant at Curie temperature is 5500. A relatively stable capacitance response in function of temperature up to 100^{0}C has been noticed in all doped samples. With higher dopant concentration (0.5 wt%) the flatness of permittivity temperature response is observed for Yb doped samples.

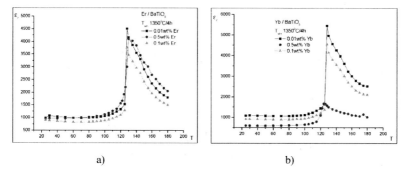

Figure 5. Dielectric constant vs. temperature for a) Er/BaTiO₃ and b) Yb/BaTiO₃.

The decrease in dielectric constant in doped samples with the increase of dopant concentration, are due to the nonhomogeneous distribution of additive throughout the specimens. The Curie temperature (T_C), determined from the maximum of the dielectric constant ε_r in the dielectric temperature characteristic, was in the range from 124 for BaTiO₃ doped with 0.5 wt% of Yb to 128°C for the samples doped with 0.5 wt% of Er. All specimens have a sharp phase transition and follow the Curie-Weiss law. Data for other specimens were omitted for clarity although they have been used to calculate the Curie constant (C) and Curie-Weiss temperature (T_0). The Curie constant (C) decreases with the increase of additive amount in both types of specimens and have an extrapolated Curie-Weiss temperature (T_0) down to lower temperature. The critical exponent γ for BaTiO₃ single crystal is 1.08 and gradually increases up to 2 for diffuse phase transformation in modified BaTiO₃. In our case the critical exponent γ is in the range from 1.12 to 1.19, and increases with the increase of additive concentration.

CONCLUSION

In this article the investigations of the influence of rare earth dopants Yb₂O₃ and Er₂O₃ on BaTiO₃ ceramics microstructure and corresponding electrical properties has been presented. Our investigations showed that for the sintering temperature of 1320°C ceramic densities varied from 72% of theoretical density (TD), for high doped samples, to 89%TD for the low doped samples, being higher for Yb doped ceramics. We have noticed that the increase of rare-earth cations content inhibits the abnormal grain growth. The average grain size in specimens doped with low content of additive (0.01-0.1 wt%) ranged between 10-30µm and that with 0.5 wt% ranged from 5-15 µm. Dielectric mesaurements showed that, in general, Er-BaTiO₃ samples exhibit greater dielectric constant compared with Yb-BaTiO₃ samples. . The dielectric constant of the investigated samples ranged from 650 to 1200 at room temperature. For 0.01 wt% Er doped BaTiO₃ dielectric constant is 1200 and for 0.5 wt%Yb-BaTiO₃ dielectric constant is 650. The decrease in dielectric constant in doped samples with the increase of dopant concentration, was explained by nonhomogeneous distribution of additive throughout the specimens.

The critical exponent γ for BaTiO$_3$ single crystal was 1.08 and gradually increased up to 2 for diffuse phase transformation in modified BaTiO$_3$. The obtained results enables further optimisation of electrical properties of barium-titanate based materials especially from the intergranular contacts point of view.

Acknowledgements: This research is a part of the project "Investigation of the relation in triad: synthesis-structure-properties for functional materials" (No.142011G). The authors gratefully acknowledge the financial support of Serbian Ministry for Science for this work.

REFERENCES

[1] S. Wang, G.O. Dayton Dielectric Properties of Fine-Grained Barium Titanate Based X7R Materials J. Am. Ceram. Soc. 82 (10), (1999) 2677–2682

[2] C.Pithan, D.Hennings, R. Waser Progress in the Synthesis of Nanocrystalline BaTiO3 Powders for MLCC International Journal of Applied Ceramic Technology 2 (1), (2005) 1–14

[3] B.D.Stojanovic, C.R.Foschini, V.Z.Pejovic, V.B.Pavlovic, J.A.Varela, "Electrical properties of screen-printed BaTiO$_3$ thick films, Journal of the European Ceramic Society 24 (2004) 1467-1471

[4] J. Nowotny, M. Rekas Positive temperature coefficient of resistance for BaTiO3-based materials Ceram. Int., 17(4) (1991) 227-41

[5] R.Wernicke The influence of kinetic process on the electrical conductivity of donor-doped BaTiO 3 ceramics Phys. Stat. Solidi (a) 47 (1978) 139

[6] V.V. Mitić, I. Mitrović, D. Mančić, "The Effect of CaZrO$_3$ Additive on Properties of BaTiO$_3$-Ceramics", Science of Sintering, Vol. 32 (3), pp. 141-147, 2000

[7] N.H.Chan, D.M.Smyth, Defect chemistry of donor-doped BaTiO3. J. Am. Ceram. Soc., 67(4) (1984) 285-8.

[8] V.V. Mitić, I. Mitrović, "The Influelnce of Nb2O5 on BaTiO$_3$.Ceramics Dielectric Properties", Journal of the European Ceramic Society, Vol. 21 (15), pp. 2693-2696, 2001

[9] H.M.Chan, M.P.Hamer, D.M.Smyth, Compensating defects in highly donor-doped BaTiO 3. J.Am. Ceram. Soc., 69(6) (1986) 507-10

[10] P.W.Rehrig, S.Park, S.Trolier-McKinstry, G.L.Messing, B.Jones, T.Shrout Piezoelectric properties of zirconium-doped barium titanate single crystals grown by templated grain growth J. Appl. Phys. Vol 86 3, (1999) 1657-1661

[11] D.Makovec, Z.Samardzija M.Drofenik Solid Solubility of Holmium, Yttrium and Dysprosium in BaTiO$_3$ J.Am.Ceram.Soc. 87 [7] 1324-1329 (2004)

[12] D. Lu, X. Sun, M. Toda Electron Spin Resonance Investigations and Compensation Mechanism of Europium-Doped Barium Titanate Ceramics Japanese Journal of Applied Physics Vol. 45, No. 11, 2006, pp. 8782-8788

DIFFUSION OF ALUMINUM INTO ALUMINUM OXIDE

Jairaj J. Payyapilly and Kathryn V. Logan.
Materials Science and Engineering Department,
Virginia Polytechnic Institute and State University,
Blacksburg, Virginia 24061.

ABSTRACT

A self-propagating high temperature synthesis (SHS) reaction involving aluminum (Al), titanium dioxide (TiO$_2$) and anhydrous boron oxide (B$_2$O$_3$) forms alpha-alumina (α-Al$_2$O$_3$) and titanium diboride (TiB$_2$) as final products. The SHS reaction progresses rapidly to the stable, equilibrium state. If the reaction rate is reduced, certain metastable products/compounds can form having unique microstructure. The present research involves a study of the interaction of Al with Al$_2$O$_3$ by simulating various SHS reaction kinetic conditions. The interfacial region between Al and Al$_2$O$_3$ is studied using transmission electron microscopy (TEM) to determine the phases formed before the stable oxide of Al is reached. The measured atomic concentration distribution of the interacting species through the interfacial region is used to understand the mechanisms of Al diffusion into Al$_2$O$_3$. An attempt has been made to draw a correlation between the phases formed at the interface and the variation in diffusivity of Al through the interface. Such correlation is used to define the kinetics of reactant interactions in SHS reactions.

INTRODUCTION

An SHS reaction involving stoichiometric amounts of aluminum, anatase and anhydrous boron oxide powders forms a composite of alumina and titanium diboride as given by Eq. (1).[1,2] Information about the formation of product phases, other than the stoichiometric reaction product as given by Eq. (1), has not been generally reported.

$$3TiO_2 + 3B_2O_3 + 10Al \rightarrow 3TiB_2 + 5Al_2O_3 \qquad ...(1)$$

Logan[3] has observed that compounds and phases that form at the SHS reaction rate (~50°C/min) are different from products that form during the slower reaction rates (~10°C/min) involving the same reactants. Compounds and phases that form at a slower heating rate have a high-aspect ratio morphology compared with products that form at a faster heating rate. Materials with high aspect ratio morphology could potentially be used for structural applications with improved properties[4] such as toughness[5] and strength.

Kinetics of the stoichiometric SHS reaction would depend on the readiness with which metallic Al would be available to reduce the oxides of Ti and B. Aluminum on the surface of the Al particles would readily oxidize to form a passive shell of Al$_2$O$_3$. Further reduction of the Ti and B oxides is restricted by the availability of metallic Al from within the Al$_2$O$_3$ shell. The difference between the coefficients of thermal expansion of Al and Al$_2$O$_3$ causes the Al$_2$O$_3$ shell to break open at high temperatures. A thicker oxide shell would require a relatively higher temperature to expose additional metallic Al from within the oxide shell on the powder. Thus SHS reaction kinetics could be dependent on the rate of formation of an aluminum oxide shell. The rate of formation of an oxide shell on the surface of aluminum is dependent on the diffusion of both Al and O ions through the Al$_2$O$_3$ layer.[6,7] Aluminum diffuses into the oxide by two mechanisms as outlined by Gall et al.[8]: a lattice diffusion mechanism and a sub-boundary mechanism. According to *Jeurgens et al.*,[9] oxidation of aluminum occurs by different possible mechanisms depending on the temperature regime, pressure and composition of the gas it is exposed to. Kinetics of aluminum oxidation has been well described by *Jeurgens L. P. H. et al.*[9] *Ciacchi L. C. and Payne M. C.*[10] have described a basic mechanism underlying the various steps and mechanism of oxidation under specific conditions.

M. Le Gall et al. describes experiences and difficulties in measuring the diffusivity of Al in an α-Al_2O_3 single crystal.[11] The results showed that aluminum and oxygen diffuse at about the same velocity in the lattice with aluminum diffusing only slightly faster than oxygen. However, in sub-boundaries, aluminum diffuses more rapidly than oxygen. Experiments by *Paladino and Kingery*[12] showed that the aluminum diffusivity values obtained by Le Gall et al. were three orders of magnitude lower than the ones obtained by *Paladino and Kingery*. It was thus considered that the diffusion rates of both aluminum and oxygen are similar. From the results obtained by *M. Le Gall et al.*[11] it is suggested that aluminum diffuses faster than oxygen in dislocations. Based on the results and analysis shown, *Le Gall et al.*[11] suggest that α-Al_2O_3 scale grows by simultaneous transport of both aluminum and oxygen.

ATOMIC CONCENTRATION MEASUREMENT AND DIFFUSIVITY CALCULATION

Processes like recrystallization, grain growth and solid state reactions are generally driven by diffusion. Knowledge of diffusion is essential for synthesis of products with enhanced microstructure for application at high temperatures.

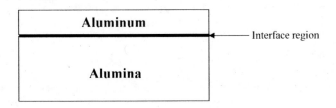

Figure 1. Schematic of diffusion couple structure to be annealing and used for diffusivity measurement.

Figure 1 shows a schematic of the film-substrate sample structure that is used for atomic concentration measurement studies for diffusivity measurement. The binary diffusion couple experiment for the Al-Al_2O_3 system is aimed at determining the intrinsic diffusivity of the diffusing species (Al) as a function of its activity (a_{Al}) in a composition gradient in the diffusion zone. Although the interdiffusion coefficient is a good measure of redistribution of components during the diffusion process, it does not give any information about relative diffusivities of the species. A more fundamental quantity to measure would be intrinsic or tracer diffusivities of the involved species which is related to the atomic fluxes (J_i) with respect to the lattice planes by Fick's law given as[13]:

$$J_{Al} = -D_{Al}\frac{\partial C_{Al}}{\partial x} = -D_{Al}\frac{V_{Al2O3}}{V_M^2}\frac{\partial N_{Al}}{\partial x} \qquad \ldots (2)$$

where D_{Al} and D_O are intrinsic diffusivities of aluminum and oxygen respectively in Al_2O_3 system. N_{Al} and N_O are the mole fractions of aluminum and oxygen respectively at the interface region where measurement is required to be done. V_{Al2O3} is the mole fraction of Al_2O_3 and V_m is the molar volume. The ratio intrinsic diffusion coefficient at the plane of interface (Kirkendall plane) between the two regions can be calculated as[13]:

$$\frac{D_{Al}}{D_O} = \frac{V_{Al}}{V_O} \left[\frac{N_{Al}^+ \int_{-\infty}^{x_k} (\frac{N_{Al} - N_{Al}^-}{V_m}) dx - N_{Al}^- \int_{x_k}^{\infty} (\frac{N_{Al}^+ - N_{Al}}{V_m}) dx}{-N_O^+ \int_{-\infty}^{x_k} (\frac{N_{Al} - N_{Al}^-}{V_m}) dx + N_O^- \int_{x_k}^{\infty} (\frac{N_{Al}^+ - N_{Al}}{V_m}) dx} \right] \quad \text{... (3)}$$

N^+_{Al} is the composition of aluminum at the aluminum end of the diffusion couple where diffusion has not occurred. Similarly N^-_{Al} is the composition of aluminum at the oxide end of the diffusion couple before the sample is annealed. V_m is the total molar volume of the oxide. V_{Al} and V_O are partial molar volumes of aluminum and oxygen respectively. The molar volume is to be considered constant since the compound is a near stoichiometric compound. It is sufficient to measure the ratio of $(D_A V_B)/(D_B V_A)$ instead of determining V_B/V_A and calculating the ratio D_A/D_B. Tracer diffusivity of oxygen in Al_2O_3 has been measured by *Nabatame et al.,*[14] which can be used to calculate the self diffusivity of Al at the Al/Al_2O_3 interface.

The intrinsic diffusion coefficient D_{Al} is associated to the tracer diffusivity D^*_{Al} via the Darken-Manning formula:[15,16,17]

$$D_{Al} = D^*_{Al} [\frac{\partial \ln a_{Al}}{\partial \ln N_{Al}}](1 + W_{Al}) \frac{V_m}{V_O} \quad \text{... (4)}$$

where a_{Al} is the chemical activity of aluminum (the term $\partial \ln a_{Al}/\partial \ln N_{Al}$ is the thermodynamic factor) and the W_{Al} is the vacancy wind factor. The measured ratio $(D_A V_B)/(D_B V_A)$ can be related to the tracer diffusion coefficients through:

$$\frac{D_{Al} V_O}{D_O V_{Al}} = \frac{D^*_{Al}(1 + W_{Al})}{D^*_O (1 - W_O)} \quad \text{... (5)}$$

The vacancy wind factor W_i formulates how the net vacancy flux influences the mobility of the diffusing species.

$$W_{Al} = \frac{2N_{Al}(D^*_{Al} - D^*_O)}{M_\circ (N_{Al} D^*_{Al} + N_O D^*_O)} \quad \text{... (6)}$$

$$W_O = \frac{2N_O(D^*_{Al} - D^*_O)}{M_\circ (N_{Al} D^*_{Al} + N_O D^*_O)} \quad \text{... (7)}$$

In the above equations, the correlation effects are considered through M_o. According to the theories presented by Manning and Ikeda[18,19] diffusion of the minor elements into the sublattice of the other species occurs via. the ordinary vacancy mechanism. Thus we could establish M_o=4.43.

EXPERIMENTAL PROCEDURE

An aluminum (99.9% pure) film about four μ thick was sputter deposited (Innovative Coatings LLC.) on Al_2O_3 substrates (α-Al_2O_3 having 99.99 % purity. CERAC Inc.), vacuum deposition grade with 10-12 mm diameter x 4-5 mm height dimensions. Surface roughness of substrates was less than one μ as determined by AFM. The surface texture allowed adequate adherence of the film to the substrate after deposition.

The film-substrate samples were treated to elevated temperature for extended periods of time before the interface region was studied using TEM. For the sake of consistency, Al-Al_2O_3 samples were treated under the same set of constraints, e.g. were heated at the same temperature for the same periods of time. Diffusivity of oxygen at 700°C into Al_2O_3 as reported by *Nabatame et al.*[14] ($1.8x10^{-23}$ m^2/second) was used to calculate the depth of diffusion of either of the diffusing species by using the well know equation:

$$l = \sqrt{Dt} \quad \dots (8)$$

where l is the depth of diffusion for the diffusing species, D is the diffusivity of the diffusing species into substrate and t is the time required for diffusion. It was calculated that the diffusing species (oxygen) would travel a depth of 50 nm, 100 nm and 200 nm by exposing the sample to 700°C for 9 hours, 36 hours and 144 hours respectively. It was decided to choose 9 hours and 36 hours as exposure times at 700°C to study the interface. The first sample was introduced in a furnace maintained at 700°C in air and the sample was soaked at 700°C for 9 hours and the second sample for 36 hours. The samples were allowed to furnace cool after which the interfaces between the film and the substrate were inspected using the TEM. Cross-section slices were prepared from the untreated samples at the interface for the TEM investigation using the cutting-dimpling-milling method. Focused ion beam milling was used to thin cross-sections at the interface of samples treated at elevated temperatures.

RESULTS AND DISCUSSIONS

The untreated and treated Al-Al_2O_3 samples were studied using a Philips EM 420 scanning transmission electron microscope (STEM). The interface between the film and the substrate is the area of interest and is probed using the TEM. There are clearly four layers observed at the interfacial region. The TEM sample was prepared initially by mounting an Al_2O_3 substrate over the Al side of the Al-Al_2O_3 sample. It was not trivial to identify the four layers as seen in Figure 2. Layers 1, 2, 3 and 4 are the alumina substrate, aluminum film, glue and the alumina substrate again respectively. The interface between the film and the substrate as shown in Figure 2 is used for comparison with the treated samples as studied using the TEM. The layers were identified using TEM micrographs and the respective selected area diffraction patterns (SADP) as shown below. Figures 3, 4, 5 and 6 are micrographs of individual layers.

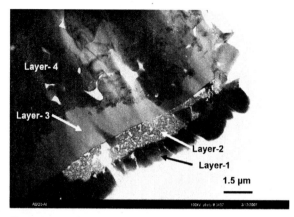

Figure 2. Overview of the interface of the untreated Al-Al$_2$O$_3$ film-substrate sample as observed using the TEM

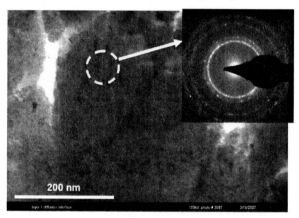

Figure 3. Layer 1 with the selective area diffraction pattern (inset) showing FCC copper.

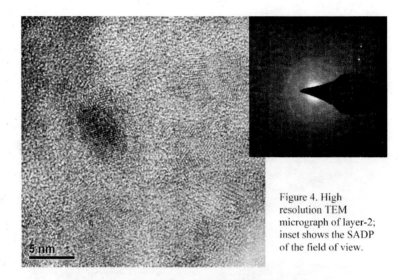

Figure 4. High resolution TEM micrograph of layer-2; inset shows the SADP of the field of view.

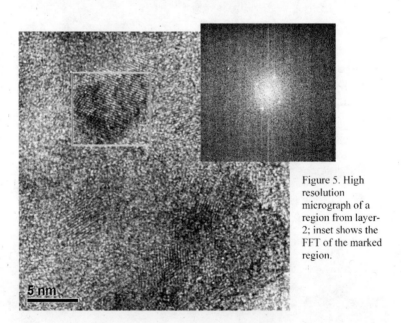

Figure 5. High resolution micrograph of a region from layer-2; inset shows the FFT of the marked region.

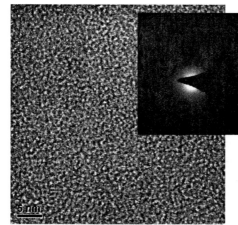

Figure 6. High resolution
micrograph of layer 3;
inset shows the SADP
having the diffused ring
pattern.

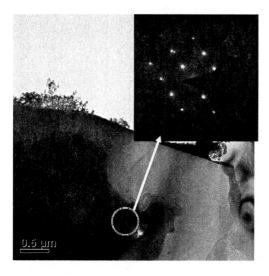

Figure 7. TEM micrograph of layer-4; inset shows SADP of the marked region
having a crystalline region.

Figure 3 shows the area over which the diffraction pattern was obtained and the inset shows the diffraction pattern of the field of view. The SADP was indexed and was found to be the face centered cubic (FCC) structure of copper. The sample was milled at the interface to get a tiny hole at the center and also to thin the region enough so as to be suitable for TEM analysis. The sample was mounted on a copper grid for sake of mechanical rigidity. This explains the presence of FCC copper on the surface of the Al-Al$_2$O$_3$ sample. Figures 4 and 5 show high magnification micrographs of regions in layer-2. The inset in Figure 4 shows the selective area diffraction pattern (SADP) over the area shown in the micrograph. The SADP shows the ring pattern indicating that the material is polycrystalline in nature. The inset in Figure 5 shows the fast Fourier Transformation (FFT) of a high resolution TEM micrograph which indicates that the material is crystalline. Neither the SADP nor the FFT was indexed but the FFT clearly indicates that the pattern is not hexagonal structure (for Al$_2$O$_3$) but the pattern could have cubic symmetry indicating that material from layer-2 could possibly be Al. Layer-3 is possibly the glue that was used to attach the Al$_2$O$_3$ substrate on the Al-Al$_2$O$_3$ sample. Figure 6 shows a high resolution image of layer 3 and the inset shows the SADP having fuzzy rings of light indicating that the material is amorphous. Layer-4 has grains with well defined grain boundaries as would be expected for the Al$_2$O$_3$ substrate. SADP of a region in layer-4 shows that the material is crystalline as is seen in the inset of the TEM microstructure in Figure 7. Layers 1 and 4 do not appear similar possibly due to the fact that copper was re-deposited over layer-1 while layer-4 was clear of copper.

TEM ANALYSIS

Figure 8 shows a micrograph of the interfacial region of the sample that was exposed to 700°C for 9 hours. The different regions are clearly marked on the micrograph. The micrograph shows the polycrystalline Al$_2$O$_3$ grains from the substrate region. The Al film is the dark region above the oxide grains. There is a clear line of separation between the Al$_2$O$_3$ region and the Al-film region. The dotted region near the Al-film side is an artifact that will not be visible when the sample is tilted about the X- or Y-axis. There is no indication of an interfacial region between the film and the substrate in this case.

Figure 9 shows a micrograph of the interfacial region of the sample that was exposed to 700°C for 36 hours. It is not a distinct line that separates the film and the substrate as in the previous sample but instead the separation is a fuzzy region. The fuzzy region is not likely to be an artifact since it is seen even when the sample is rotated about the various possible axes on the double-tilt sample holder in the TEM. Figure 10 shows the interfacial region at different locations along the region separating the film from the substrate. The strip of interfacial region between the film and the substrate is not continuous. The reason for the discontinuous nature of the interfacial strip is not yet known.

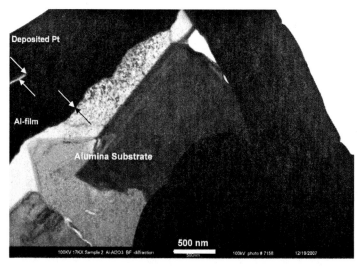

Figure 8. TEM micrograph showing the interface between the film and substrate for the Al-Al$_2$O$_3$ sample soaked at 700°C for 9 hours.

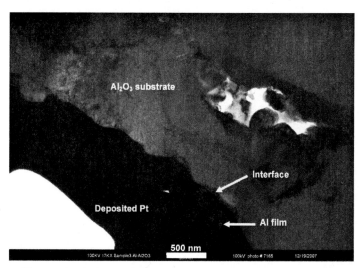

Figure 9. TEM micrograph of film-substrate of Al-Al$_2$O$_3$ sample exposed to 700°C for 36 hours.

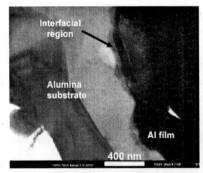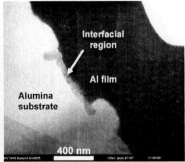

Figure 10. TEM micrographs of the region between the Al film and Al_2O_3 substrate showing the interfacial region after the sample was exposed to 700°C for 36 hours.

DIFFUSIVITY CALCULATIONS

The molar concentrations of Al and O atoms (N_{Al} and N_O respectively) in Al_2O_3 are to be measured and used in Eq. (3), (6) and (7). An X-ray photospectrometer could be used to measure the concentration of the diffusing species from one end (Al - end) of the diffusion couple (semi infinite diffusion couple) through the interface to the other end (Al_2O_3 - end). The molar concentration of Al at the Al_2O_3 end (where the diffusion of Al from the Al side has not reached) would have a certain residual value N_{Al}^-. Similarly N_O^+ is the residual molar concentration of O atoms on the Al_2O_3 side. N_{Al} and N_O are measured at various points from one end of the semi-infinite diffusion couple through the interface to the other end. The Al concentration profile is to be plotted as 'percentage of Al' versus the 'distance of the measurement from one end' of the diffusion couple to the other end.

The concentration profile could be one of several ways as explained below. The concentration profile could be a step function of 'distance' with ($n+2$) steps when there are n phases formed at the diffusion interface. In the event of a single phase at the interface, a three-step concentration profile would be simple and diffusivity calculations would be minimized. However, with multiple phases at the interface, diffusivity at each phase would have to be measured and calculated separately. The step (new phase/compound region or the diffusion region) on the concentration curve could either have a non-zero slope or a zero slope depending on whether the new phase/compound varies in concentration. A positive slope would mean that the new phase/compound does not have a fixed concentration but steadily varies from one end to the other. It could also mean that the compound could have variable stoichiometry similar to many oxide systems. The step would be flat when the phase/compound formed would be stoichiometric or the concentration of any of the species that form the phase or compound does not vary. In both the above mentioned cases, the value of $(D_A V_B)/(D_B V_A)$ would be non-zero, allowing a feasible calculation to determine the diffusivity. In case the diffusion region comprises of two phases, one on either side of the Kirkendall plane, and that the Kirkendall plane is not easily visible due to the irregular nature of diffusion of the two diffusing species into each other, it can still be roughly calculated. The contours on both sides of the interface, in the diffusion region, would be fitted with a polinom. The software fitted polinom would help determine the area on both sides of the original plane of interface. The average thickness of the diffusion layer would be calculated by summing the two areas mentioned above. The position of the Kirkendall plane can be calculated by taking the ratio of the two areas measured.

The value $(D_{Al}V_O)/(D_OV_{Al})$ can be calculated from Eq. (3), by using the measured value of N_{Al}. Substituting for values of W_{Al} and W_O from Eq. (6) and (7) into Eq (5) and using the calculated value of $(D_{Al}V_O)/(D_OV_{Al})$, it is possible to calculate the tracer diffusivity of Al into Al_2O_3.

For simplification let the measured and calculated value of

$$\frac{D_{Al}V_O}{D_OV_{Al}} = P \quad \dots (9)$$

Also as mentioned above,

$$P = \frac{D_{Al}^*\left[1 + \dfrac{2N_{Al}\left(D_{Al}^* - D_O^*\right)}{M_\circ(N_{Al}D_{Al}^* + N_O D_O^*)}\right]}{D_O^*\left[1 - \dfrac{2N_O\left(D_{Al}^* - D_O^*\right)}{M_\circ(N_{Al}D_{Al}^* + N_O D_O^*)}\right]} \quad \dots (10)$$

Simplifying equation 8 in terms of the tracer diffusivity D_{Al}^*,

$$D_{Al}^*\big)^2 N_{Al}(M_\circ + 2) + D_{Al}^*[M_\circ N_O D_O^* - 2N_{Al}D_O^* - P(D_O^*)^2 M_\circ N_{Al} + 2P(D_O^*)^2 N_O] = P(D_O^*)^3 N_O(M_\circ + 2) \dots$$
$$(10)$$

The tracer diffusivity of Al can thus be determined by solving for D_{Al}^* in the above equation (Eq. 10).

SUMMARY AND CONCLUSIONS

Results of the interaction between $Al-Al_2O_3$, components of the SHS reactants are summarized. An interfacial region is formed between Al and Al_2O_3 when exposed to an elevated temperature for extended periods of time. The interfacial region is not continuous along the $Al-Al_2O_3$ interface. Determination of the diffusivities of Al and O is possible. The diffusivity values will clearly determine the mechanism of oxidation of Al in the presence of Al_2O_3. Al oxidation and the interaction of Al with its oxide could be a significant parameter for controlling the kinetics of the SHS reaction.

ACKNOWLEDGEMENTS

The authors thank Dr. William Reynolds Jr. for valuable discussions and help with TEM characterization. Thanks to the Multifunctional Materials Research Group for constructively critiquing this article. Part of this work was carried out using instruments at the Nanoscale Characterization and Fabrication Laboratory, a Virginia Tech facility partially supported by the Institute for Critical Technology and Applied Science. The authors gratefully acknowledge the financial support from the National Institute of Aerospace under the Grant Award # VT-03-01.

REFERENCES

1. K. V. Logan and J. D. Walton, 'TiB$_2$ Formation using Thermite Ignition,' *Ceramic Engineering and Science Proceedings*, W. J. Smothers, ed., The American Ceramic Society, Columbus, Ohio, No. 7-8, pp. 712-738 (1984).
2. I. K. Lloyd, K. J. Doherty and G. A. Gilde, 'Phase Equilibrium Studies in Al$_2$O$_3$-TiB$_2$,' *Ceramic Armor Materials by Design*, Ceramic Transactions, Vol. 134, pp. 623-628 (2002).
3. K. V. Logan, Private communication.
4. K. V. Logan, *U.S. Patent No. 6,090,* 321 (July 18, 2000).
5. H. Teshima, K. Hirao, M. Toriyama and S. Kanzaki, 'Fabrication and Mechanical Properties of Silicon Nitride Ceramics with Unidirectionally Oriented Rod-like Grains,' *Journal of the Ceramic Society of Japan*, Vol. 10, No. 12, pp. 1216-1220 (December 1999).
6. C. Wagner, 'Passivity and Inhibition During the Oxidation of Metals at Elevated Temperatures', Corrosion Science, Vol. 5, pp. 751-764, (1965).
7. G. J. Yurek, in Corrosion Mechanisms, edited by F. Mansfield (Marcel Dekker, Inc., New York, NY), pp. 397-446, (1987).
8. M. Le Gall, B. Lesage and J. Bernardini, 'Self-diffusion in α–Al$_2$O$_3$, I. Aluminum Diffusion in Single Crystals', *Philosophical Magazine A*, Vol. 70, No. 5, pp. 761-773, (1994).
9. L. P. H. Jeurgens, W. G. Sloof, F. D. Tichelaar and E. J. Mittemeijer, 'Growth Kinetics and Mechanisms of Aluminum-Oxide Films Formed by Thermal Oxidation of Aluminum,' *Journal of Applied Physics*, Vol. 92, No. 3, pp 1649-1656, (August 2002).
10. Ciacchi L. C. and Payne M. C., 'Hot-Atom O$_2$ Dissociation and Oxide Nucleation on Al(111),' *Physical Review Letters*, Vol. 92, No. 17, pp 176104-1 – 176104-4, (30 April 2004).
11. M. Le Gall, B. Lesage and J. Bernardini, 'Self-Diffusion in α-Al$_2$O$_3$ I. Aluminum Diffusion in Single Crystals,' *Philosophical Magazine A*, Vol. 70, No. 5, pp. 761-773 (1994).
12. P. E. Paladino and W. D. Kingery, 'Aluminum Ion Diffusion in Aluminum Oxide,' *The Journal of Chemical Physics*, Vol. 37, p. 957-962 (1962).
13. C. Cserhati, A. Paul, A. A. Kodentsov, M. J. H. van Dal and F. J. J. van Loo, 'Intrinsic Diffusion in Ni$_3$Al System,' *Intermetallics*, Vol. 11, pp. 291-297, (2003).
14. T. Nabatame, T. Yasuda, M. Nishizawa, M. Ikeda, T. Horikawa and A. Toriumi, 'Comparative Studies on Oxygen Diffusion Coefficients for Amorphous and γ-Al$_2$O$_3$ Films using ^{18}O Isotope,' *Japanese Journal of Applied Physics*, Vol. 42, 7205-7208, (2003).
15. P Shewmon, 'Diffusion In Solids' Second Edition, a publication of *The Minerals, Metals & Materials Society*, Warrendale, Pennsylvania 15086.
16. D. J. Shmartz, H. Domian and H. I. Aaronson, *Journal of Applied Physics*, No. 37, 1741, (1966).
17. N. R. Iorio, M. A. Dayananda and R. E. Grace, *Metallurgical Transactions*, No. 4, 1339, (1973).
18. T. Ikeda, A. Almazouzi, H. Numakura, M. Koiwa, W. Sprengel and H. Nakajima, *Acta Materialia*, No. 46, 5396, (1998).
19. M. Koiwa and S. Ishioka, Phil. Mag. A, No. 48, 1, (1983).

Science of
Ceramic Interfaces

EVALUATION OF THE INTERFACIAL BONDING BETWEEN CUBIC BN AND GLASS

Chris Y. Fang, Hoikwan Lee, Alfonso Mendoza, David J. Green, Carlo G. Pantano
Materials Research Institute, The Pennsylvania State University
University Park, Pennsylvania, USA

ABSTRACT

The goal of this study is to evaluate the mechanical strength of cubic BN-glass interfaces. To measure interfacial bonding strength, when only sub-mm sized cubic boron nitride (cBN) crystals are available, a glass-cBN-glass sandwich test structure was developed. Samples were prepared by joining two glass rods end-to-end with an intervening layer of 120–140 mesh (115±10 μm) cBN particles. The sandwich structures were heat treated to fuse the glass to the cBN particles with negligible glass-to-glass contact. Four-point bend tests were performed on these samples to obtain the nominal failure stress. The cBN-glass bonding area and nature of the de-bonded surfaces was observed using scanning electron microscopy (SEM). The interfacial bonding strength was then estimated to be 22 MPa from the integrated bonding area and the four-point bend strength.

1. INTRODUCTION

Cubic boron nitride (cBN) is an attractive material for abrasive applications due to its high hardness, thermal and chemical stability.[1,2] Glass bonded cBN grinding tools offer many benefits such as high mechanical strength, high heat resistance, sharp cutting edges, low wear rate, longer-life dressing and lower cost per part.[3] They are considered ideal abrasive tools for high speed, high efficiency, and high precision grinding with lower cost and less environment pollution.[4-6]

The mechanical strength is a performance property for glass-bonded cBN grinding products. The chemistry and morphology of the surface of the cBN particles, the composition of the glass, and the processing conditions all influence the strength through interfacial bonding of the cBN and glass matrix.[3] The objective of this work was to measure and characterize the cBN-glass interface strength directly. Large cBN crystals and dense hot-pressed cBN ceramics are not available for the creation of macroscopic interfaces for testing and thus, an alternate method has been developed, which is described here. The test is based on measurement of the bending strength of two butt-joined glass rods held together by a monolayer of cBN particles that are fused to the two glass rods. The challenges are to reproducibly create fused cBN-particle/glass interfaces (with negligible glass-to-glass bonding) and to quantitatively determine the actual cBN/glass interface area after the test. The glass used in this study is a commercial borosilicate; it has an expansion coefficient that is fairly close to cBN (Pyrex 3.3 ppm/°C[7]; cBN 3.5 ppm/°C[8]) and can be softened to fuse to the cBN at 850-900°C,[9] which is a temperature range where there is no degradation of the cBN. In particular, oxidation of cBN must be avoided. Fortunately, this does not take place below 1060°C according to the study by Li et al.[3] who found that cBN abrasives begin to oxide in air at 1063°C and approach a maximum oxidation rate at 1177 °C.

2. EXPERIMENTAL PROCEDURE

2. 1. Sample Preparation

2.1.1. Glass Rods

Pyrex glass rods (Pyrex, Corning Inc, Corning, NY) of 10 mm diameter were cut into 30 mm long cylindrical pieces on a low speed diamond saw. The alignment of the rod during cutting is critical to ensure the cut surface to be flat, smooth, and orthogonal to the length.

2.1.2. Production of cBN Monolayer

The cBN monolayer was created with the help of an organic adhesive (later burned out during the cBN/glass fusion). The cut rods were wrapped with paper with only the ends exposed for adhesive spray (bottom of Figure 1). The end to be joined was sprayed with the adhesive (Krylon[TM], Sherwin-Williams Company, Solon, OH) from a distance of about 6 inches, and then the paper was removed. This end of the rod was then dipped into the cBN powder (115±10 μm CBN 400, Diamond Innovations, Worthington, OH) with slight force to pick up a monolayer of cBN particles. The other rod was then butt-joined onto the cBN monolayer in a right-angle fixture with enough force to ensure good alignment. Appropriate force was applied to push the two rods against each other to join the two rods as shown in the fixture (top of Figure 1).

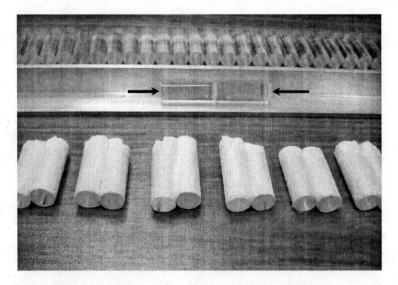

Figure 1. Preparation of glass rods for joining to cBN particles.

2.1.3. Fusing the samples

The glued sample pairs were carefully loaded into the holes of a die constructed from fibrous ceramic insulation (Fiberfrax Duraboard[®] 3000, Unifrax Corp, Niagara Falls, New York). The exposed sample ends were covered with a rectangular ceramic block with uniform thickness to produce a uniaxial stress of 10 kPa during fusion (Figure 2).

The die with samples loaded was heated to 900°C for 60 min, ramping at 5°C/min to 500°C, holding 60 min at 500°C to burn out the glue and then heating at 5°C/min to 900°C for an isothermal hold (Figure 3). The final temperature was chosen according to the thermal behavior of Pyrex glass and a matrix of preliminary experiments. At the firing temperature, the glass fused to the cBN particles without direct glass-glass bonding. After cooling, the samples were removed carefully from the die and inspected visually to ensure they were joined properly without any visible defects.

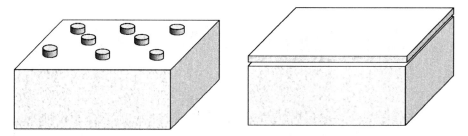

Figure 2. Joining glass pairs loaded in the insulation die for thermal fusion. A ceramic plate was used to place the samples under a small compressive stress.

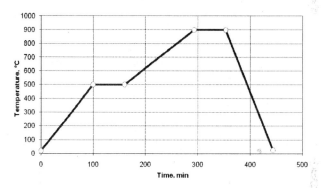

Figure 3. Temperature profile for joining samples.

2.2. Four-Point Bend Test

The four-point bend test was performed on a testing machine (Model 4206, Instron Corp., Norwood, MA). The sample fixture was made of steel with a 40 mm outer support span and 20 mm inner span, as schematically shown in Figure 4. The loading speed of 0.1 mm/min was used in the test, which led to an average failure time of 18 s. The load was recorded versus time till the sample failed (Figure 5). In the first 5 seconds or so, the loading curve was non-linear. The loading curve became linear afterwards (Figure 5). The sample failed in the joint at the maximum load. The non-linear behavior in the first few seconds is believed to be due to the sample and fixture alignment. The data sampling rate interval was 0.5 seconds. The bend strength, σ_f, was determined using [10]

$$\sigma_f = \frac{2F_c a}{\pi r^3} \qquad (1)$$

where F_c is the failure load, a is the moment arm and r is the radius of the sample rod.

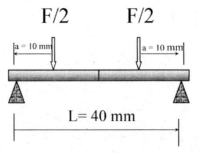

Figure 4. Schematic showing the configuration of the four-point bend test.

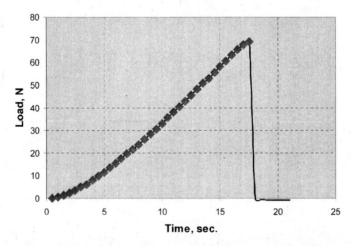

Figure 5. A typical loading curve in the 4-point bend test of the butt-joined sample.

2.3. Determination of bonding area

Assuming that only cBN-glass interface exists, and given that only a certain percentage of the nominal joint area is bonded by the cBN monolayer, a direct measurement of the actual bonded area is required to quantitatively interpret the measured strength. In this way, the interfacial strength can be determined from the measured strength divided by the fraction of bonded interfacial area. The determination of the bonding area was made on representative SEM images of the failed samples at 50 times magnification. The bonded areas could be easily identified because both the free surface of the glass and any un-bonded cBN particles appear smooth, whereas the de-bonded interfaces appear rough (as shown in Figure 6). The de-bonded areas were outlined manually, and the total de-bonded area was found by summing the individual de-bonded areas. The sum of the individual de-bonded areas was found by a paper cutting and weighing method; i.e., the enlarged SEM images were printed on paper, the weight of the whole image SEM image, w_0, was measured, then the highlighted de-bonded areas

de-bonded glass   de-bonded cBN

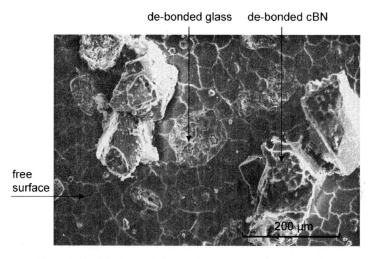

free
surface

200 µm

Figure 6. Identification of the interfacial de-bonded areas on the failed surface.

were cut off and weighed as w_1. The percentage, $(w_1/w_0)\%$, or fraction of the de-bonded areas, was taken as the bonded "interface" area. It was not practical to measure this total interface area on each failed sample. After each test, the high, low and intermediate strength samples were selected as "representative". At least 3 images from each representative sample were used to get an average value for the percentage-bonded area, which was then used to determine the interfacial strength for the entire sample set.

3. RESULTS AND DISCUSSION

Figure 7 shows the four-point bend strength data of a representative set of 23 samples. The average tensile strength of the set was 2.74±0.5 MPa, with a minimum value at 1.65 MPa, and the maximum at 3.72 MPa. Figure 8a shows one example of an SEM image used for bonding area determination. The de-bonded areas were highlighted as shown in Fig. 8b. It is to be noted that some particles were very smooth and showed no evidence of interfacial failure, and so it was concluded these surfaces were not bonded and thus, were excluded. The bonded area of the image in Fig. 8 was found to be 13.7%. From the four-point bend strength of the corresponding sample (2.85 MPa), and the percentage bonded area, the interfacial bonding strength between cBN and glass can be calculated to be 20.8 MPa (i.e. 2.85 MPa/13.7% = 20.8 MPa) for this individual sample.

In order to substantiate the method, numerous sets of samples were tested and directly evaluated for bonding area. In general, it was found that the measured bonding area was directly proportional to the failure load in the four-point bend test. Figure 9 presents this relationship for several complete sets of samples. Although the nominal cross section is the same for all the samples prepared under the same condition, the bonded areas varied with four-point bend strength. A weaker sample always shows a smaller bonded area than a stronger one. The relationship between the bonding area and the load was found to follow a linear relationship:

$$y = 0.23x \qquad\qquad (2)$$

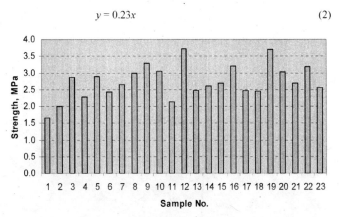

Figure 7. The four-point bend strength data of a typical set of 23 butt-joined samples.

where y is the bonded area, x is the load. From this relationship, the area could also be estimated from the load in the four-point bend test.

Interface strength is a specific property of the material system and processing and should be a constant independent of the bonded area for any one specimen. Figure 10 shows the statistical distribution of the interfacial bonding strength as a function of the bonding area for each of the specimens in Figure 9. Clearly, the interfacial bonding strength converges to 22 MPa. But, it can also be seen that when the bonding area is very low (<15-20%), the standard deviation of the data is large, whereas when the bonding area is high, the deviation is substantially smaller. This is probably a difficulty associated with the statistics of counting "small" bonding areas and/or of measuring small breaking loads. It leads to the conclusion that this test is not very precise or accurate when the bonded area is less than 15-20%.

It is fully appreciated that this proposed test procedure is approximate. In reality, the cBN-glass bonding is more or less 3-dimensional, for example, the embedded edges of the cBN particles. Fortunately, intrusion of the cBN particles into the glass is small compared to the particle size. Thus, it is reasonable to take the 2-dimensional bonding as an approximation, i.e., using the projection of the bonding area as the estimate of actual bonding area. Of course, this greatly simplifies the procedure. It remains to be determined, in the practice of this test, whether the relationship demonstrated in Figure 9 is sufficiently reproducible and universal. A primary application of this test is to evaluate the effects of matrix composition, particle surface treatments, coatings, and processing on interface strength. It is possible that the introduction of nanoscale roughness or interface crystallization (due to particle surface or matrix effects) could change the relationship between 2D geometric interface area and the bend strength of the joined glass rods.

CONCLUSIONS
A test was developed for estimating the interfacial bonding strength of cBN and Pyrex glass based upon the measuring a set of four point bend strengths for a unique sample design, and then normalizing those strengths to a measured or calibrated interfacial area. The sample consists of two glass rods joined end to end by a monolayer of cBN particles fused at 900°C for 60 min under a stress of 10 kPa

(1.5 psi.) applied in the longitudinal direction. The fused samples were broken using a four-point bending geometry. In this work, the percent bonding area was obtained by direct examination

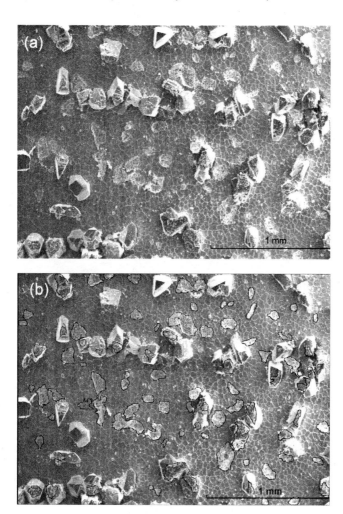

Figure 8. Example of an SEM image used for determining the bonded area: (a) original SEM image of the failure surface, and (b), bonded areas outlined and counted to yield the percentage-bonded area of 13.7%.

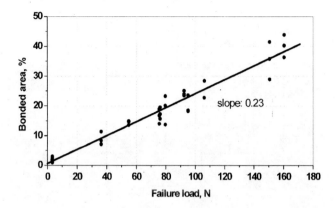

Figure 9. The bonding area (%) increases linearly with failure load.

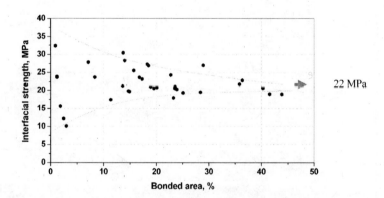

Figure 10. Statistical distribution of the calculated interfacial bonding strength versus bonded area (%).

and quantitative SEM image analysis of numerous failure surfaces (i.e., the fracture surface created by the four-point bend testing of the butt-joined glass rods). The interfacial bonding strength of the cBN-Pyrex glass was estimated to be 22 MPa.

ACKNOWLEDGEMENTS
This work was supported by Diamond Innovations. The authors would like to thank I-Kang Chen and Matthew Fimiano for technical assistance in the mechanical tests.

REFERENCES

[1]C. F. Gardiner, *Am. Ceram. Soc. Bull.*, **67**, 1006 (1988).
[2]A. C. Carius, *Tooling & Production*, **66**, 45 (2000).
[3]Z. H. Li, Y. H. Zhang, Y. M. Zhu and Z. F. Yang, Sintering of Vitrified Bond CBN Grinding Tool, *Key Engineering Materials,* **280-283**, 1391-1394 (2005).
[4]N. P. Navarro, *Machine and Tool Blue Book*, **81**, 51 (1986).
[5]H. O. Juchem, B. A. Cooley, *Industrial Diamond Reviewer*, **43,** 310 (1983).
[6]X. Chen, W.B. Rowe and R. Cai, *Inter. J. Machine Tools Manufact.,* **42**, 585 (2002).
[7]J-H. Jean, T-H. Kuan and T. K. Gupta, Crystallization inhibitors during sintering of Pyrex borosilicate glass, *J. Mat. Sci. Lett.*, **14**, 1068-1070 (1995).
[8]Z. H. Li, Y. M. Zhu, X. W. Wu, Q.M. Yuan and Z.F. Yang, Investigation on the Preparation of Vitrified Bond CBN Grinding Tools, *Key Engineering Materials* **259-260,** 33-36 (2004).
[9]E. B. Shand, Engineering Glass, in *Modern Materials*, **6**, 262 (1968).
[10]D. J. Green, An introduction to the mechanical properties of ceramics, Cambridge University Press (1998).

OXIDATION BEHAVIOUR OF HETERO-MODULUS CERAMICS BASED ON TITANIUM CARBIDE

Igor L. Shabalin
Institute for Materials Research, University of Salford
Salford, Greater Manchester, M5 4WT, UK

ABSTRACT
    Recent studies on oxidation of the TiC - 7 vol.% C (graphite) hetero-modulus ceramics, at temperatures of 400–1000 °C and oxygen pressures of 0.13-65 kPa, led to the discovery of a temperature-pressure-dependent phenomenon called 'ridge effect'. The oxidation rate of material rises rapidly to a maximum at ridge values of oxygen pressure (p_{O2}) or temperature (T), but then declines with subsequent growth of p_{O2} or T. The thermogravimetric (TGA) data and carbon analysis of oxidized samples were applied in modeling for carbon burn-off and oxygen consumption processes. The apparent activation energy Q and order of reaction (exponent) m were determined for different oxidation parameters (p_{O2}, T) and steps of the process, described using a linear-paralinear model. The ridge values of p_{O2} and T mark a change in the prevailing mechanism, as while traversing the ridge parameter the values of Q or m change their sign. The meanings of ridge parameters, which are served as boundaries between $p_{O2} - T$ regions with different oxidation mechanisms, are considered and explained on the basis of X-ray diffraction (XRD), microanalysis by electron probe (EMPA), energy-dispersive X-ray spectroscopy (EDX), optical and scanning electron microscopy (SEM) analyses. The oxidation mechanisms, essentially different within the ranges of parameters, are identified according to the developed ridge-effect approach, which can be applied probably to a variety of ceramic materials.

INTRODUCTION

    The subclass of hetero-modulus ceramics (HMC) presents the combination of ceramic matrix having high Young's modulus (300–600 GPa) with particles and/or fibers of a phase having significantly lower modulus (15–20 GPa) such as sp^2-structured graphite or boron nitride. Correspondingly it becomes more efficient to use the refractory compounds, such as carbides, oxides, nitrides etc., which possess the highest melting points, in modern high-temperature installations. The generally low thermal shock resistance of these brittle materials can be greatly improved by the addition of low-modulus phases. Similar materials, successfully applied in rocket design, were referred to as "high-E – low-E composites" in the USA[1-2] and "hetero-modulus ceramics" in the USSR, Russia and Ukraine[3-5], and also known as "soft ceramics" to emphasize another great advantage of HMC, namely the remarkable machinability by conventional tools[5] that is not normally feasible with conventional ceramics. The experience gained through the application of HMC was subsequently shifted from space and nuclear technologies to metallurgy and machinery as HMC provide significant opportunities. One of the types of HMC, in particular the refractory transition-metal carbide – carbon composites[1,3], are prospective materials for a number of high-temperature applications: as thermally stressed components of rocket motors, elements of thermal protection for re-entry spacecraft, diaphragms for casting metallurgical equipment, highly loaded brake-shoes in aviation and automobile production, high-temperature lining and heating elements and others, including TiC – carbon composites, which are real candidates for thermonuclear fusion reactors as a plasma facing material[6].

    If the scales, formed on the surface of HMC during gas exposure at elevated temperatures, possess protective properties and prevent corrosion propagation, the resistance to corrosion of carbide – carbon HMC in chemically active media, will become significantly higher than those for carbon – carbon composites, which are currently widely used at high temperatures in different technologies. Therefore, advanced HMC can provide an opportunity to increase the high-temperature strength, working temperatures and/or operational life times for components and structures. However, some previous physico-chemical studies have shown that in general the formation of oxide scales on the transition-metal carbides as a result of surface oxidation does not generate an effective barrier to inhibit the bulk oxidation of material[7-8]. The process on the carbide surface during oxygen exposure seems to be highly sensitive to temperature and gas pressure. Hence, knowledge of the oxidation behaviour, chemical kinetics and mechanism of this process can help us to prepare the corrosion-resistant scales. Coating the working surfaces of carbide – carbon HMC components, by the means of preliminary oxidation under strictly controlled conditions, could protect them from damage and erosion in high-speed and high-enthalpy gas and plasma flows. Similar technique is also of great relevance to manufacturing of functionally graded materials developed over the last decade.

EXPERIMENTAL

The carbothermic TiC powders with a particle size 1–4 μm and the additive of natural graphite with mineral remains less than 0.1 % were used for the fabrication of the TiC – 7 vol.% C (graphite) HMC in this study. The composite blanks were prepared by hot-pressing in graphite moulds at 2700 ^{0}C; all other details of the manufacturing method are given in our recent paper[3]. The physico-chemical characteristics of the HMC material are reported in Tables I-II.

Table I. Physico-chemical characteristics of TiC–7 vol.% C (graphite) HMC

Material code	C/Ti	Composition vol.%			Chemical analysis wt.%					Density	
		TiC	C	Ti	C^a	O	N	W	Fe +Co	g/cm^3	%^{b}
TKU	1.17	93	7	74.7	21.9	0.6	0.5	0.6	0.3	4.60	97.5

^{a} Total carbon content.
^{b} Relative density calculated from XRD measurements for the components.

Table II. Physico-chemical characteristics of phases in TiC–7 vol.% C (graphite) HMC

Material code	C/Ti	Phase analysis	Titanium carbide			Graphite^a	
			Lattice parameter a, nm	Micro-hardness HV, GPa	Average grain size L, μm	Lattice parameter a, nm	Lattice parameter c, nm
TKU	1.17	TiC$_{1-x}$, C (graphite)	0.4326±0.0001	24±1	75±10	0.246±0.001	0.670±0.001

^{a} Lattice parameters before high-temperature hot-pressing: a = 0.246 nm and c = 0.677 nm.

The samples for oxidation were shaped into plates with dimensions of 7×7×1 mm and weight of about 200 mg by cutting the hot-pressed blanks and subsequent polishing the surfaces with diamond tools. The thermogravimetric analysis (TGA) of the samples was carried out using a modernized version of the installation described by Afonin et al.[9] It allowed to stabilize the gas pressures in the range of 0.1-100 kPa with mass rate of oxygen flow of 0.06-0.16 g min^{-1}. The determination of carbon content in the oxidized samples was carried out using a coulometric method with a CuO – Pb flux employing a special analyzer with the relative accuracy of measurement at about 0.2%. The samples were investigated by the means of X-ray diffraction (XRD), optical and scanning electron microscopy (SEM) analyses as well as microanalysis by electron probe (EMPA) and energy-dispersive X-ray spectroscopy (EDX).

RESULTS AND DISCUSSION

The isobaric-isothermal oxidation of the HMC material was performed at temperatures of 400-1000 ^{0}C and pressures of 0.13-65 kPa in pure oxygen with mass flow 0.12±0.02 g/min. The notable weight gain was observed only at temperature of 500 ^{0}C and above it. The change of sample weight was normalized to the initial surface area of the sample and presented via plots showing weight gain per unit surface, w vs. time, t; some typical results of the TGA obtained at different temperatures and oxygen pressures are shown in Fig. 1.

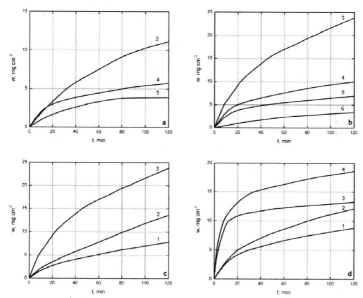

Fig. 1. The isobaric-isothermal oxidation of the TiC - 7 vol.% C (graphite) HMC at different oxygen pressures, p_{o2}: (a) 0.13 kPa; (b-c) 1.3 kPa; (d) 26 kPa and temperatures, T: (1) 500 $^{\circ}$C; (2) 600 $^{\circ}$C; (3) 700 $^{\circ}$C; (4) 800 $^{\circ}$C; (5) 900 $^{\circ}$C; (6) 1000 $^{\circ}$C.

The mathematical processing showed that the TGA curves could be divided into three (or four) steps. In the initial step the oxidation was almost linear with time. Subsequently, in the second step the oxidation rate, w is best described by a parabolic function; for the case of oxidation at higher temperatures / higher oxygen pressures ($T \geq 700$ ^{0}C and $p_{O2} \geq 13$ kPa), this step was followed by another parabolic step with different kinetics constants. The parabolic steps of the process take up the majority of the current time with w changing from 1-5 up to 25 mg/cm^{2}. In the final oxidation step the process returns to a linear form similar to that found in the first step, but with significantly decreased values for constants. The analogous behaviour was observed across the entire range of temperatures and pressures applied. Thus, for example, to describe the process at higher temperatures / higher oxygen pressures, one would apply two linear rate constants: k_{l1} and k_{l2}, and two parabolic rate constants: k_{p1} and k_{p2}, however, for other regions, defined by higher or lower values of oxidation parameters, the constants, which completely describe the oxidation behaviour of the material, were: k_{l1}, k_p and k_{l2}. Hence, the assembly of observed TGA curves can be characterized comprehensively as a linear-paralinear model, whose main properties are in the alternation of linear and parabolic steps rather differentiated for the higher temperatures / higher oxygen pressures range.

It should be pointed out especially that there was no steady increase of weight gain from the lowest temperatures and oxygen pressures to the highest ones. Indeed, there is a clearly defined maximum, which corresponds to particular values of the oxidation parameters, which were not the maximal temperature or pressure exposed. Previously, Stewart and Cutler[10] have reported a maximum in the oxidation rates for TiC, Voitovich and Pugach[11] found the highest values of the rate at 800 ^{0}C, when studying oxidation of TiC in air at 600-1000 ^{0}C. Shimada and Kozeki[12] and Gozzi et al.[13] reported about decrease of oxidation rate with growing of oxygen pressure. The obtained model for the oxidation behaviour can be better visualized by the use of 3D plots (w-surfaces). The typical $w - t - T$ (temperature) diagram is presented in Fig. 2a; the TGA curve at 700 ^{0}C appears as a ridge on the w-surface corresponding to an oxygen pressure of 1.3 kPa. The analogous situation is also observed in the $w - t - lgp_{O2}$ (oxygen pressure) diagrams, for example, at 800 ^{0}C a ridge lies at a pressure of about 13 kPa (Fig. 2b). A ridge violates the monotonic property of the kinetics model. These ridge temperatures of about 700-800 ^{0}C and/or ridge oxygen pressures of about 1.3-13 kPa, which were revealed experimentally by TGA, served as clearly visible boundaries between areas with different behavior of the material and/or prevailing oxidation mechanisms.

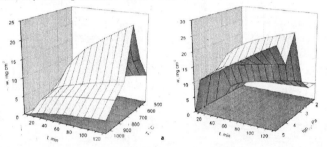

Fig. 2. 'Ridge effect' plots showing the influence of temperature T and oxygen pressure p_{O2} on the oxidation kinetics $w=f(t)$ at different values of these parameters: (a) $p_{O2} = 1.3$ kPa; (b) $T = 800$ ^{0}C.

The obtained TGA curves for the studied composite possess some kind of quantitative uncertainty in relation to real chemical reactions, as they are influenced by two concurrent processes with opposite effects such as formation of oxygen containing scale, accompanied by a weight gain, and carbon burn-off, accompanied by a weight loss. Similar situations are often observed during the oxidation of ceramic matrix composites[14]. By means of a carbon analysis in the samples exposed to different oxidation conditions, it became possible to establish that the carbon burn-off relationship between the normalized to surface area weight of burnt-off carbon $\Delta m_c/s$ and exposure time, t effectively follows the parabolic rate rule. The determination of weight loss due to the formation of carbon oxides allowed converting the results of TGA into the data of real oxygen consumption by solids. Some examples of the calculated kinetic curves for oxygen solid-state consumption (plots of $\Delta m_o/s$ vs. t) are presented jointly with experimental curves for carbon burn-off in Fig. 3.

The effects of gas pressure on the solid – gas interaction kinetics is one of the most important factors, which are necessary for clear understanding of the reaction mechanism. Some examples of the plots of lgk_i vs. lgp_{O2} are shown in Fig.4 to illustrate the influence of oxygen pressure on oxidation rate. The calculated values of exponent m (the apparent order of reaction) across the studied range of oxidation parameters have been collected together in Table III.

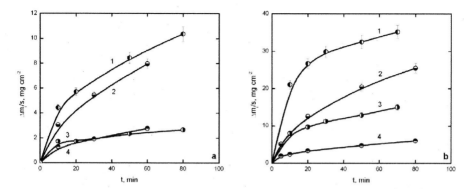

Fig. 3. The oxygen consumption (1, 2) and carbon-off (3, 4) kinetics during the oxidation at temperatures, T: (a) 500 ^{0}C; (b) 700 ^{0}C and oxygen pressures, p_{O2}: (2,4) 1.3 kPa; (1,3) 13 kPa.

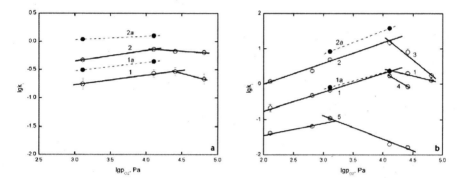

Fig. 4. Relationship between lgk_i for rate constants (1) k_{l1}, (2) k_p, (3) k_{p1}, (4) k_{p2}, (5) k_{l2} and oxygen pressures lgp_{O2} during the oxidation at different temperatures, T: (a) 500 ^{0}C; (b) 700 ^{0}C (curves with indices 1a and 2a relate to oxygen consumption kinetics).

Table III. The calculated values of apparent reaction order m at various oxidation parameters [a]

m	0.13-0.65 kPa	0.65-1.3 kPa	1.3-13 kPa	13-26 kPa	26-65 kPa
500 ^{0}C	–		1/6 / 1/5-1/6 / 0	1/6 / -1/10 / 0	-1/3 / -1/10 / 0
600 ^{0}C	–	1/5-1/6 / 1/3 / 1/6-1/10		1/5-1/6 / -1/2 / 1/6-1/10	
700 ^{0}C	1/2 / 1/2 / 1/3		1/2 / 1/2 / -(1/2-2/3)	-(1/2-1/3) / -1 / -(1/2-2/3)	
800 ^{0}C		1/2 / 1/2 / 1/3		1/2 / -1-0 / 1/3	
900 ^{0}C			1/2-1/3 / 1/2 / 1/3	–	
1000 ^{0}C		–		1 / 1 / 1/2	

[a] Light grey areas provide the values of m and the dark grey regions represent negative m values for k_{l1} / k_p / k_{l2} in the relationships $k_i \sim p_{O2}{}^m$; white boxes represent regions where no data has been taken.

The negative values for m were revealed in the range of lower temperatures / higher pressures for the most part during steps of linear-paralinear oxidation. This is an interesting result given that after a ridge oxygen pressure has been reached, the oxidation rates are decreased by further increase in oxygen pressure. Another key characteristic of this process, observed for both linear and parabolic steps, is the gradual increase of the absolute value of m with an increase of oxidation temperature from $m = 0$-$\frac{1}{3}$ at 500 ^{0}C and $m = 1/10$-$1/2$ at 600 ^{0}C up to $m = \frac{1}{3}$-$\frac{1}{2}$ at 900 ^{0}C and $m = \frac{1}{2}$-1 at 1000 ^{0}C, as for pure TiC powder Stewart and Cutler[10] reported about $m = 1/6$ at 600-800 ^{0}C and $m = \frac{1}{4}$ at 800-900 ^{0}C. It is relevant to note that the application of the data on real oxygen consumption by solids, at the same temperatures and oxygen pressures, gives approximately the same values for m as those determined for the TGA curves.

Preliminary inspection of the experimentally obtained w vs. t plots suggests a clear dependence of oxidation rate on temperature, although this rate tends to rise to a maximum at the ridge temperature before falling with further temperature increases. The Arrhenius plots of lnk_i vs. $1/T$ for the different steps of the process at oxygen pressures from 0.13 up to 13 kPa are shown in Fig. 5.

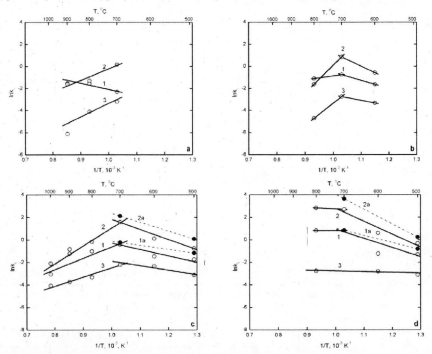

Fig. 5. Arrhenius plots of lnk_i for rate constants (1) k_{l1}, (2) k_p, (3) k_{l2} versus $1/T$ for the oxidation at different oxygen pressures, lgp_{O2}: (a) 0.13 kPa; (b) 0.65 kPa; (c) 1.3 kPa; (d) 13 kPa (curves with indices 1a and 2a relate to oxygen consumption kinetics).

For temperatures below the ridge temperature of 700 ^{0}C, the Arrhenius plot behaviour is typical, in that the rate of oxidation of the material increases with temperature. Within the range of higher temperatures, the same behaviour was found at the lowest oxygen pressure, but only for the initial step of oxidation (see Fig. 5a, curve 1). For low oxygen pressures in the temperature interval from 700 to 1000 °C, the rate of oxidation for the material is generally slowed with an increase of temperature. At the ridge temperature in the plot of lnk_i vs. $1/T$ there is an inflexion point (see Fig. 5b-d), at which the activation energy Q changes sign as was found for the inflection point at the ridge oxygen pressure, where the order of reaction m also changed its sign in the same manner. Comparison of the lnk_i vs. $1/T$ plots, obtained from the data on real oxygen consumption by solids, with those from the TGA, as demonstrated by Fig. 5c-d, shows that these two types of data are in good agreement with each other. At oxygen pressures from 26 to 65 kPa the situation is similar; as the Arrhenius plots show a change in gradient at the ridge temperature, but only without a change of the sign for Q. After this ridge temperature, the values of the activation energy Q dramatically drop, as the gradient occasionally falls to near zero. The calculated values of the activation energy Q for the different steps and parameters of oxidation are presented in Table IV.

Table IV. The calculated values of apparent activation energy Q at various oxidation parameters [a]

Q, kJ/mol	500-600 °C	600-700 °C	700-800 °C	800-900 °C	900-1000 °C
0.13 kPa			46±8 / -(87±2) / -(102±6)		–
		61±8 / 99±6 /	-(30±8) /		
0.65 kPa	–	41±5	-(210±10) /		
			-(170±10)		
1.3 kPa	47±3 / 75±1 / 33±1		-(84±2) / -(132±2) / -(68±2)		
13 kPa	67±5 / 96±5 / 5±1		0 / 11±6 / 5±1	–	
26 kPa	61±3 / 79±5 / 5±1		61±3 / 0 / 5±1		
65 kPa	70±3 / 84±1 / 9±1		0 / 22±6 / 15±6 / 9±1 [b]		

[a] Light grey areas provide the values of Q and the dark grey regions represent negative Q values for k_{l1} / k_p / k_{l2}; white boxes represent regions where no data has been taken.

[b] For k_{l1} / k_{p1} / k_{p2} / k_{l2}.

In summarizing the kinetics data, a number of observations on the process can be made:
– for the initial linear step, the activation energy Q is 46-70 kJ/mol throughout the range of oxidation parameters, with the exception of 700-1000 °C at oxygen pressures from 0.65 to 1.3 kPa, where Q is negative and has values between 30 to 84 kJ/mol;
– for the parabolic steps, the activation energy Q is 75-100 kJ/mol at 500-700 °C (independent of oxygen pressure) and 11-22 kJ/mol at 700-1000 °C over the range of oxygen pressures from 13 to 65 kPa, except for the higher temperature / lower oxygen pressure range, where Q is negative and has values between 87 to 220 kJ/mol;
– for the final linear step, the activation energy Q is 33-41 kJ/mol at 500-700 °C for oxygen pressures of 0.65-1.3 kPa and 5-9 kJ/mol at higher oxygen pressures (13-65 kPa), except for the higher temperature / lower oxygen pressure range, where Q is negative and has values 68-170 kJ/mol.

As a result of this kinetic analysis, it becomes possible to divide the studied range of temperatures and oxygen pressures into four main regions; each region defined by high or low values of oxidation parameters (Fig. 6). The ridge temperature and ridge oxygen pressure serve as boundaries for these regions and mark a change in the prevailing oxidation mechanism.

The evolution of weight gain during the oxidation process was accompanied with a change of the chemical composition of the surface layers. The XRD patterns taken from the oxidized surface at different temperatures are shown in Fig. 7. No XRD peaks corresponding to the newly formed phases were found on the surface for $w < 8$ mg cm^{-2}. However, for all samples oxidized under different conditions, the lattice parameter of the carbide phase changed slightly; for example, at the temperature

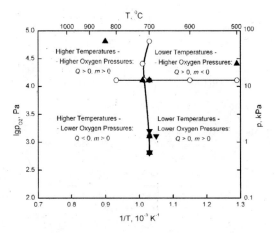

Fig. 6. An oxidation parameter plot of lgp_{O2} versus $1/T$; each inflexion point, at ridge temperatures / ridge oxygen pressures for different kinetics stages, has been marked as: ○ - parabolic (k_p, k_{pl}); ▲ - first linear (k_{l1}); ▼ - second (final) linear (k_{l2}).

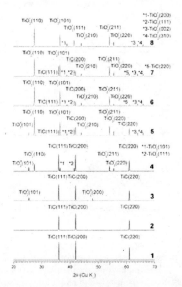

Fig. 7. XRD patterns of the initial composition (1) and samples oxidised at: (2) $T = 500\ ^{\circ}C$, $p_{O2} = 1.3$ kPa; (3) $T = 600\ ^{\circ}C$, $p_{O2} = 1.3$ kPa; (4) $T = 700\ ^{\circ}C$, $p_{O2} = 1.3$ kPa; (5) $T = 800\ ^{\circ}C$, $p_{O2} = 1.3$ kPa; (6) $T = 900\ ^{\circ}C$, $p_{O2} = 13$ kPa; (7) $T = 1000\ ^{\circ}C$, $p_{O2} = 1.3$ kPa; (8) $T = 1000\ ^{\circ}C$, $p_{O2} = 65$ kPa (indices relate to different polymorphic crystalline modifications of titanium dioxide: a - anatase; r - rutile).

of 500 ^{0}C and oxygen pressure of 1.3 kPa, the lattice parameter reduced from the initial 0.4326 ± 0.0001nm to 0.429±0.002 nm. This decrease of lattice parameter was accompanied by a significant broadening of all XRD peaks. However, it proved to be impossible to find any correlations between this change in lattice parameter and values of the oxidation temperature and pressure, as the decrease of lattice parameter was approximately equal for all XRD patterns obtained. The first oxide phase appears at 600 ^{0}C, and it is represented on the pattern by small broad peaks of TiO_2 (anatase). At 700 ^{0}C, the oxide scale consists just of two crystalline structures of TiO_2 – anatase and rutile. The latter phase becomes predominant at 800 ^{0}C when compared with carbide and anatase, and at $T \geq 900$ ^{0}C and $p_{O2} \geq 13$ kPa, the rutile phase completely replaces the anatase in scale. The peaks of the carbide disappeared only at the highest values of temperature and oxygen pressure.

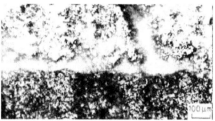

Fig. 8. Microstructures of oxide scale and oxide – composite interface (polarized light) for the sample oxidised at temperature $T = 700$ ^{0}C and oxygen pressure $p_{O2} = 1.3$ kPa for 2h.

It seems likely that the common layered structure of oxide scale, e.g. observed during the oxidation of metals, breaks down in the case of the HMC (Fig. 8), probably, because of the wide development of short-circuit diffusion processes in this type of material[3]. At earlier stages of oxidation, employing SEM/EDX, it became possible to determine the atomic ratio of O/Ti = 0.28±0.05, which characterizes the material and corresponds to the incubation of the oxide on the solid – gas interface.

The obtained experimental results suggest that oxidation of the TiC – 7 vol.% C (graphite) HMC is very sensitive to both temperature and oxygen pressure and that the mechanism changes significantly with transfer from lower to higher values of oxidation parameters. The process itself can be divided into two sub-processes characterized by the different physico-chemical transformations:
– the formation of titanium oxides due to oxidation of the carbide phase:

$$TiC_{1-x} + O_2 \rightarrow TiC_{1-x-y}O_z + yC\,(CO, CO_2) + O_2 \rightarrow TiO_n + C\,(CO, CO_2) \text{ and} \qquad (1)$$

– the oxidation of carbon with formation of carbon oxides:

$$C + O_2 \rightarrow CO, CO_2 \qquad (2)$$

Here, it should be noted that Eq. (2) describes the burn-off for both the graphite phase and "secondary" carbon formed by a reaction in Eq. (1). Although the relationships between these sub-processes may be complex and dependent on oxidation conditions, it appears that the pressure of the same gaseous products determines the reaction equilibrium for both sub-processes, in which carbon itself represents not only a reagent, but also a solid product.

At the lower values of oxidation parameters, 500-700 ^{0}C and 0.65-13 kPa, it was found that the initial linear step in oxidation kinetics is characterized by $Q = 45$-70 kJ/mol with m subsequently increasing from 1/6 at 500 ^{0}C to ½ at 700 ^{0}C. On the basis of information derived from the XRD, SEM and EPMA analyses, it appears reasonable to connect this step with the dissolution of oxygen in the carbide and formation of the oxycarbide phase $TiC_{1-x-y}O_z$, as described by Eq. (1) for the initial stage of reaction. At temperatures of about 500 ^{0}C and oxygen pressures of 3.9-16 kPa, Shimada and Kozeki[12] reported a very similar value of Q for the oxidation of TiC powders, but for parabolic kinetics with m = 0.2-0.6. They proposed that the diffusion of oxygen through the oxycarbide phase should be responsible for such low value for Q. However, according to Schuhmacher and Eveno[15], the activation energy of oxygen diffusion in $TiC_{0.97}$ is close to that of carbon, with a value of 383±8 kJ/mol. From this we can conclude that it is more likely that the decomposition of oxygen-oversaturated oxycarbide[16], followed by the formation of oxide nuclei according to reaction:

$$TiC_{1-x-y}O_z \rightarrow TiC_{1-x-y-a}O_{z-b} + aC\ (CO,\ CO_2) \rightarrow TiO_{2-d}C_d + kC\ (CO,\ CO_2) \rightarrow TiO_{2-u} + C\ (CO,\ CO_2),\ (3)$$

acts as the governing stage for the oxidation of carbide material in this case. It was clear that the deviations of lattice parameter for the carbide phase were approximately the same for samples oxidized under different conditions. Hence, according to data available in literature[16], the experimentally obtained lattice parameter of 0.429±0.002 nm corresponds to the oxycarbide with composition in the range of $TiC_{0.3-0.8}O_{0.1-0.8}$. However, SEM, EDX and EMPA measurements allowed determining the composition of oxycarbide (coexisting with rare particles of oxide on the surface of material) as following $TiC_{0.80±0.15}O_{0.28±0.08}$. Shimada[17] observed similar oxidation behaviour for TiC, attributing his findings to a structure with lattice parameter of 0.4327±0.0001 nm and composition of $TiC_{>0.5}O_{<0.5}$.

The subsequent transition to the parabolic kinetic step leads to a value of Q = 75-105 kJ/mol and exponent m, which increases with temperature increase as observed previously. The variations of m in the interval from 1/6 to ½ reflect the solid-state diffusion character of the process and connect with the concentration of oxygen vacancies, as it is in good agreement with data on defect structure of non-stoichiometric p-conductive TiO_2[18]. During this step, it would appear that the process transits to the solid-state diffusion regime and the most probable rate determining stage in this range of temperatures and pressures is oxygen diffusion through the anatase $TiO_{2-d}C_d$ scale. The formation of carbon-doped anatase phase $TiO_{2-d}C_d$ with d = 0.02÷0.05, as an initial product of TiC oxidation, was confirmed recently by Irie et al.[19] as well as Shen et al.[20] at lower temperatures in oxygen and air. There are available data in the literature on the activation energy for the parabolic oxidation of single crystals and polycrystalline TiC by oxygen[10,21] with values close to 200 kJ/mol; e.g. Voitovich and Lavrenko[22] determined the value of Q in the interval from 184 to 217.5 kJ/mol, to be dependent on the purity of TiC. All of these values are in good agreement with the activation energy of oxidation for metal Ti, which is connected with the process of oxygen lattice diffusion[23]. Nevertheless, in the earlier works it was suggested that, at low temperatures and low oxygen pressures (500 ^{0}C, 13 Pa), a high percentage of the oxide area consists of paths of low diffusion resistance, and hence, grain boundary diffusion plays an overwhelming role in the oxidation of Ti compared to the negligible role played by lattice diffusion[24]; additionally, relatively low values of Q obtained in this work are clearly connected with the structure, characterized by a widely developed grain boundary network[3]. The short-circuit diffusion along grain boundaries and dislocations in similar materials[3] plays a more important role in general than it would with single phase, usually – hypostoichiometric, refractory carbides. However, the activation energy for grain boundary diffusion is smaller than that for lattice diffusion and the value for the former accounts for about a half of the latter in metal oxidation processes[23], although Rothschild et al.[25] determined this ratio for non-stoichiometric TiO_2 films at lower temperatures as ¾ and the value of Q for grain boundary diffusion of about 50 kJ/mol. Yet the dispersion of carbon inclusions in the initial structure of the material, is not the only factor contributing to this dominant role for short-circuit diffusion. Other factors, such as the lower temperatures of oxidation ($T \leq 0.5$ of the melting point of TiO_2) and the fine dispersal of residual carbon throughout the oxide scale, also contribute significantly to a predominantly solid state transport mechanism along low resistance diffusion paths.

The residual carbon dispersed throughout the scale is a result of non-stoichiometric oxidation of the carbide and the composite in a whole, as the fractions of Ti and C oxidized during the gas exposure differ significantly. The total (integral) ratio of C/Ti involved in the oxidation process (Fig. 9) was calculated by only taking into account the formation of such oxides as TiO_2 and CO/CO_2, as the small contribution of the formation of oxycarbide to the overall process could be safely neglected and no other products were identified in the oxidized samples. However, the analysis showed that at lower oxidation parameters the process of TiO_2 formation generally dominates over the carbon – oxygen interaction. An illustration of this is provided by comparison of the number of Ti atoms entering into reaction with oxygen at 700 °C and 1.3 kPa over an hour with those of carbon; the former is about two times more than the latter, showing a clear dominance of the metal oxide scale producing mechanisms. Therefore it is understandable that the grain growth of oxide scale is retarded, not only by initial carbon inclusions, but also by the presence of secondary carbon formed because of the non-stoichiometric oxidation of the combined carbon in the carbide phase.

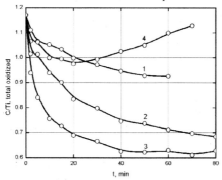

Fig. 9. Evolution of the current value of C/Ti total (integral) oxidised ratio during the oxidation process at different temperatures, T: (1, 2) 500 °C; (3, 4) 700 °C and oxygen pressures, p_{O_2}: (1, 3) 1.3 kPa; (2, 4) 13 kPa (calculated by taking into account the formation of TiO_2, CO and CO_2 only).

In general, the shift of the reaction mechanism to short-circuit diffusion is mainly due to the character of carbon oxidation, which lags behind the oxidation of Ti atoms. The removal of carbon oxides probably forms the specific micro- and nano-porous structure of the scale. In keeping with the laminar-nature of most oxide scales, the microstructure is also characterized by some "lateral elements". These easy diffusion paths for oxygen, afterwards transformed into the grain boundaries of scale, are formed initially in decomposing oxycarbide by precipitated carbon like those observed by Voitovich and Lavrenko[22] in the oxidized TiC.

It seems obviously that the final period of the oxidation at lower oxidation parameters is connected with the initiation and further development of micro- and nano-channels, which are formed by gases escaping the bulk sample. The transfer of the rate determining stage of the process from solid state diffusion to predominantly gas diffusion in the final step is also evidence of the priority of short-circuit diffusion during the initial propagation of the oxidation process. This step is described by linear kinetics with Q = 30-40 kJ/mol. The exponent m, growing from 0 at 500 °C up to ⅓ at 700 °C, also supports the suggestion concerning the transition of the prevalent mechanism to gas processes. With reference to these observations, and on the basis of microstructure studies, the most likely rate determining stage in this case is the transition regime of carbon gasification between the chemical reaction (carbon oxidation) and gas diffusion stages. This conclusion seems credible, as the obtained value for Q is very close to those for the processes controlled by the gas diffusion stage for oxidation of carbonaceous materials[26-27], while m is of the same order as those for the processes dominated by the chemical reaction stage of the carbon oxidation[28].

The increase of oxygen pressure at 500-700 °C up to 13-65 kPa results in the initial linear step with Q = 60-75 kJ/mol and negative values of $m = -(\frac{1}{2}-\frac{1}{3})$, so it is suggested that controlling stage of the process shifts from the solid state diffusion to the transition regime between solid state and gas diffusion processes. This regime is realized due to the development of sintering and grain growth processes in the scale. The rate of oxidation, being limited by the gas diffusion process, obviously correlates with the characteristics of porosity of the oxide scales because of the influence of these characteristics on gas diffusion (Knudsen flow, surface diffusion, adsorption-desorption processes) through the scale[29]. At the same time, the structure characteristics, such as gas permeability, is dependent directly on values of the solid-state diffusion parameters, by which sintering process and grain growth of porous scale are governed. Although to date there have been no attempts to evaluate quantitatively the relationship between permeability of scale and its recrystallisation temperature, a similar situation with the influence of scale sintering on the reactivity of solids seems to be typical for the oxidation of both individual metal carbides and carbide composites and very important for analysis of the process[8]. At moderate temperatures the solid-state diffusion of carbon is significantly retarded, so a key point is the contribution of the gas diffusion processes, necessary for the removal of carbon gasification products, to the overall process of oxidation. The difficulties associated with the removal of CO/CO_2 can noticeably influence the rate of oxidation for carbides in general and appear to be especially important for oxidation of carbide – carbon HMC.

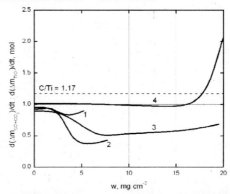

Fig. 10. Plots of $d(\Delta m_{CO/CO_2})/dt$: $d(\Delta m_{TiO_2})/dt$ vs. w for the oxidation process at different temperatures, T: (1, 2) 500 °C; (3, 4) 700 °C and oxygen pressures, p_{O_2}: (1, 3) 1.3 kPa; (2, 4) 13 kPa.

The transition in the kinetics to the parabolic step at lower temperatures / higher oxygen pressures is characterized by a slight increase of Q to 75-100 kJ/mol; while m varies widely, it remains negative and increases in absolute value from $1/10$ at 500 °C to 1 at 700 °C. The oxidation mechanism does not change essentially and remains in the same transition regime as in the previous step, but with a greater contribution from the gas diffusion mechanism, which dominates completely during the subsequent final linear step. This step differs from those previously by an accelerating carbon burn-off. At the ridge temperature and/or oxygen pressure the process of carbon oxidation is "catching up" with the oxidation of Ti atoms (Fig. 10). Over the range from 13 to 65 kPa, the final step is characterized by the same value of Q = 5-10 kJ/mol and slightly variation of $m = \frac{1}{2}-\frac{1}{3}$, sufficient grounds to identify this step as being regulated by the interdiffusion of O_2 and CO/CO_2 gas flows through the porous scale.

The oxidation behaviour of the material at the lowest level of oxygen pressure convincingly demonstrated the large influence that sintering of the scales has on the oxidation process; the unusual Arrhenius plots at this pressure become straightforward to interpret in terms of this mechanism. It was found that, at 700-900 ^{0}C and 0.13 kPa, the initial linear step does not significantly differ from those at lower temperatures, 500-600 ^{0}C and oxygen pressures, 0.65-13 kPa. However, the subsequent steps at this pressure are realized by a completely different mechanism, characterized by negative values of Q, which increase insignificantly (in absolute value) from 85-90 kJ/mol to 95-110 kJ/mol for the following parabolic and linear kinetics steps respectively, while m fluctuates between ½ and ⅓. So it is quite possible to conclude that the scale, which was nucleated and formed during the first linear step, sinters and subsequently recrystallizes, with an apparent enlargement of the contribution from lattice diffusion, during the further development of the oxidation process. These peculiarities shift the process to a gas diffusion regime, which is governed by the permeability of the scale; hence this regime is dependent on the processes relative to recrystallisation (grain growth, evolution of porosity) of oxide – carbon and/or oxide scale. However, these processes, in turn, are determined by solid-state diffusion, as was mentioned above. Due to the greater segregation of oxide nuclei in the volume of oxycarbide, especially in carbide – carbon HMC, the contribution of the scale sintering to the oxidation process is more significant, and observed at lower temperatures than was found for the oxidation of Ti metal[23].

Negative values of Q were found in all steps of the linear-paralinear model throughout the quadrant of higher temperatures / lower oxygen pressures at $0.65 \leq p_{O2} \leq 13$ kPa. In the initial linear step, Q is 20-85 kJ/mol; afterwards, in the parabolic step, Q increases to 85-220 kJ/mol and then slightly decreases in the final linear step. These values are in good agreement with those gained for the activation energy of the densification process for TiO_2 during the initial stage of sintering, corresponding to the plastic flow mechanism in the kinetics of sintering[30]. It is also significant that the highest absolute values for Q are in excellent agreement with those for the oxygen lattice diffusion coefficient, calculated from TGA data for the oxidation of Ti metal[23]. At 700-900 ^{0}C and 0.65-13 kPa, m = ½ for the first two steps, being independent of oxidation parameters, so it is suggesting that the defect structure of sintering oxide does not significantly change. The different behaviour of the scale before and after the ridge temperature reflects the intrinsic change in the solid products of oxidation. The presence of carbon within the scale during the oxidation of transition-metal carbides very often leads to the formation and stabilization of oxides typically unstable in this range of temperatures[8,11]. A similar situation was revealed in this study, as the anatase phase, contained in the oxide scale, was observed up to 800 ^{0}C, although, according to XRD analysis data in the recent paper by Zhang et al.[31], anatase – rutile polymorphic transition begins at 500 ^{0}C and reaches about 95% at 700 ^{0}C. Another key observation is that the anatase phase, in contact with carbon, is more chemically stable; this is confirmed by noting that the carbothermic reduction of anatase starts at higher temperatures when compared with those for the rutile phase[32].

From the obtained experimental data, a convincing argument would be that, during the oxidation of TiC composite, the anatase – rutile transition is directly connected with the ridge temperature, as the intensive polymorphic transformation is concurrent with deeper carbon burn-off in the scale. Presumably, the preferential formation of anatase at lower temperatures, as the result of the oxycarbide decomposition, is connected with a deficiency of energy (exposed temperature) and reagents (oxygen pressure) as the body-centered tetragonal structure of anatase originates from the simpler, face-centered cubic structure[33], which is inherent to the carbide/oxycarbide phases. Hence, this structural correspondence simplifies the nucleation of oxide phase in the initial oxycarbide. The lower density of the anatase (~9% less than that of rutile) provides more opportunities for the dissolution of carbon (as well as for diffusion of oxygen) than are available to the rutile structure. During the polymorphic transition, the scale provides much lower protection from oxidation than normal, as the diffusion resistance of the structure is reduced. The effect of ridge temperature, when the oxidation rate of materials rises rapidly to a maximum and then declines with subsequent growth of temperature, is caused by the accumulated impact of interconnected processes occurring in the scale, such as the accelerated release of carbon oxides and the polymorphic transition, from low to high density phases, of TiO_2. The eventual rise of oxidation temperature leads to a steady decrease of carbon content coupled with a steady increase of rutile fraction in the scale. These factors bring about a significant change in both gas and solid diffusion processes, due to the development of recrystallisation and consequent reduction in gas permeability through the scale. The kinetics reflects the mechanism, as at the ridge temperature the sign of Q changes and further temperature increase corresponds to a decrease of the oxidation rate, although the absolute value of Q does not change significantly as it describes a similar solid-state diffusion process, but with the opposite effect on the rate.

The increase of oxygen pressure above the ridge value at 700-900 ^{0}C brings about a change in the oxidation mechanism, as Q varies dramatically in both sign and magnitude. The initial step is characterized by a negligibly small value of Q, whereas afterwards the value fluctuates: 15-30 kJ/mol for the first parabolic step and 5-10 kJ/mol – for the final, second linear step. All this allows us to classify the current mechanism as a process governed by the different stages of gas interdiffusion through the porous scale with structure that varies with the evolution of oxidation. These rate determining stages of the oxidation process can include gas adsorption/desorption, diffusion of oxygen to the apparent carbide – oxide interface and/or counter diffusion of gas products back through the porous medium. A noticeable characteristic of the oxidation process at the highest studied temperature of 1000 ^{0}C is the value of exponent m, which is practically independent of oxygen pressure and reaches a maximum value for all kinetics steps of $m = 1$, probably due to the higher rate of scale sintering, inherent to the material oxidized at this temperature.

CONCLUSION

The ridge effect, when the oxidation rate of solid rises rapidly to a maximum and then declines with subsequent growth of oxidation parameters (p_{O2}, T), was revealed during the oxidation of TiC-based HMC. This phenomenon is caused by the accumulated impact of interconnected processes occurring in the scale, such as the accelerated carbon burn-off and the anatase – rutile polymorphic transition characterized by the apparent stabilization of anatase by carbon. These processes bring about an essential change in both gas and solid diffusion in the formed scale, due to development of its recrystallisation and a consequent reduction in gas permeability through the scale. In evolution of the oxidation process prevailing mechanisms change, with increasing p_{O2} and/or T, in the following subsequence: solution of oxygen in carbide phase (oxycarbide formation) $\rightarrow$ decomposition of oversaturated oxycarbide $\rightarrow$ formation of oxide/oxide-carbon nuclei $\rightarrow$ solid-state, preferably short-circuit, diffusion $\rightarrow$ chemical stage of carbon gasification $\rightarrow$ gas diffusion governed by permeability (sintering) of the formed scale $\rightarrow$ gas interdiffusion through micro- and nano-porous scale.

The ridge effect is obviously reflected in the linear-paralinear kinetics model, as at the ridge oxidation parameter, the sign of the apparent order of reaction m or activation energy Q changes and further increase in parameters corresponds to a decrease of the oxidation rate. However, the absolute value of Q sometimes does not change significantly as it describes a similar solid-state diffusion process, but with the opposite effect on the oxidation rate. It should be emphasized that mobility of atoms, directly increasing with the rise of temperature, is not directly connected with chemical reactivity of solids, as in some cases the mobility can suppress the reactivity on the interaction surface due to formation of products with specific physico-chemical properties.

It is proposed that the newly developed model with the positive/negative values for m and Q can be transferred from the Ti – C – O system to variety of HMC materials, which are involved in solid-state gas-exchange reactions, and use of the model can contribute significantly to the design of advanced functionally gradient HMC for ultra-high temperature application.

REFERENCES

[1]D. P. H. Hasselman, P. F. Becher and K. S. Mazdiyasni, Analysis of the Resistance of High-E, Low-E Brittle Composites to Failure by Thermal Shock, *Z. Werkstofftech.*, **11**(3), 82-92 (1980).

[2]K. S. Mazdiyasni and R. Ruh, High/Low Modulus Si_3N_4 – BN Composite for Improved Electrical and Thermal Shock Behavior, *J. Am. Ceram. Soc.*, **64**(7), 415-9 (1981).

[3]I. L. Shabalin, D. M. Tomkinson and L. I. Shabalin, High-Temperature Hot-Pressing of Titanium Carbide – Graphite Hetero-Modulus Ceramics, *J. Eur. Ceram. Soc.*, **27**(5), 2171-81 (2007).

[4]Ya. L Grushevskii, V. F. Frolov, I. L. Shabalin and A. V. Cheboryukov, Mechanical Behavior of the Ceramics Containing Boron Nitride, *Sov. Powder Metall. Met. Ceram.*, **30**(4), 338-41 (1991).

[5]Yu. N. Zhukov, A. V. Cherepanov, A. R. Beketov and I. L. Shabalin, Inspection of the Nonuniformity of a Ceramic Compact for Machining by Hardness Measurement, *Sov. Powder Metall. Met. Ceram.*, **30**(4), 349-51 (1991).

[6]M.Araki, M. Sasaki, S. Kim, S. Suzuki, K. Nakamura and M.Akiba Thermal Response Experiments of SiC/C and TiC/C Functionally Gradient Materials as Plasma Facing Materials for Fusion Application, *J. Nucl. Mater.*, **212-215**, 1329-34 (1994).

[7]V. A. Lavrenko, L. A. Glebov, A. P. Pomitkin, V. G. Chuprina and T. G. Protsenko, High-Temperature Oxidation of Titanium Carbide in Oxygen, *Oxid. Metals*, **9**(2), 171-9 (1975).

[8]S. G. Kuptsov, V. G. Vlasov, A. R. Beketov, I. L. Shabalin, O. V. Fedorenko and Yu. P. Pykhteev, Low-Temperature Oxidation Mechanism of Zirconium Carbide, *Kinetics and Mechanism of Reactions in Solid State*, Minsk, Byelorussian State University, 182-3 (1975).

[9]Yu. D. Afonin, Thermogravimetric Unit for Studies of Gas – Solid Body Interaction Processes, *J. Phys. Chem. USSR*, **50**(8), 2156-7 (1976).

[10]R. W. Stewart and I. V. Cutler, Effect of Temperature and Oxygen Partial Pressure on the Oxidation of Titanium Carbide, *J. Am. Ceram. Soc.*, **50**(4), 176-80 (1967).

[11]R. F. Voitovich and E. A. Pugach, High-Temperature Oxidation of Titanium Carbide, *Sov. Powder Metall. Met. Ceram.*, **11**(2), 132-6 (1972).

[12]S.Shimada and M. Kozeki, Oxidation of TiC at Low Temperatures, *J. Mater. Sci.*, **27**(7), 1869-75 (1992).

[13]D.Gozzi, M. Montozzi and P. L. Cignini, Oxidation Kinetics of Refractory Carbides at Low Oxygen Partial Pressures, *Solid State Ionics*, **123**(1), 11-8 (1999).

[14]K. G. Nickel, Ceramic Matrix Composite Models, *J. Eur. Ceram. Soc.*, **25**(10), 1699-704, (2005).

[15]M. Schuhmacher and P. Eveno, Oxygen Diffusion in Titanium Carbide, *Solid State Ionics*, **12**, 263-70 (1984).

[16]S. I. Alyamovsky, Y. G. Zainulin and G. P. Shveikin, *Oxycarbides and Oxynitrides of IVa and Va Subgroups Metals*, Moscow, Nauka (1981).

[17]S. Shimada, A Thermoanalytical Study of Oxidation of TiC by Simultaneous TGA – DTA – MS Analysis, *J. Mater. Sci.* **31**(3), 673-7 (1996).

[18]R.Haul and G. Dümgen, Sauerstoff-Selbstdiffusion in Rutilkristallen, *J. Phys. Chem. Solids*, **26**, 1-10 (1965).

[19]H. Irie, Y. Watanabe and K. Hashimoto, Carbon-Doped Anatase TiO_2 Powders as a Visible-Light Sensitive Photocatalyst, *Chem. Lett.*, **32**(8), 772-3, (2003).

[20]M.Shen, Z. Wu, H. Huang, Y. Du, Z. Zou and P. Yang, Carbon-Doped Anatase TiO_2 Obtained from TiC for Photocatalysis under Visible Light Irradiation, *Mater. Lett.*, **60**(5), 693-7, (2006).

[21]M. Reichle and J. J. Nickl, Untersuchungen über die Hochtemperaturoxidation von Titankarbid, *J. Less-Common Metals*, **27**, 213-36 (1972).

[22]V. B. Voitovich and V. A. Lavrenko, Oxidation of Titanium Carbide of Different Purity, *Sov. Powder Metall. Met. Ceram.*, **30**(11), 927-32 (1991).

[23]P. Kofstad, *High Temperature Corrosion*, New York, Elsevier (1988).

[24]J. Markali, An Electron Microscopic Contribution to the Oxidation of Titanium at Intermediate Temperatures, *Proc. 5th Int. Cong. on Electron Microscopy*, New York, Academic Press, C4 (1962).

[25]A. Rothschild, Y. Komem and F. Cosandey, The Impact of Grain Boundary Diffusion on the Low Temperature Reoxidation Mechanism in Nanocrystalline $TiO_{2-\delta}$ Films, *Interface Sci.*, **9**(3-4), 157-62 (2001).

[26]D. Gozzi, G. Guzzardi and A. Salleo, High Temperature Reactivity of Different Forms of Carbon at Low Oxygen Fugacity, *Solid State Ionics*, **83**(3-4), 177-89 (1996).

[27]L. E. Cascarini de Torre, J. L. Llanos and E. J. Bottani, Graphite Oxidation in Air at Different Temperatures, *Carbon*, **29**(7), 1051-2 (1991).

[28]J. M. Thomas, Microscopic Study of Graphite Oxidation, *Chemistry and Physics of Carbon*, New York, Marcel Dekker, Vol. 1, 135-68 (1965).

[29]R. M. Barrer, Surface and Volume Flow in Porous Media, *The Solid – Gas Interface*, New York, Marcel Dekker, Vol. 2, 557-609 (1967).

[30]L. A. Perez-Maqueda, J. M. Criado and C. Real, Kinetics of the Initial Stage of Sintering from Shrinkage Data: Simultaneous Determination of Activation Energy and Kinetic Model from a Single Nonisothermal Experiment, *J. Am. Ceram. Soc.*, **85**(4), 763-8 (2002).

[31] J. Zhang, M. Li, Z. Feng, J. Chen and C. Li, UV Raman Spectroscopic Study on TiO_2. I. Phase Transformation at the Surface and in the Bulk, *J. Phys. Chem. B.*, **110**(2), 927-35 (2006).

[32]N. Setoudeh, A. Saidi and N. J. Welham, Carbothermic Reduction of Anatase and Rutile, *J. Alloys Compd.*, **390**(1-2), 138-43 (2005).

[33]G. Cangiani, Ab-Initio Study of the Properties of TiO_2 Rutile and Anatase polytypes, *These N 2667 Pour L'Obtention du Grade de Docteur es Sciences*, Lausanne, Ecole Polytechnique Federale de Lausanne (2003).

Materials for
Solid State Lighting

A POTENTIAL RED-EMITTING PHOSPHOR FOR UV-WHITE LED AND FLUORESCENT LAMP

K. U. Kim[1,2], S. H. Choi[1,*], H.-K. Jung[1] and S. Nahm[2]

[1]Advanced Materials Division, Korea Research Institute of Chemical Technology, P.O. Box 107, Yuseong, Daejeon 305-600, Republic of Korea
[2]Department of Materials Engineering, Korea University, Seoul 136-701, Republic of Korea

ABSTRACT

Luminescence properties of $NaLnGeO_4$:Eu^{3+} (Ln=rare earth) phosphors were investigated. It has a strong red emission under mercury discharge and deep blue/near-UV excitations. With the proper atomic ratio of rare earth ions, Y/Gd, and activator concentration, the emission intensity of Eu^{3+}-activated germanate compound is comparable to that of commercial red-emitting phosphor, Y_2O_3:Eu^{3+}, under 254 nm excitation with the CIE coordinates are x=0.655, y=0.339. Moreover, the maximum emission intensity is about four times higher than that of SrS:Eu^{3+} under GaN-based blue light excitations (λ_{ex}=400-465 nm).

INTRODUCTION

Highly efficient and power saving back light units (BLUs) are essential for the application of flat-panel displays, especially liquid crystal displays (LCDs), due to strong demands in excellent color gamut with enhanced brightness. Thus, the selection of the proper red-, green- and blue-emitting (RGB) backlight phosphor becomes an important factor determining the overall optical property of LCD devices.[1]

At present, lighting elements based on white light emitting diodes (LEDs) have made remarkable breakthroughs in the display fields. Conventional LEDs using GaN-based semiconductors as the excitation source converts bright violet-blue light to longer wavelength via the light-conversion phosphors. Nakamura *et al.* invented the first white light LEDs combining an InGaN blue LED emitting at 465 nm with a broad-band yellow phosphor, e.g. Ce-doped yttrium aluminate.[2,3] To improve the color rendering property of the above mentioned blue LED/yellow phosphor approach, white light can be ideally generated by combining discrete RGB phosphors with a deep blue/UV LED or laser diode (LD). Unfortunately, the efficiency of the sulfide-based red phosphors is much lower than that of the green and blue phosphors, there is an urgent need to make superior red phosphors that absorb in the UV and emit in the red.

The white light generated by fluorescent lamp is another option for adopting as a back lighting in LCDs. Even though the available lamp phosphors are already well known for these applications, the highly efficient red-emitting phosphors with pure color gamut are also demanded in fluorescent lamps for LCD-BLUs.[4,5]

Eu^{3+}-doped olivine type $NaLnGeO_4$ (Ln=rare earth) phosphors are suggested as an possible candidate for red-emitting elements under vacuum ultra violet conditions.[6-8] Germanates are well considered as host materials of phosphors in the search for luminescent materials with reasonable conductivity due to the smaller electronegativity difference between Ge and O other than silicates and aluminates compounds.[9]

In this work, we examines the luminescence properties of a Eu^{3+}-activated $NaLnGeO_4$ (Ln=Y, Gd) red-emitting phosphor for sophisticated BLU elements. The obtained results reveal its suitability for application in a novel phosphor LCDs.

279

EXPERIMENTAL

Eu^{3+}-doped $Na(Y,Gd)GeO_4$ phosphors were prepared by conventional high temperature solid-state reaction method. The reagent grade raw materials, Na_2CO_3 (99%), Y_2O_3 (99.99%), Gd_2O_3 (99.99%), Eu_2O_3 (99.9%) and GeO_2 (99.999%), were properly mixed together with a mortal. After thoroughly mixed using acetone and then dried in room temperature, they were calcined at 1100□ for 12h in air. After firing, samples were gradually cooled to room temperature. Phase analysis of the prepared powders was investigated by X-ray diffraction (XRD) analysis using with Cu Kα (λ=1.5406) radiation (Rikaku D/MAX-33 X-ray diffractometer). Photoluminescence (PL) emission, excitation (PLE) spectra and color coordinates were recorded using a fluorescence spectrophotometer at room temperature with a xenon lamp as an excitation source. For the reliable estimation, the relative PL and PLE intensity were measured with a commercialized Y_2O_3:Eu^{3+} and SrS:Eu^{3+} phosphors.

RESULTS AND DISCUSSION

Fig. 1 shows the XRD patterns of the $Na(Y,Gd)GeO_4$:Eu^{3+} prepared by the solid state reaction. Although we have synthesized $Na(Y,Gd)GeO_4$:Eu^{3+} phosphor with various compositions, we will focus on $Na(Y_{0.3}Gd_{0.2}Eu_{0.5})GeO_4$ since it has the highest emission efficiency among other samples. Compared with the standard $NaLnGeO_4$ (Ln=Y, Gd) diffraction patterns, all prepared samples are assigned to olivine phase without any impurity or secondary phases during a high temperature process. The phases of $NaYGeO_4$ and $NaGdGeO_4$ are iso-structure so they can form continuous solid solutions in the whole range (x=0-1.0) for the system $NaY_{1-x}Gd_xGeO_4$. Moreover, all diffraction peaks were in good agreement with the known reference data with larger activator concentrations which means that the Eu^{3+} incorporation does not change the lattice constant of $Na(Y,Gd)GeO_4$ much.

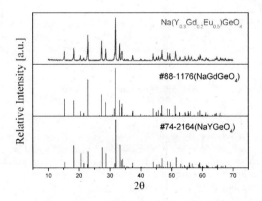

Figure 1. XRD patterns of the Eu^{3+}-doped $Na(Y,Gd)GeO_4$.

Fig. 2 presents the PLE spectra of $Na(Y_{0.3}Gd_{0.2}Eu_{0.5})GeO_4$ comparing with a commercial red-emittingY_2O_3:Eu^{3+} under UV region. The excitation spectrum of $Na(Y_{0.3}Gd_{0.2}Eu_{0.5})GeO_4$ mainly consist of two parts; the first one is a broad band centered at 256 nm, which were attributed to the charge-transfer transition between Eu^{3+} and O^{2-}, $i.e.$, an electron transfer from the ligand O^{2-} ($2p^6$)

orbital to empty states of $4f^6$ for Eu^{3+} configuration and the other one is extended peaks ranged from 350 to 425 nm owing to the intra-configuration $4f^6$ excitation peaks of Eu^{3+}.[9,10]

PL emission spectra of $Na(Y_{0.3}Gd_{0.2}Eu_{0.5})GeO_4$ and the dependence of emission intensities on Eu^{3+} content are also shown in Fig. 3. As shown in Fig. 3 (a), the emission spectrum of $Na(Y_{0.3}Gd_{0.2}Eu_{0.5})GeO_4$ is mainly dominated by a intense peak appearing around 610-625 nm, attributed to the electric dipole transition of $^5D_0$-$^7F_2$, while the several weak emissions arising from the magnetic dipole transition and $^5D_0$-$^7F_{3,4}$ transition. This indicates that the red emission of $Na(Y,Gd)GeO_4$:Eu^{3+} is mainly induced by the asymmetrical positioned Eu^{3+} with the lack of the inversion symmetry.[10] Phosphors with various activator concentration and Y/Gd atomic ratio were also examined and their emission spectra were similar to those in Fig. 3 (a). Moreover in comparison with a commercial Y_2O_3:Eu^{3+} phosphor, we could get a more pronounced color rendering property in our samples since the quenched emission in the vicinity of 630 nm. The CIE color coordinates are shown in Fig. 4 (b) under various excitation wavelengths. The dependence of PL intensities on Eu^{3+} content is shown in Fig. 3 (b). Maximum emission intensity is about 99% of commercialized Y_2O_3:Eu^{3+} under the mercury discharge condition, 254 nm, with Eu^{3+} content at x=0.5. The emission intensity gradually enhanced with increasing Eu^{3+} concentration up to 0.5 mol % and then finally decreased. In general, the emission intensity is proportional to the concentration of activator within limited ranges and begins to decrease beyond the critical concentration level. It is well known as a "concentration quenching effect" which leads activators to non-radiative emission process via energy transfer between neighboring activators. The high critical concentration may be due to low possibility of the energy transfer among the emission centers.[11] As we can see from the dependence of emission intensities on activator content, Fig. 3 (b), $Na(Y,Gd)GeO_4$ has a wide activator saturation range.

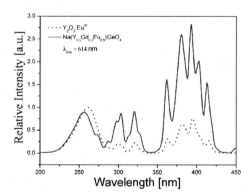

Figure 2.   PLE spectra of the $Na(Y,Gd)GeO_4$:Eu (solid line) and commercial Y_2O_3:Eu (dot line).

In some rare earth activated complex ion type phosphors, such as phosphates and borates, the extent of activator incorporation is relatively higher than an appropriate value (usually several wt%). We can explain these phenomena based on the characteristic crystal structure of olivine type compound. An ordered olivine structure has a quasi-two dimensional rare earth sublattice and each Eu^{3+} ion is effectively isolated by the surrounding sublattice group without non-radiative sites, even if the activators are highly incorporated to the host compound. So the excitation energy transfer would take place within the planes because of long separation among the emission centers under UV excitation.[8,12]

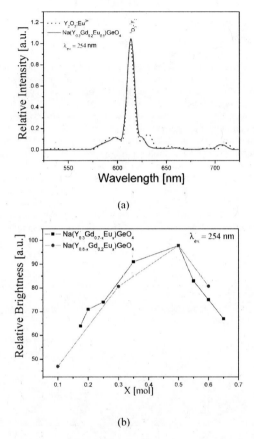

(a)

(b)

Figure 3. (a) PL emission spectra Na(Y,Gd)GeO$_4$:Eu (solid line) and commercial Y$_2$O$_3$:Eu (dot line) and (b) dependency of PL emission intensity with various Eu concentrations under 254 nm excitation.

Fig. 4 exhibits the emission property of Na(Y,Gd)GeO$_4$:Eu under the excitation of blue light wavelengths (λ_{ex}=405, 465 nm) for white light LEDs application. From Fig. 4 (a), it can be seen that the overall emission spectra of Na(Y,Gd)GeO$_4$:Eu^{3+} phosphor shows strong red emission under blue-LEDs conditions compared with that of commercialized Y$_2$O$_3$:Eu^{3+} under 254 nm. K. Toda et $al.$ reported that the different emission process of NaLnGeO$_4$ (Ln=Y, Gd) between VUV and UV excitation with proper activator contents.[6] Under UV excitation, it has a high critical activator concentration (x=0.75) similar to our results. Therefore, we can conclude that our suggested chemical composition is suitable for UV-LEDs red-emitting phosphors. Fig. 4 (b) summarizes the relative brightness and CIE chromaticity coordinates of $Na(Y_{0.3}Gd_{0.2}Eu_{0.5})GeO_4$ together with that of Y_2O_3:Eu^{3+} as a comparison. The resultant CIE chromaticity coordinates, x=0.652 and y=0.336, are superior to Eu^{3+}-activated sulfides under blue-LEDs condition and close to National Television Standard Committee (NTSC) standard values.

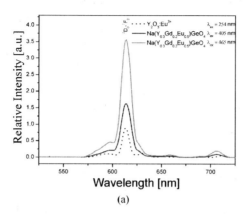

(a)

Phosphors	Excitation Wavelength (nm)	CIE chromaticity coordinates		Relative Brightness (%)
		x	y	
Y_2O_3:Eu	254	0.647	0.344	100
$Na(Y_{0.3}Gd_{0.2}\ Eu_{0.5})GeO_4$	254	0.655	0.339	99
	405	0.652	0.336	217
	465	0.658	0.340	461

(b)

Figure 4. (a) UV-PL emission spectra of $Na(Y_{0.3}Gd_{0.2}Eu_{0.5})GeO_4$ and Y_2O_3:Eu under various excitation conditions and (b) Summary of the CIE chromaticity coordinates and relative brightness of these phosphors

CONCLUSION

Red-emitting Eu^{3+}-doped $NaLnGeO_4$ (Ln=Y, Gd) phosphors were synthesized and their luminescence properties have been investigated. With the efficient photoluminescence excitation induced by both CTS band and internal f-f transition of Eu^{3+} ions, these phosphors has an improved CIE color coordinates (x=0.658, y=0.340) under mercury discharge and blue-LEDs condition close to the NTSC standard values. Among other various compositions, the $Na(Y_{0.3}Gd_{0.2}Eu_{0.5})GeO_4$ gives stronger emission intensity under near UV excitation, especially for that excited by 465 nm which is of 4.6 times stronger than Y_2O_3:Eu^{3+} (λ_{ex}=254 nm). In conclusion, Eu^{3+}-activated olivine type rare earth germanates phosphor, $NaLnGeO_4$ (Ln=Y, Gd), might be one of the potential candidates for fluorescent lamp phosphors and red-emitting elements for white light LEDs.

ACKNOWLEDGEMENTS
This work was supported by a grant from Information Display R&D Center, one of the 21st Century Frontier R&D Program funded by the Ministry of Commerce, Industry and Energy of Korea government.

FOOTNOTES
*Corresponding author; Advanced Materials Division, Korea Research Institute of Chemical Technology, P.O. Box 107, Yuseong, Daejeon 305-600, Republic of Korea.
Tel: +82-42-860-7372; fax: +82-42-861-4245
 E-mail address: shochoi@krict.re.kr

REFERENCES
[1] K. Kakinuma, Technology of Wide Color Gamut Backlight with Light-Emitting Diode for Liquid Crystal Display Television, Jpn. J. Appl. Phys., 45, 4330 (2006).
[2] R. Mueller-Mach, G.O. Mueller, M.R. Krames, and T. Trottier, High-power phosphor-converted light-emitting diodes based on III-nitrides, IEEE J. Sel. Top. Quantum Electron., 8, 339 (2002).
[3] S. Nakamura, M. Senoh, and T. Mukai, High-power InGaN/GaN double-heterostructure violet light emitting diodes, Appl. Phys. Lett., 62, 2390 (1993).
[4] S. Neeraj, N. Kijima, and A.K. Cheetham, Novel red phosphors for solid-state lighting: the system NaM(WO$_4$)$_{2-x}$(MoO$_4$)$_x$:Eu^{3+} (M=Gd, Y, Bi), Chem. Phys. Lett., 387, 2 (2004).
[5] X. Wang, J. Wang, J. Shi, Q. Su, and M. Gong, Intense red-emitting phosphors for LED solid-state lighting, Mat. Res. Bull., 42, 1669 (2007).
[6] K. Toda, Y. Imanari, T. Nonogawa, K. Uematsu, and M. Sato, New VUV phosphor, NaLnGeO$_4$:Eu^{3+} (Ln=Rare Earth), Chem. Lett. 32, 346 (2003).
[7] A. P. Dudka, A.A. Kaminskii, and V.I. Simonov, Refinement of NaGdGeO$_4$, NaYGeO$_4$, and NaLuGeO$_4$ single-crystal structures, Phys. Stat. Sol.(a) 93, 495 (1986).
[8] G. Blasse, A. Bril, Structure and Eu^{3+}-fluorescence of lithium and sodium lanthanide silicates and germanates J. Inorg. Nucl. Chem., 29, 2231 (1967).
[9] Y. Li, Y. Chang, B. Tsai, Y. Chen, and Y. Lin, Luminescent properties of Eu-doped germanate apatite Sr$_2$La$_8$(GeO$_4$)O$_2$ J. Alloys Compd., 416, 199 (2006).
[10] S. Shionoya, W. M. Yen, Phosphor Handbook, CRC press, 183 (1999).
[11] D.L. Dexter, Theory of concentration quenching in inorganic phosphors, J. Chem., Phys., 22, 1063 (1954).
[12] X. X. Li, Y. H. Wang, Synthesis and photoluminescence properties of (Y,Gd)Al$_3$(BO$_3$)$_4$:Tb^{3+} under VUV excitation, Mater. Chem. Phys., 101, 191 (2007).

COPRECIPITATION AND HYDROTHERMAL SYNTHESIS OF PRASEODYMIUM DOPED CALCIUM TITANATE PHOSPHORS

James Ovenstone[1], Jacob Otero Romani[2], Dominic Davies[2], Scott Misture[1] and Jack Silver[3]
[1]NYSCC Alfred University, Alfred , New York.
[2]School of Chemical and Life Sciences, University of Greenwich, Island Site, Woolwich, London.
[3]Wolfson Centre for Materials Processing, Brunel University, Uxbridge, Middlesex UB8 3PH, UK

ABSTRACT
The synthesis of the cathodoluminescent phosphor material praseodymium doped calcium titanate (CAT) using co-precipitation and hydrothermal techniques has been investigated. For the co-precipitation method the type of base used has been found to have a strong influence on both the microstructure and the luminance. Co-precipitation failed, however to produce phase pure product, but instead produced a mixture of rutile and calcium titanate after calcinations. Hydrothermal processing also failed to make the phase pure material due to the strong kinetic drive towards anatase formation under hydrothermal conditions. Despite this, the phase mixture formed contained a significantly higher proportion of calcium titanate, and manifested significantly brighter luminance than the co-precipitated material under both PL and CL conditions.

INTRODUCTION
Field emission displays (FEDs) are one of the many potential candidates for advanced flat panel display applications which are becoming increasingly important as the world demands better and better multimedia information access[1]. Potentially they will be able to provide high contrast and brightness, but with low weight and low power consumption. They work on a similar principle to traditional cathode ray tube (CRT) displays, in that the phosphor pixels are excited by an electron stream, but whereas the CRT unit uses a single large electron gun, in a FED display, each individual pixel is addressed by its own microtip cathode. FEDs must operate at significantly lower voltages than CRTs ($\leq$5 kV) and higher current densities (10-100 μA/cm^2), thus requiring the phosphors to operate with high efficiency at low voltages, with high resistance to current saturation, with equal or better chromaticity than CRT phosphors[2,3].

Calcium titanate doped with praseodymium (CAT) was first reported as a potential red emission phosphor for use in low voltage and in particular FEDs by Vecht et. al. in 1994[4]. Sung et. al., in 1996, further improved the properties of the phosphor through improved synthesis techniques[5]. The red emission is attributed to a $^1D_2 \rightarrow {}^3H_4$ transition in the Pr^{3+} dopant ion, and has Commission International d'Eclairage (CIE) coordinates x=0.680 and y=0.311, which is very close to the National Television Standards Committee (NTSC) 'ideal red'[6]. Significant work has since been carried out in the optimisation of CAT, in terms of synthesis[6,7], and investigation of emission mechanisms[8]. Others have attempted to form composites in order to better control the morphology of the material. It has been recognised that small spherical particles with a narrow size distribution and no agglomeration offer potential advantages in terms of packing, brightness, high definition, and low scattering[9]. Liu et. al. in 2006, successfully developed core-shell structured CAT on silica cores in order to control the morphology[10]. Boutinaud et. al., on the other hand investigated the potential for depositing CAT thjin films using radio frequency sputtering[11]. There has also been much work carried out investigating the potential for further additives to the material which may improve its emission properties further. Addition of aluminium has been found to significantly improve the emission of CAT[12-14]. Zhang et. al. in 2007 reported the enhancement of the red emission of Cat by addition of the rare earth oxides Lu,

La, and Gd. In this work, the PL emission of the CAT almost tripled with the addition of 5 mol% Lu_2O_3[15].

Hydrothermal reactions take place in aqueous media at temperatures greater than 100 °C, and at pressures greater than 0.1 MPa. Hydrothermal synthesis has been shown to be a highly effective synthetic method for producing, directly, perovskite ceramic compounds such as calcium doped lanthanum chromite without calcination[16]. Less extreme hydrothermal conditions (<200 °C) have also been shown to significantly lower the calcination temperatures required to form solid solutions in perovskite powders[17, 18]. The enhanced reactivity is attributed to the high ionic mobility and increased solubility of species under hydrothermal conditions. Hydrothermal treatment has also been shown to produce highly sinteractive powders, with small particle size and good morphology control. Despite the advantages of hydrothermal processing reviewed by Dawson, the technique has not been widely used in the production of phosphor materials[19]. In the current work we report the application of co-precipitation and hydrothermal processing techniques to the production of calcium titanate doped with praseodymium, in an attempt to produce phosphor particles with sub-micron, spherical morphology and a tight size distribution.

EXPERIMENTAL

The precursor for all the syntheses of CAT:PR was amorphous titania powder. Since the reactivity of amorphous titania has been shown to be strongly dependent on the synthesis method, the amorphous powder was prepared by four different methods.

(a) $Ti(OPr)_4$ (10 ml) was diluted in a beaker with 100 ml of water, and then 2 ml of ammonia solution (conc.) was added with stirring. The white precipitate formed was vacuum filtered and dried in an oven at 80 °C for two hours. The white powder produced was amorphous titania (AP1).

(b) 200 ml of ammonia solution (40 ml NH_3 7.3 M + 160 ml of H_2O) was added dropwise into 50 ml of a $TiCl_4$ solution (20 ml $TiCl_4$, 0.182 mol + 200 ml of ice cold water) with mixing, resulting in the formation of a white precipitate. The white precipitate formed was vacuum filtered and dried in an oven at 80 °C for two hours.
The white powder produced was amorphous titania (AP2).

(c) 50 ml of a $TiCl_4$ solution (20 ml $TiCl_4$, 0.182 mol + 200 ml of ice cold water) was added dropwise into 200 ml of ammonia solution (40 ml NH_3 7.3 M + 160 ml of H_2O) with mixing, resulting in the formation of a white precipitate. The white precipitate formed was vacuum filtered and dried in an oven at 80 °C for two hours. The white powder was amorphous titania (AP3).

Having prepared the amorphous powders, CAT:PR was then prepared by two co-precipitation methods.

Method 1

2 g amorphous titania (AP1) was dissolved in 25 ml HCl (conc.). 4.1 g $Ca(NO_3)$, and $Pr(NO_3)_3$ (0.1 M) and 60 g of urea were added with stirring. The solution was heated to 90 °C for 60 minutes until precipitation ceased. The amorphous precipitate was vacuum filtered and dried in an oven at 80 °C for 24 hours. This preparation was repeated to produce a range of phosphors from 0.1 mol% Pr to 2 mol% Pr (CP1).

Method 2

2 g amorphous titania (AP1) was dissolved in 25 ml HCl (conc.). 4.1 g $Ca(NO_3)_2$, and $Pr(NO_3)_3$ (0.1 M) were added, and the mixture added dropwise to 200 ml ammonia solution (7.3 M). The amorphous precipitate formed was vacuum filtered and dried in an oven at 80 °C for 24 hours. This preparation was repeated to produce a range of phosphors from 0.1 mol% Pr to 2 mol% Pr (CP2).

For the hydrothermal treatments 0.5 g of amorphous titania (or anatase supplied by Aldrich) was added to 1.025 g $Ca(NO_3)_2$, 0.3125 ml $Pr(NO_3)_3$ (0.1 M) and 1.825 M ammonia solution (40 ml, large excess). The concentrations of base and $Ca(NO_3)_2$ were varied between reactions. Hydrothermal processing was carried out using PTFE lined polycarbonate autoclaves. The vessels were heated using microwaves (CEM MDS-2100 microwave autoclave system) to a range of temperatures from 120 °C to 200°C. Higher temperatures could not be attained due to the development of excessive pressure within the vessels. All reactions were carried out for 1 hour. The precipitates were all vacuum filtered, and dried in an oven at 80°C for 24 hours. All products were calcined for 1 hour in a furnace at 1100 °C.

X-ray powder diffraction (XRD) was carried out using a Philips pw1710 diffractometer. A step size of 0.02° 2-theta was used with a scan speed of 0.5° per minute. XRD spectroscopy showed that the initial titania powders were indeed amorphous, and free from any crystalline phases, as no peaks were observed. Scanning electron microscopy (SEM) studies were carried out using a Cambridge Instruments Stereoscan 90 scanning electron microscope.

Photoluminescence brightness measurements were made using a Pritchard PR880 photometer. The powders were packed into a glass holder approximately 0.5mm deep, and with an area approximately 1.5 cm by 1 cm. The surface was smoothed using a glass slide, and then illuminated with a 100W U.V. lamp emitting light at 366nm.

Emission and excitation spectra were measured using a customised system. Excitation wavelengths were selected from a Xenon Lamp (Bentham, type IL7D), using a Bentham M300BA monochromator. Samples were placed in a Bentham SC-FLUOR fluorescence chamber, and the emission wavelengths measured using another M300BA monochromator. The emission spectra were measured for an excitation wavelength of 366 nm.

Cathodoluminescence (CL) measurements were made on a range of samples. Cathodoluminescent excitation was achieved under vacuum at around 10^{-6} Torr and using a Kimball Physics Inc. (Walton, NH) model EGPS-7 electron gun. The e-beam used was 1.44 mm in diameter and the current was 8.5 μA. Measurements were made from 400 V to 3000 V at a current of 50 μA. The samples were prepared by electrophoretic deposition onto aluminium stubs.

RESULTS AND DISCUSSION

The SEM micrographs in Figure 1 show the typical microstructures for the calcined co-precipitated powders. From these pictures we can see that method 1 produces smaller more spherical powder particles (around 500 nm), when compared to method 2 (1-3 nm). In these experiments, AP1 was the amorphous titania precursor.

From the XRD spectra in Figure 2, we can see that the unfired powders made by method 1 (outlined above) manifest peaks corresponding to anatase. In addition, the samples with 0.5 mol% and 1 mol% praseodymium also manifest other peaks at $2\theta = 22.4, 29.5, 30.7$, and 31.9°. The peaks could not, however, be identified, but are suspected to be complex oxides of titanium and calcium. After firing, however, all the samples produced a phase mixture of rutile and calcium titanate (Figure 2b).

There was no significant difference in the XRD patterns of the fired materials, regardless of the method used.

Figure 3a shows a graph of the PL brightness for the phosphors made by method 1 with different praseodymium concentrations. The amorphous titania precursor in this instance is AP1. There is a clear increase in PL brightness as the praseodymium concentration is increased from 0.1 mol% to 0.5 mol%. Beyond this concentration, the PL emission brightness falls significantly as concentration quenching occurs. This type of quenching is commonly observed in phosphors, and occurs as the result of quantum mechanical resonant energy exchange between neighbouring activator ions, resulting in non-radiative relaxation of the excited states. A similar pattern of behaviour is seen for the brightness under cathodoluminescence conditions (Figure 3b). As expected, the phosphors all manifested a linear increase in PL brightness with increasing voltage up to 3000V, with the exceptions of the 1% and 2% Pr doped materials, which did not exhibit any cathodoluminescence at all, again due to concentration quenching. Having established the optimum activator concentration for maximum brightness in the system, all further preparations used a 0.5 mol% praseodymium concentration.

Figure 4a shows the PL emission spectrum for the material made using method 1 with 0.5wt% Pr doping. The PL spectrum manifests the characteristic red luminescence of the material. Due to experimental constraints the CL spectrum of the material could not be measured. In Figure 4b method 1 and method 2 are further compared in terms of the brightness of luminescence for the 0.5wt% Pr samples under cathodoluminescence conditions. Method 1 clearly outperforms method 2 in terms of brightness, particularly at higher voltages (>2500V).

Despite the apparent attractiveness of using urea as the base, for the microwave hydrothermal synthesis reactions to obtain small spherical particle morphology, urea could not be used, as this results in the production of calcium carbonate rather than calcium titanate (CAT), due to the extremely low solubility of calcium carbonate under hydrothermal conditions. The polymorph of calcium carbonate produced was dependent on the hydrothermal conditions, and was found to be topotactically templated by the titania precursor. This is the subject of another paper. As a result, only ammonia was used as the base in these reactions.

The SEM micrographs in Figure 5 shows the typical morphologies of the calcined powders produced hydrothermally at 120°C. The particles produced from precursor AP1 (Figure 5a) have a spherical morphology with a very small particle size of around 300nm. Figures 5b and 5c show the morphologies for the powders produced using the AP1 and AP2 precursors respectively. These precursors again produced small particle sizes (sub 1 micron), however the particle size distribution is not so narrow, and larger particles above 1 micron were also observed. Figure 5d shows powder produced using $TiCl_4$ (stoichiometric with the calcium nitrate) instead of amorphous titania. The particle size is significantly larger, and particle shapes more irregular.

Figure 6a shows the XRD spectra for the as-produced products for the hydrothermal reactions at temperatures between 120 °C and 200 °C for 1 hour for the AP1 powder. Calcium titanate has not formed directly in the autoclave, however, there is some evidence for the crystallisation of anatase from the amorphous titania powder (the broad band centred around 2 theta = 25 °. This suggests that the formation of anatase is kinetically favourable compared to the formation of CAT under these conditions. Clearly, increasing the reaction temperature increases the rate of crystallisation of anatase (as would be expected), but has seemingly no effect on the crystallisation of CAT. All three amorphous titania precursors gave very similar results, as did the $TiCl_4$. In the latter case, however, hydrothermal treatments could only be carried out at 120°C, as higher temperatures resulted in failure of the pressure vessels due to the high pressures generated by the decomposing precursors. A typical XRD spectrum of the calcined products from a hydrothermal treatment at 120°C is shown in Figure 7. In all cases, after firing, mixture of rutile (the thermodynamically stable phase of titania) and CAT is formed. Again, as for the hydrothermal product, it is the formation of the titania phase, which has the kinetic advantage, and so forms first. Similar results were observed for all the amorphous precursors and the TiCl$_4$. It should be noted, however, that the proportion of calcium titanate produced relative to rutile was significantly higher for the hydrothermally produced powder when compared to the co-precipitated material. When rutile or anatase was used as starting materials, the only crystalline products remained as rutile and anatase respectively after hydrothermal processing. In both cases CAT and rutile were the calcined products.

PL and CL luminance were measured for the powders prepared from theAP1, AP2, AP3 and TiCl$_4$ precursors, with a Pr dopant level of 0.5wt%, but in all cases the AP1 precursor produced the brightest emission. As a result further experiments were carried out only with the AP1 precursor. Figure 8a shows how the PL brightness varies with the hydrothermal processing temperature for the AP1 starting material, while figure 8b shows the variation in CL brightness with hydrothermal processing temperature and voltage. As might be expected, the best materials were those made at low temperature, since these did not suffer so much from anatase crystallisation in the autoclaves. The CL measurements showed a close approximation to a linear relationship between brightness and voltage used.

Since raising the reaction temperature did not improve the formation of CAT, the concentrations of base and calcium nitrate were altered, maintaining a reaction temperature of 120 °C to limit the formation of anatase. In all cases, the as-produced products were shown to be amorphous by XRD, and to be mixtures of CAT and rutile after calcination (Figure 9). Surprisingly, no change in the relative intensities of the rutile and CAT were observed for any of the combinations tested, even when large excesses of calcium nitrate were used.

Figures 10a and 10b (fig24 and 25) show the relationship between the base concentration and the PL brightness or CL brightness respectively. There is clearly an optimum base concentration, which results in improved PL brightness. The increase in brightness with base concentration may well be due to the hindrance of anatase crystallisation under strongly basic conditions. This occurs due to the immediate solubilisation of anatase nuclei as they form, preventing further growth. The same phenomenon can be expected for CAT nuclei, at higher base concentrations, hence the drop in PL brightness at the highest base concentration used.

Figures 11a and 11b show the effect of changing the calcium nitrate concentration on PL and CL brightness respectively. In both cases, increasing the calcium nitrate concentration to a large excess, increases the brightness significantly. One explanation is that the large excess of starting material drove the reaction to completion. XRD, however, did not show any changes in the ratio of rutile to CAT in the final products, indicating that the amount of CAT formed has not increased. The reason for this improvement in luminance, and the limiting concentration that gives maximum brightness will be the subject of further study at a later date.

CONCLUSIONS

Calcium titanate doped with praseodymium has been prepared by a co-precipitation route and by hydrothermal processing. A doping level of 0.5 % has been found to be the optimum for maximising luminescence. The base used for precipitation has a strong influence over the morphology of the final product, with urea giving small spherical particles, while ammonia gives larger more irregular shaped particles. The brightness of the phosphors were also strongly influenced by the base, with the urea giving the brighter product. In all cases a phase mixture of CAT and rutile was formed after calcinations. The hydrothermal route has been found to produce brighter material than the co-precipitation method, although the product is again not phase-pure CAT, but a mixture of CAT and rutile. The proportion of CAT compared to rutile after calcinations is, however, significantly larger for the hydrothermally produced powders compared to the co-precipitated powders. Low temperature hydrothermal processing has been found to be advantageous in avoiding anatase formation, as has a large excess of calcium nitrate, and a medium-high base concentration. Longer reaction times may well allow formation of phase pure CAT:Pr, however, such experiments are not possible using the equipment available to the authors at this time.

ACKNOWLEDGEMENTS
The authors would like to thank the TCD (Program No. 2912) for their support of JO. Mr. Geoff Cooper is also thanked for his assistance with the electron microscope.

REFERENCES
[1] J.C. Park, H.K. Moon, D.K. Kim, K.S. Suh, Appl. Phys. Lett. 77, 2162, (2000).
[2] T. Justel, H. Nikol, and C. Ronda, Angew. Chem., Int. Ed. 37, 3084, (1998).
[3] Y.D. Jing, F. Zhang, C.J. Summers, and Z.L. Wang, Appl. Phys. Lett. 74, 1677, (1999).
[4] A. Vecht, D.W. Smith, S.S. Chadha, and C.S. Gibbon, J. Vac. Sci. Technol. B12, 781, (1994).
[5] S.H. Cho, J.S. Yoo, and J.D. Lee, J. Electrochem. Soc. 143, L231, (1996).
[6] E. Pinel, P. Boutinaud, G. Bertrand, C. Caperaa, J. Cellier, and R. Mahiou, J. Alloys and Compounds 374, 202, (2004).
[7] Y. Pan, Q. Su, H. Xu, T. Chen, W. Ge, C. Yang, and M. Wu, J. Sol. State Chem. 174, 69, (2003).
[8] E. Pinel, P. Boutinaud, and R. Mahiou, J. Alloys and Compounds 380, 225, (2004).
[9] M.I. Martinez-Rubio, T.G. Ireland, G.R. Fern, J. Silver, and M.J. Snowden, Langmuir 17, 7145, (2001).
[10] X. Liu, P. Jia, J. Lin, and G. Li, J. App. Phys. 99, 124902, (2006).
[11] P. Boutinaud, E. Tomasella, A. Ennajdaoui, and R. Mahiou, Thin Sol. Films 515, 2316, (2006).
[12] S. Okamoto, and H. Yamamoto, J. Appl. Phys. 15, 5492, (1999).
[13] P.T. Diallo, K. Jeanlouis, P. Boutinaud, R. Mahiou, and J.C. Cousseins, J. Alloys and Compounds 323, 218, (2001).
[14] J. Tang, X. yu, L. Yang, C. Zhou, and X. Peng, Mat. Lett. 60, 326 (2006).
[15] X X. Zhang, J. Zhang, X. Zhang, L. Chen, Y. Luo, and X. Wang, Chem. Phys. Lett. 434 237, (2007).
[16] M. Yoshimura, S. Song, and S. Somiya, in Ferrites: Proceedings of the International Conference Sept-Oct 1980 Japan.
[17] J. Ovenstone and C.B. Ponton, British Ceramics Proceedings 58, 155, (1998).
[18] J. Ovenstone, K.C. Chan, and C.B. Ponton, J. Mat. Sci. 37, 3315, (2002)
[19] W.J. Dawson, Am. Ceram. Soc. Bull. 67 [10], 1673 (1988).

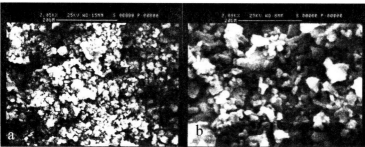

Figure 1   SEM micrographs of calcined powders from a) method 1, and b) method 2, showing the smaller particle sizes and more uniform size distribution achievable using urea as base during co-precipitation.

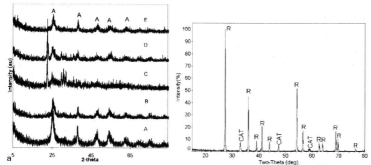

Figure 2   X-ray diffraction spectra for a) unfired precipitates prepared using method 1 A) 0.1 mol% Pr; B) 0.25 mol% Pr; C) 0.5 mol% Pr; D) 1 mol% Pr; and E) 2 mol% Pr. b) Calcined product.

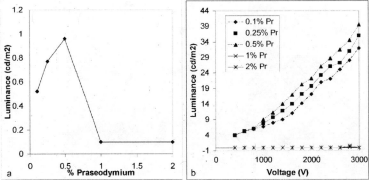

Figure 3　　Luminescence as a function of Pr concentration for calcined powder prepared by method 1 using AP1 precursor. a) PL luminescence, and b) CL luminescence.

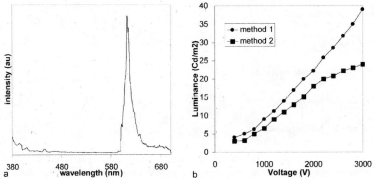

Figure 4　　a) PL emission spectra for the 0.5 mol% Pr sample made by method 1 excited using 366 nm U.V. light. b) Cathodoluminescence for the $CaTiO_3$ with 0.5 wt% Pr prepared by method 1 and method 2.

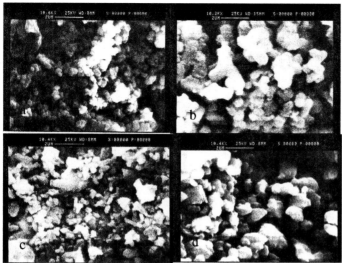

Figure 5   SEM micrograph of calcined, hydrothermally produced CAT:Pr with a Pr concentration of 0.5 wt%. Hydrothermal treatment carried out at 120°C using a) AP1 precursor and b) AP2 precursor, c) AP3 precursor and d) TiCl₄.

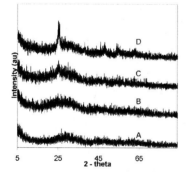

Figure 6   XRD spectra of material prepared by hydrothermal synthesis, using amorphous titania AP1 and ammonia, at different temperatures. A) 120 °C; B) 150 °C; C) 175 °C; and 200°C.

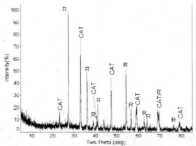

Figure 7   Typical XRD spectrum of the hydrothermal product produced at 120°C using AP1 precursor, and then calcined.

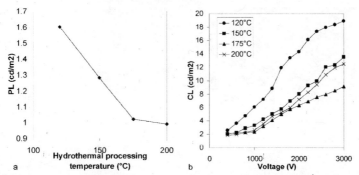

a   b

Figure 8   a) PL as a function of hydrothermal processing temperature for 0.5wt%Pr doped CAT.
b) CL as a function of voltage for 0.5wt% doped CAT produced using different hydrothermal processing temperatures.

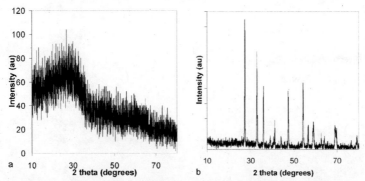

a   b

Figure 9   a) Typical X-ray diffraction pattern for hydrothermal product at 120°C. b) Typical X-ray diffraction spectrum for hydrothermal product at 120°C after calcination.

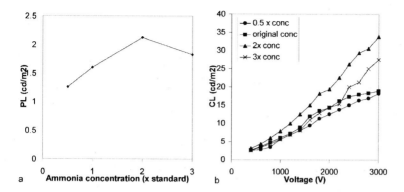

Figure 10   Luminescence for calcined products of hydrothermal runs at 120°C with varying base concentration: a) PL and b) CL.

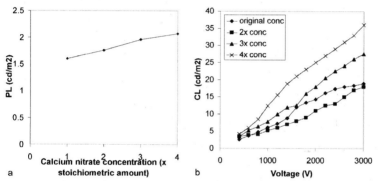

Figure 11   Luminescence for calcined products of hydrothermal runs at 120°C with varying calcium nitrate concentration: a) PL and b) CL.

Author Index